Handbook of Mango Fruit

Handbook of Mango Fruit:

Production, Postharvest Science, Processing Technology and Nutrition

Editor

Muhammad Siddiq
Michigan State University, USA

Associate Editors

Jeffrey K. Brecht
University of Florida, USA

Jiwan S. Sidhu
Kuwait University, Kuwait

WILEY Blackwell

This edition first published 2017

Registered Offices
John Wiley & Sons Ltd, The Atrium, Southern Gate, Chichester, West Sussex, PO19 8SQ, UK

Editorial Office
9600 Garsington Road, Oxford, OX4 2DQ, UK
The Atrium, Southern Gate, Chichester, West Sussex, PO19 8SQ, UK

For details of our global editorial offices, customer services, and more information about Wiley products visit us at www.wiley.com.

Wiley also publishes its books in a variety of electronic formats and by print-on-demand. Some content that appears in standard print versions of this book may not be available in other formats.

Library of Congress Cataloging-in-Publication Data
Names: Siddiq, Muhammad, 1957- editor. | Brecht, Jeffrey K., 1942- editor. | Sidhu, Jiwan S., editor.
Title: Handbook of mango fruit : production, postharvest science, processing technology and nutrition / edited by Muhammad Siddiq, Michigan State University, USA, Jeffrey K. Brecht, University of Florida, USA, Jiwan S. Sidhu, Kuwait University, Kuwait
Description: Oxford, UK : Wiley, 2018. | Includes bibliographical references and index. | Description based on print version record and CIP data provided by publisher; resource not viewed.
Identifiers: LCCN 2017011132 (print) | LCCN 2017011768 (ebook) | ISBN 9781119014386 (pdf) | ISBN 9781119014379 (epub) | ISBN 9781119014355 (cloth)
Subjects: LCSH: Mango. | BISAC: TECHNOLOGY & ENGINEERING / Food Science.
Classification: LCC SB379.M2 (ebook) | LCC SB379.M2 H36 2018 (print) | DDC 634/.44--dc23
LC record available at https://lccn.loc.gov/2017011132

Cover Design: Wiley
Cover Images: (left to right) Courtesy of Dr. Jeffrey K. Brecht; © nine_far/iStockphoto; © RUJITOP/iStockphoto

Set in 10/12pt Warnock Pro by SPi Global, Chennai, India

10 9 8 7 6 5 4 3 2 1

Contents

Preface

Mango is one of the most consumed fresh fruits in the world, with production occurring in more than one hundred countries. Global production of the fruit has more than doubled in the last three decades. There are over one thousand different varieties of mangoes, giving rise to various shapes, sizes, colors, textures and nutritional properties. The mango market and trade has grown considerably since the 1990s, with the two major import markets being the United States of America and European Union countries. Since the mango fruit has made its way into mainstream market outlets in most of the developed countries and is available year-round, it is no longer classified as exotic. The year-round availability of the fruit is attributed to several factors, including the fact that the fruit is grown under diverse climatic conditions, which allows harvesting throughout the year, and improvements in transportation, market access, pre-harvest production practices, and postharvest treatment allow the crop to be shipped long distances relatively free of any pests and diseases. Demand for mangoes is also on the rise, as more health conscious consumers incorporate the fruit into their diets, based on its nutritional value and unique flavor. Along with the increased consumption of this nutrient-rich fruit, the processed mango market has also seen similar growth, especially in fresh-cut fruit, juice and beverages, and shelf-stable dried products.

This book provides a contemporary source of information that brings together current knowledge and practices in the value-chain of mango production, postharvest handling, and processing. This value-chain approach to the topic is the unique feature of this book, with an in-depth coverage on a wide variety of pertinent topics: production, harvesting and GAPs, postharvest entomology and pathology, postharvest physiology and storage, packaging technologies, processing and processed products, innovative processing technologies, nutritional profile and health benefits, bioactive and phytochemical compounds, and value-added utilization of mango by-products. An experienced team of over 30 contributors from Asia, North America, and South America has contributed to this book. These contributors come from a field of diverse disciplines, including horticulture, crop sciences, plant pathology and entomology, food science and technology, food biochemistry, food engineering, nutritional sciences, and agricultural economics.

The editors acknowledge many individuals for their support from conception through to final development of this book. Foremost is our sincere thanks and gratitude to all authors for their contributions and for bearing with us during the review and finalization process of their chapters. We are grateful to our family members for their understanding and support, enabling us to complete this work. We dedicate this work to the worthy contributions of the numerous researchers and students throughout the world, for their decades long devoted efforts to improve the quality and utilization of fresh mango and its processed products.

Muhammad Siddiq
Jeffrey K. Brecht
Jiwan S. Sidhu

List of Contributors

Poonam Aggarwal
Department of Food Science
and Technology
Punjab Agricultural University
Ludhiana, Punjab
India

Ender Arserim
Department of Food Science, Rutgers
The State University of New Jersey
New Brunswick, New Jersey
USA

Fredy H. Ballen
Center for Tropical Agriculture
University of Florida, IFAS
Homestead, Florida
USA

Suresh Bhise
Department of Food Science &
Technology
Punjab Agricultural University
Ludhiana, Punjab
India

Andrea Birke
Instituto de Ecología A.C.
Congregación El Haya
Veracruz
Mexico

Jeffrey K. Brecht
Horticultural Sciences Department
University of Florida
Gainesville, Florida
USA

Daniel Carrillo
Tropical Research & Education Center
University of Florida
Homestead, Florida
USA

Edward A. Evans
Center for Tropical Agriculture
University of Florida, IFAS
Homestead, Florida
USA

Antonio de Goes
Department of Plant Pathology
Faculty of Agricultural and
Veterinarian Sciences
Sao Paulo State University (UNESP)
Jaboticabal, São Paulo
Brazil

Larissa Guillen
Instituto de Ecología A.C.
Congregación El Haya
Veracruz
Mexico

Rajinder P. Gupta
BIS Institute of Science and Technology, Gagra
Moga, Punjab
India

Yearul Kabir
Department of Biochemistry and Molecular Biology
University of Dhaka
Dhaka
Bangladesh

Anu Kalia
Electron Microscopy and Nanoscience Laboratory, College of Agriculture
Punjab Agricultural University
Ludhiana, Punjab
India

Mukund Karwe
Department of Food Science, Rutgers
The State University of New Jersey
New Brunswick, New Jersey
USA

Amarjeet Kaur
Department of Food Science & Technology
Punjab Agricultural University
Ludhiana, Punjab
India

Maria Gloria Lobo
Department of Tropical Fruits
Instituto Canario de Investigaciones Agrarias, Valle de Guerra
La Laguna, Tenerife, Canary Islands
Spain

Aman Ullah Malik
Institute of Horticultural Sciences
University of Agriculture, Faisalabad
Pakistan

Olga Martin-Belloso
Departament de Tecnologia d'Aliments
Universitat de Lleida
Lleida
Spain

Robert E. Paull
Tropical Plant & Soil Sciences
College of Tropical Agriculture and Human Resources
University of Hawaii at Manoa
Honolulu, Hawaii
USA

J.E. Peña
Tropical Research and Education Center
University of Florida
Homestead, Florida
USA

Sunisa Roidoung
Department of Food Technology and Nutrition
Maha Sarakham University,
Maha Sarakham
Thailand

Blanca Salinas-Roca
Departament de Tecnologia d'Aliments
Universitat de Lleida
Lleida
Spain

Deepti Salvi
Department of Food Science, Rutgers
The State University of New Jersey
New Brunswick, New Jersey
USA

Hossain Uddin Shekhar
Department of Biochemistry and Molecular Biology
University of Dhaka, Dhaka
Bangladesh

Farihah Siddiq
Food Packaging Associate
East Lansing
Michigan
USA

Muhammad Siddiq
Department of Food Science &
Human Nutrition
Michigan State University
East Lansing, Michigan
USA

Jiwan S. Sidhu
Department of Food Science and
Nutrition, College of Life Sciences
Kuwait University, Safat
Kuwait

Dalbir S. Sogi
Department of Food Science and
Technology
Guru Nanak Dev University
Amritsar, Punjab
India

Robert Soliva-Fortuny
Departament de Tecnologia d'Aliments
Universitat de Lleida
Lleida
Spain

Andressa de Souza-Pollo
Laboratory of Molecular Epidemiology
Sao Paulo State University (UNESP)
Faculty of Agricultural and Veterinarian
Sciences, Jaboticabal, São Paulo
Brazil

Chantalak Tiyayon
Department of Plant and Soil Sciences
Faculty of Agriculture
Chiang Mai University, Chiang Mai
Thailand

Jorge Welti-Chanes
Tecnológico de Monterrey, Escuela
de Ingeniería y Ciencias, Centro de
Biotecnología FEMSA, Monterrey
Mexico

Elhadi M. Yahia
Horticultural Sciences Department
Universidad Autonoma de Queretaro
Queretaro
Mexico

Tasleem A. Zafar
Department of Food Science
and Nutrition,
College of Life Sciences
Kuwait University, Safat
Kuwait

1

Mango Production, Global Trade, Consumption Trends, and Postharvest Processing and Nutrition

Edward A. Evans[1], *Fredy H. Ballen*[1] *and Muhammad Siddiq*[2]

[1] *Center for Tropical Agriculture, University of Florida, IFAS, Homestead, Florida, USA*
[2] *Department of Food Science and Human Nutrition, Michigan State University, East Lansing, Michigan, USA*

1.1 Introduction

In terms of production, mangoes, pineapples, papayas, and avocados are considered the major tropical fruits, while lychees, durian, rambutan, guavas, and passion fruits are among the minor ones. Out of global production of the major tropical fruits, mangoes accounted for more than half of the total production in 2012, followed at a distance by pineapples, papayas, and avocados, with shares of 28.36, 15.08, and 5.33%, respectively (FAO 2015). It is to be noted that the FAO reports mango, mangosteen, and guava production and trade as one aggregated category. There are over a thousand different varieties of mangoes, giving rise to various shapes, sizes, colors, texture, and nutritional properties. The fruit is considered to be one of the most consumed fresh fruits in the world, with production taking place in more than 100 countries. However, despite its nutritional value, popularity in developed countries, and tremendous potential for export revenues in many developing countries, only a relatively small portion of the produce enters into international trade; the bulk of production is consumed in the producing country. In 2013, for example, less than 4% of global mango production reached international markets. The USA and Europe are the leading importers of mangoes and have shown steady increases in the volume of fruit absorbed over the past few years.

In recent years, since the mango fruit has made its way into mainstream market outlets in most of the developed countries and is available all year round, it is no longer classified as exotic. The year-round availability of the fruit is attributed to several factors. These include the fact that the fruit is grown under diverse climatic conditions, which allows harvesting throughout the year, and improvements in transportation, market access, preharvest production practices, and postharvest treatment allow the crop to be shipped over long distances, while remaining relatively free of pests and diseases. Demand for the fruit is also on the rise, as more health conscious consumers incorporate the fruit in their diet based on its nutritional value.

Although mangoes no longer command the high prices associated with exotic fruits and with its past scarcity on the international market, prices have not declined substantially. A likely explanation for the relative buoyancy of prices is due to promotional and

Handbook of Mango Fruit: Production, Postharvest Science, Processing Technology and Nutrition, First Edition. Edited by Muhammad Siddiq, Jeffrey K. Brecht and Jiwan S. Sidhu.

consumer educational activities, which have had the effect of increasing the overall demand for the fruit despite increases in supply. Notwithstanding, prices of the fruit vary considerably depending on variety, source of origin, and time of the year.

This chapter gives an overview of mango fruit world production, exports, and imports, consumption trends, postharvest storage, processing, nutritional quality, and food safety aspects.

1.2 Mango Production, Trade, and Consumption

1.2.1 Area Harvested and World Production

The global area of mangoes harvested in 2013 was 5.41 million hectares, an increase of about 52% as compared to that in 2000 (Figure 1.1). Given the expansion in the harvested area, global production of mangoes have also increased, growing from 24.71 million metric tons (MMT) in 2000 to 42.66 MMT in 2013, which represented an increase of about 73%. Since 2010, the area harvested and production have shown an increase of 8.95% and 14.62%, respectively. It is noted that the majority of the increase in production since 2010 has been due to corresponding increase in area harvested (52%) versus improvements in yield per unit area (14%).

Mangoes are grown under very diverse climatic conditions, in tropical and subtropical regions. Asia is the dominant producing region, with over 77.17%, based on the 2013 production figures, followed by the Americas and the Caribbean (12.22%), Africa (10.50%), and Oceania (0.11%) (FAO 2015). Because of the wide distribution of the growing regions, these fruits are produced in 115 countries, with the vast majority being developing economies. Table 1.1 shows the leading mango producing countries for selected years from 2000 to 2013. Although the fruit is produced in over 100 countries, the data provided in Table 1.1 indicate that the bulk of production is highly concentrated, with 10 countries accounting for 82.06% of the total production in 2013, and since 2000 this share has remained fairly stable at or above 80%. India is by far the

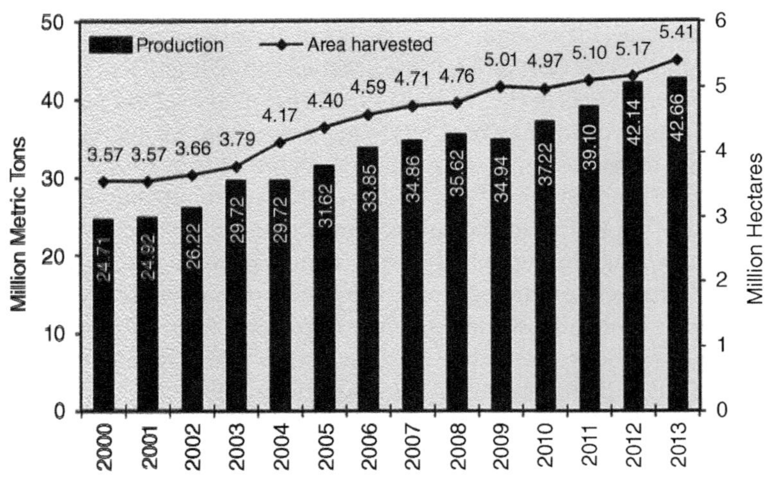

Figure 1.1 World mango production and area harvested (2000–2013). *Source:* FAO (2015).

Table 1.1 World's ten major mango[1] producing countries for selected years since 2000 (million metric tons).

Country	2000	2005	2010	2013	Percent Change 2000–13	Percent Change 2010–13
India	10.50	11.83	15.03	18.00	71.39	19.80
China	3.00	4.10	4.00	4.45	48.33	11.25
Thailand	1.62	1.80	2.55	3.14	93.57	23.18
Indonesia	0.88	1.41	1.29	2.06	134.99	59.92
Mexico	1.56	1.68	1.63	1.90	21.97	16.49
Pakistan	0.94	1.67	1.85	1.66	76.87	−10.13
Brazil	0.54	1.00	1.19	1.16	116.05	−2.24
Bangladesh	0.19	0.62	1.05	0.95	408.02	−9.34
Nigeria	0.73	0.73	0.85	0.85	16.44	0.00
Egypt	0.30	0.42	0.51	0.83	179.22	65.01
Others	*4.46*	*6.36*	*7.21*	*7.66*	71.51	6.24
World Total	24.72	31.63	37.14	42.67	72.61	14.87

[1]FAO reports mango, mangosteen, and guava data as one aggregated category.
Source: FAO (2015).

largest producer, accounting for 42.19% (18.00 MMT) of global output. Next are China and Thailand, with shares of 10.43% (4.45 MMT) and 7.36% (3.14 MMT), respectively. India's dominance as a producer is evident from the fact that in the top 10 producers, the rest of the 9 countries combined had a global share of 39.86%. Other important mango producing countries, with their share of world production include Indonesia (4.82%), Mexico (4.44%), Pakistan (3.89%), Brazil (2.73%), Bangladesh (2.23%), Nigeria (1.99%), and Egypt (1.96%).

1.2.2 Global trade – Exports and Imports

Despite the consistent growth in mango exports, the bulk of the production is still consumed in domestic markets. In 2013, only 1.65 MMT, or 3.85% of the global production, was exported, valued at about $1.69 billion (US dollars) (FAO 2016). Exports of mangoes showed a 165% increase over the 2000–2013 period, from 0.62 to 1.65 MMT. The situation was even more striking in value terms, with total exports increasing from US$ 386 million to US$ 1.69 billion over the same period, representing a 26% annual average growth rate. The higher growth rate observed for value of exports reflected a combination of increased demand for the mangoes as well as improvements in the quality of the fruit shipped. The leading mango exporting countries are shown in Table 1.2. The top 10 countries have an 85.24% share of global exports, with the top 3 countries (Mexico, India, and Thailand), accounting for 52% of the trade. Historically, Mexico and India have dominated the mango export market. However, over the 2000–2013 time period, the largest increase in exports was recorded by Thailand, from 8.76 thousand

Table 1.2 World's ten major mango[1] exporting countries for selected years since 2000 ('000' metric tons).

Country	2000	2005	2010	2013	Percent Change 2000–13	Percent Change 2010–13
Mexico	206.78	195.21	275.37	338.17	63.54	22.81
India	39.27	222.62	260.48	263.92	571.99	1.32
Thailand	8.76	1.52	144.57	252.90	2788.68	74.94
Peru	21.07	57.62	96.94	126.82	501.87	30.82
Brazil	67.17	113.88	124.38	122.18	81.89	−1.77
Netherlands	34.48	68.79	107.02	110.89	221.63	3.62
Pakistan	48.45	48.86	85.92	98.93	104.17	15.13
Ecuador	25.50	39.97	39.98	61.31	140.41	53.36
Yemen	3.57	11.64	20.37	35.25	886.87	73.03
Philippines	40.03	32.44	23.74	20.88	−47.84	−12.05
Others	126.73	149.65	170.70	216.95	71.19	27.10
World Total	621.82	942.19	1,349.46	1,648.19	165.06	22.14

[1]FAO reports mango, mangosteen, and guava data as one aggregated category.
Not in top-10: 2000, 2005 (Thailand, Yemen).
Source: FAO (2015).

metric tons (TMT) to 252.90 TMT in 2013, which represented a 29-fold increase. Other major mango exporting countries include Peru, The Netherlands (re-export), Pakistan, Ecuador, Yemen, and the Philippines.

It is noteworthy that three of the top five exporting countries, namely Mexico, Brazil, and Peru, are not in the top five mango producing countries, with Peru not even included in the top ten. This reflects the fact that these countries have made a conscious decision to target the international mango markets and, as mentioned earlier, have implemented programs and made high-cost investments in support of this decision. Overall, the noticeable rise in exports can be attributed to a combination of factors, including a conscious decision in many of the exporting countries to boost exports by improving cultural practices, such as paying more attention to phytosanitary conditions, investing in high technology production and marketing systems, and focusing exports on a few varieties that have the greatest international appeal.

With respect to imports, Table 1.3 lists the top 10 mango importing countries. As can be seen, world imports have grown by 91.23%, from 0.62 MMT in 2000 to 1.19 MMT in 2013; value-wise, this translates from US$ 513 million to US$ 1.65 billion, respectively (FAO 2015). Although over 150 countries import the fruit, the trade is dominated by a handful of countries, with the top three accounting for 52.92% of the volume traded in 2013. The USA is by far the largest importer, absorbing over a third of the trade (35.70%). Next in line is The Netherlands (12.36%), followed by Saudi Arabia (4.87%). The most popular varieties of mango traded are Tommy Atkins, Kent, Keitt, Haden, Francis, Ataulfo, Alphonso, Kesar, Edward, and Manila (NMB 2015). Other varieties which are gaining in importance in the trade include Sindhri, Badami, Glenn, Valencia Pride, and Nam Dok Mai.

Table 1.3 World's ten major mango[1] importing countries for selected years since 2000 ('000' metric tons).

					Percent Change	
Country	**2000**	**2005**	**2010**	**2013**	**2000–13**	**2010–13**
USA	235.08	260.84	320.59	424.45	80.56	32.40
Netherlands	61.86	98.04	142.55	146.99	137.63	3.12
Saudi Arabia	28.33	50.63	58.25	57.86	104.26	−0.67
United Kingdom	22.02	46.92	47.58	56.23	155.40	18.18
Germany	23.32	37.14	48.45	54.61	134.15	12.71
Malaysia	20.32	18.88	42.02	48.68	139.53	15.85
United Arab Emirates	38.90	31.62	46.49	48.00	23.40	3.24
France	26.26	34.94	32.27	35.15	33.85	8.94
Spain	9.19	13.72	32.23	27.79	202.49	−13.77
Belgium	16.12	12.20	16.42	26.43	64.00	61.01
Others	140.42	149.58	231.73	262.91	87.23	13.46
World Total	621.81	754.52	1018.57	1189.10	91.23	16.74

[1]FAO reports mango, mangosteen, and guava data as one aggregated category.
Not in top-10: 2000, 2005 (Belgium, Spain), 2010 (Belgium).
Source: FAO (2015).

1.2.3 US Production

The mango was successfully introduced into Miami, Florida, in 1863; the fruit was the focus of intense research that resulted in the development of many well-known cultivars such as Tommy Atkins, Haden, Keitt, and Kent. These cultivars are the most widely grown in the Western Hemisphere for export to the US market (Campbell and Zill 2009).

Despite its early involvement in mango production and cultivar development, the USA is not ranked among the top producers; because of climatic requirements, mango cultivation is restricted to specific regions of the US states of Florida, Hawaii, Texas, and California. Florida is the main mango producing state; in 2007, the planted area was about 490 hectares (408 bearing and 82 non-bearing hectares), but has since more than doubled to an estimated 1,042 hectares in 2012 (925 bearing and 117 non-bearing hectares) (USDA-NASS 2014).

1.2.4 US Imports and Exports

As noted earlier, the USA is the largest import market for mango, absorbing more than 32% of the global import trade in 2011. As can be seen in Figure 1.2, total mango imports (fresh and processed) increased by 74.34% from 305,918 metric tons (MT) in 2004 to 533,339 MT in 2014. Fresh mango imports accounted for 85% of the total imports, and grew at an annual rate of 5.3%, reaching 424,451 MT in 2013. Mexico is by far the largest supplier of fresh mangoes to the USA, with a market share of 65.58% for 2011–2013,

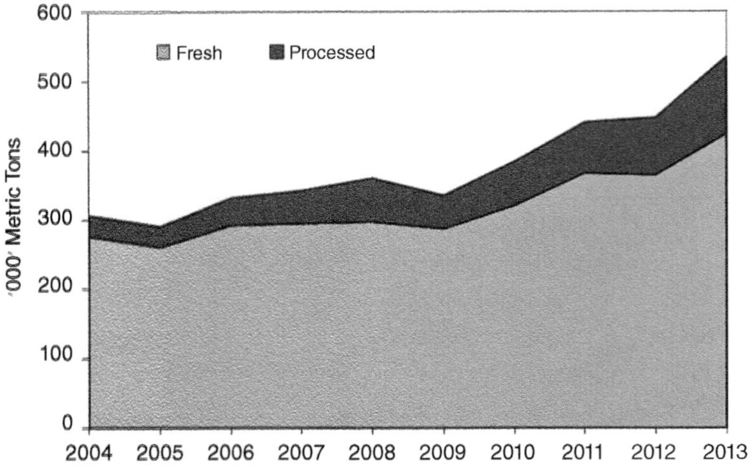

Figure 1.2 US fresh and frozen mango imports over ten year period (2004–2013). *Source:* USDA/FAS (2014).

followed by Ecuador (9.85%), Peru (9.77%), Brazil (6.31%), and Guatemala (4.46%) (Figure 1.3). Fresh mangoes are imported all year round; however, the peak of the importing season occurs during the summer months (June, July, and August). For 2011–2013, approximately 38% of the fresh fruit imports took place during these months (USDA/FAS 2014).

Data for US organic fresh mango imports are available since January 2013; this category represented about 28% of the total fresh mango imports for that year. Mexico and Peru are the main suppliers of organic mangoes to the US market, with a share of the fruit imports at 69% and 11.6%, respectively.

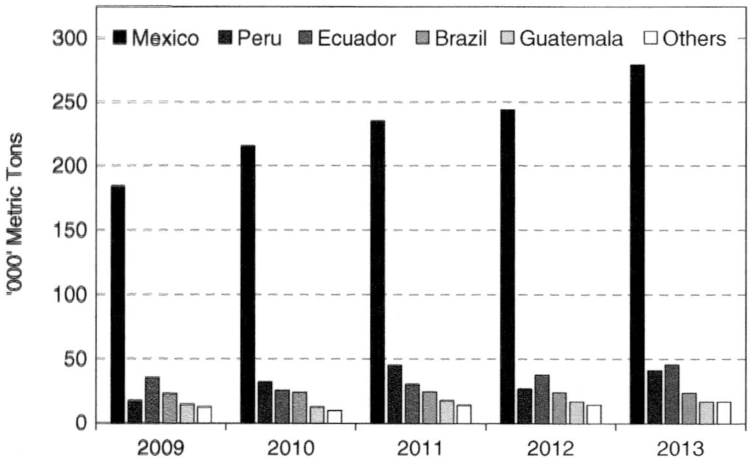

Figure 1.3 US total mango imports by country of origin, 2009–2013. *Source:* USDA-FAS (2014).

In 2013, total fresh mango imports were valued at $380.35 million; conventional fresh mango imports reached $279.65 million, while organic fresh mango imports totaled $100.7 million (USDA/FAS 2014). Imports of processed mangoes rose markedly (268%) over the period of 2004 to 2013, increasing from 29,573 MT to 108,887 MT. Within this category, frozen mango accounted for 52.75% during 2011–2013, followed by preserved mangoes (38.83%), and dried mangoes (7.77%), respectively. Frozen mangoes are marketed in the USA as concentrate, pulp, and chunks, which are used for toppings, baking, fruit salads, or snacks. Mexico and Peru are the main suppliers of frozen mango, with an import share of 70.91% and 26%, respectively (USDA-FAS 2014).

India, Thailand, and Pakistan are the more recent suppliers of fresh mangoes to the USA. In April 2007, after 18 years, the USA lifted its ban on the import of Indian mangoes, including the very popular Alphonso and Kessar varieties (USDA-APHIS 2007a). Indian mango exports to the USA have grown slightly, from 179 MT in 2007 to 275 MT in 2013; in the medium term, Indian mango exports are not expected to increase significantly, as they are not cost competitive. The average Indian mango import value in 2013 was $5.53 per kilogram (kg), which made them very expensive as compared to mango import values from Mexico ($0.87/kg) or Brazil ($1.30/kg) (USDA/FAS 2014). In July 2007, mangoes from Thailand were granted access to the US market when irradiated at low doses (USDA-APHIS 2007b). However, because the fruit is shipped by air, the cost increases significantly; the average import value for mangoes from Thailand in 2013 was $6.32/kg.

In August 2010, the USDA Animal and Plant Health Inspection Service (APHIS) formally allowed the import of Pakistani mangoes, particularly the Chaunsa variety, which rivals the Indian variety, Alphonso. It was ruled that the only authorized port of entry for Pakistani mangoes would be O'Hare International Airport in Chicago, Illinois. Furthermore, the fruit has to be irradiated in a facility at Sioux City, Iowa, before going to the market (USDA-APHIS 2011c). Because Pakistani fresh mangoes may only be imported via commercial air cargo carriers, it has served to increase the landed price and reduce its competitiveness. Most recently, approval has been given allowing imports of Australian mangoes, negotiated under the US–Australia Free Trade Agreement. Such US imports should not have a significant market impact, as Australian growers produced 45,000 metric tons (MT) of mangoes in 2012-13, with exports valued at about $13 million. Likewise, approval has been given for imports of mango from Jamaica. The annual quantity that Jamaica expects to export to the USA is about 261 MT, which represents less than 0.08% of US mango imports. As a consequence, US imports from Jamaica will not have any noticeable effect on the market (USDA-APHIS 2014).

Mangoes from Mexico, Central America, and South America are shipped to the USA at different times of the year, which means that fresh mangoes are available year round. For example, Mexican mangoes are shipped from late February until September; Peruvian mangoes are shipped from mid-November until April; Ecuadorian mangoes are shipped from late September until December; and Brazilian mangoes are shipped from late September until December. Compared with Mexican, Central American, and other South American mango exporters, Brazil has the costliest ocean freight to the USA, so it concentrates more on the European Union market.

Table 1.4 shows the annual average price of mango imports to the USA from selected countries during 2009–2013. The average annual import value has fluctuated from a low of $0.79/kg in 2010 to a high of $0.99/kg in 2013. Of the five countries shown in

Table 1.4 Average annual mango import values from the top five exporters to the US market, 2009–2013 (US $/kg).

Origin	2009	2010	2011	2012	2013
Mexico	0.68	0.75	0.71	0.79	0.87
Peru	1.27	0.91	0.99	1.23	0.98
Brazil	0.95	0.96	0.96	0.92	1.3
Ecuador	0.63	0.7	0.75	0.97	0.95
Guatemala	0.61	0.66	0.83	0.69	0.85
Average	*0.82*	*0.79*	*0.84*	*0.92*	*0.99*

Source: USDA-FAS (2014).

Table 1.4, Guatemala is the lowest-cost supplier of mangoes, while Peru appears to be the costliest supplier to the US market. Mexican mango import values have increased gradually during 2009 to 2013. The largest increase in per kg import value occurred in 2013, when it was $0.87/kg more than in 2012.

1.2.5 US Consumption

Several mango varieties are marketed in the USA, with the most popular being Tommy Atkins, Kent, Keitt, Haden, Ataulfo, Francis, and Francine. Per-capita consumption of mango has increased steadily from 0.92 kg in 2004 to 1.31 kg in 2013 (Figure 1.4), which is equivalent to a modest annual growth rate of 4.24% (USDA/ERS 2014). Consumption of mangoes in the USA is still low compared to other tropical and temperate fruits, such as pineapple (3.06 kg), bananas (12.77 kg), and apples (7.87 kg) (USDA/ERS 2014).

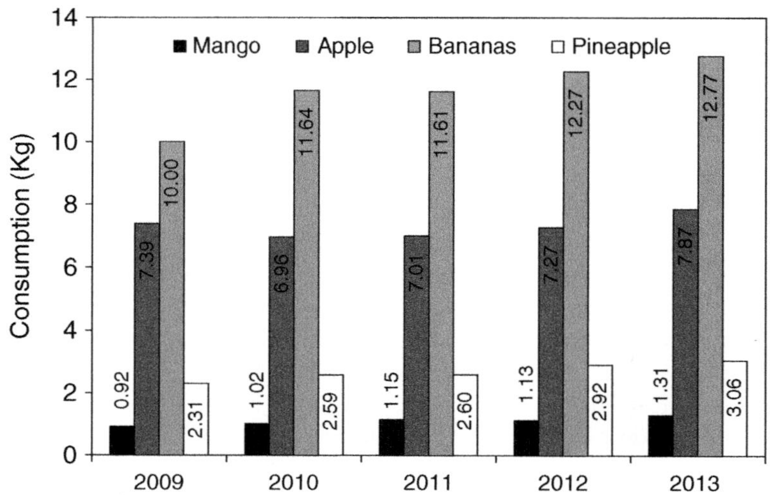

Figure 1.4 US per-capita consumption of selected fruits, 2009–2013. *Source:* USDA-ERS (2014).

Advertising and promotion programs established by agricultural commodity groups have played an important role in the increase in demand for these products. The National Mango Board (NMB), which was established in 2005 as a national promotion and research organization, with the mission to increase awareness and consumption of fresh mangoes in the USA, has been credited with much of the success with respect to rising per-capita consumption of the fruit. Price and product appearance are important attributes for consumers to decide on when purchasing fruit. The NMB conducted a study in 2011 to better understand the factors behind the decision to purchase mangoes. Findings indicated that US consumers ranked price (17.1%), ripeness (15.9%), and appearance (10.6%) as the three top reasons for purchasing mangoes (Ward 2011). This price sensitivity is not surprising given that the fruit competes with several other fruits and is not regarded as a staple. Ethnicity was the most important demographic factor influencing mango purchases; Asian consumers were the most likely ethnic group to buy the fruit, followed by African Americans, Hispanics, and whites, respectively. Finally, it was found that about 6% of the surveyed shoppers from all income levels were more likely to purchase mangoes; indicating that the effects of income on consumption of the fruit were minimal (Ward 2011).

1.2.6 European Union Market

European Union (EU) mango imports rose from 164,077 MT in 2004 to 260,845 MT in 2013, an increase of almost 60%. The more recent information by country within the EU shows that the top three mango import markets in 2011 were the United Kingdom and Germany (each with about 50,000 MT), and France (32,000 MT) (CBI 2014).

The most popular varieties sold are Tommy Atkins, Kent, Haden, and Keitt. Of these varieties, Haden commands the highest price, followed by Kent and Keitt traded at similar prices and then Tommy Atkins. Among the trade there is a definite preference for Tommy Atkins due to its ability to withstand long transportation and postharvest treatments, and its longer shelf life. At the same time, the organoleptic properties are considered by many buyers to be less desirable when compared to other varieties. The latter factor may explain the reason why some of the other varieties are gaining market shares at the expense of the Tommy Atkins (Pinheiro and Lopez 2012).

1.2.7 Market Outlook

Over the last decade, global production and trade of mangoes have increased substantially, with the growth in value exceeding that of quantity, suggesting that international demand outstripped that of supplies and/or increases in prices due to the general improvement in quality of fruit shipped. Despite the noticeable rise in volume of mangoes traded internationally and increased supplies in importing countries, prices in international markets have remained fairly competitive. As a response to these market opportunities, many governments of developing countries have taken steps to actively promote the production and trade of the fruit to increase their export earnings and improve socio-economic conditions in specific rural areas. Countries such as Mexico and Peru, for example, have significantly increased their export-oriented mango production based on a combination of government and private sector involvement. Because adjustments in output and changes in production areas take years to respond to market signals, exporting countries might consider how they can join with agencies/organization in the importing countries to continue to grow the demand for mangoes and stave off downward price pressure.

Two recent developments are worth noting. The first is the rapid expansion in production and trade of organic versus conventionally produced fresh mangoes. As pointed out earlier, US imports of organic mangoes accounted for 28% of the total mango imports in 2013. This could imply that conventional fresh mango exports may have limited expansion opportunities in international markets due to health and environmental considerations becoming more important to consumers. The second development is the observable rise in the trade of processed mangoes. Unlike organic mangoes, processed mangoes represent a better market alternative for countries wishing to diversify their mango industry and add value to their exports. Processed mangoes, especially frozen mango exports, also have good prospects, since the product is ready to eat, and can be used in different culinary applications.

Although prospects for mangoes appear encouraging, it is worthwhile to sound a word of caution. Like most fruits, the demand for mangoes is elastic, meaning that the quantity demanded is responsive to a change in price. That is, an increase in price, in the absence of significant growth in the demand for the product, is likely to result in a decrease in the quantity consumed.

1.3 Postharvest Handling and Packaging

The ripening behavior and quality of mangoes are greatly influenced by the stage of harvest. The characteristic taste and flavor of the mango varieties cannot be attained unless the fruit is harvested at the appropriate stage of maturity. However, demand and marketing pressure often undermines this criterion and farmers go for pre-mature harvest followed by forced ripening using chemicals. The storage potential, marketable life, and quality of mango fruits depend on the stage of maturity at which it is harvested. Poor quality and uneven ripening are caused by early harvesting while late harvesting results in extremely poor shelf life (Narayana *et al.* 2012).

Fruit maturity is typically correlated with various physical characteristic like skin color, shape, size, and specific gravity. A number of chemical parameters are also used to assess maturity, which include total soluble solids (TSS), acidity, starch, phenolic compounds, and carotenoids. It is to be noted that the TSS, sugars, carotenoid pigments, and pH are directly proportional to fruit's specific gravity, both at harvest and during ripening (Narayana *et al.* 2012). Grading mangoes after harvest is important from a marketing point of view. Harvesting fruit at stages beyond mature green will reduce their shelf stability and shorten their fresh market life. The rate of respiration and ripening, development of pigments, flavor compounds, phenolic compounds, sugars, fruit quality, and postharvest diseases are all affected by handling procedures following harvest (Baldwin *et al.* 1999; Narayana *et al.* 2012; Roy and Joshi 1989).

Mango is a climacteric fruit that is harvested at a physiologically mature green stage and allowed to ripen for fresh market. The ripening of mango fruit involves many chemical and physiological changes as the climacteric peak of respiration is reached. Unripe fruits are characterized by their hard texture, high starch content, low carotenoids, high organic acid concentrations, and subsequent low pH (Narayana *et al.* 2012). The postharvest losses in mangoes have been estimated to be in the range of 25–40% from harvesting until they reach consumers. The climacteric nature of mango makes it ripen quickly after harvest. Disease susceptibility, sensitivity to low storage temperatures, and perishability due to ripening and softening, are serious causes of postharvest losses in mango, limiting its handling, storage, and transport potential. Good handling practices during harvesting and postharvest stages can minimize mechanical damage and reduce subsequent wastage due to microbial attack (Narayana *et al.* 2012). Storage under ambient or higher refrigerated temperature leads to substantial postharvest losses, mainly due to moisture loss and/or microbial activity.

The postharvest shelf life of mangoes can be increased by using different storage and packaging techniques, such as:

- evaporative cool storage
- low temperature storage/cold storage
- modified atmosphere packaging (MAP)
- controlled atmosphere (CA) storage
- low pressure or hypobaric storage.

There are several ways of packaging mangoes using different types of packaging material. Depending on the cultivar to be packed, ventilated lugs, fiberboard boxes, and corrugated cartons of different dimensions have been used in India, Trinidad, Philippines, Florida, and Jamaica. Roy and Joshi (1989) reported that corrugated fiber board boxes with a ventilated partition were the best for transportation of "Alphonso" mangoes, because of less physiological loss in weight and low levels of spoilage and ripening.

Kader (2003) recommended maintaining cold chain through all marketing channels to maintain perishable commodities' quality and minimize postharvest losses:

- *Harvest*: Protect the product from the hot sun and transport quickly to the packinghouse;
- *Cooling*: Minimize delays before cooling and use efficient cooling techniques;
- *Temporary storage*: Store the product at its optimum temperature and practice "first-in-first-out" rotation, ship to market as soon as possible;
- *Transport to market*: Use refrigerated loading area, cool truck before loading, put insulating plastic strips inside door or reefer if the truck makes multiple stops, and avoid delays during transport;
- *Handling at retail market*: Use a refrigerated unloading area, monitor product temperature carefully, move product quickly to proper storage area, and display at proper temperature range;
- *Handling at destination*: Store in refrigerator at home or food service establishment.

Shelf life of mangoes can be extended significantly by implementing this cold chain concept. Continued technological developments in the future, to provide CA during transport and storage at reasonable cost (positive benefit/cost ratio), are essential to expanding their application on fresh tropical and subtropical fruits (Kader and Siddiq 2012).

1.4 Processed Products

The commonly processed mango products are puree/pulp, nectar, juice, juice concentrate, and dried/dehydrated mangoes (Figure 1.5). Besides these common products, there are a number of traditional products which are processed commercially in major mango producing countries, which include pickles, sweet or sour chutney (a tomato ketchup-type product), *amchoor* or dried powder, mango leather, and a variety of soft drinks and beverages (Siddiq *et al.* 2012). Ready-to-eat, fresh-cut products have been one of the fastest growing segments of the food industry in the USA in recent years. Mangoes, due to their unique flavor, are increasingly processed into fresh-cut or "minimally processed" products. Generally, fresh-cut mango products are consumed in the developed countries due to their higher prices. The other processed products are more popular in the countries where mangoes are a major fruit or in ethnic populations in North American and European countries.

Besides traditional processing technologies (canning, drying, juicing), research on the use of innovative technologies has been reported in the recent years. These technologies include high-pressure processing (HPP), pulsed electric field (PEF) processing, Ohmic heating, microwave heating, radio frequency heating, ultraviolet (UV) light, ionizing radiation, pulse light technology, ultrasound, and ozone treatments (Ahmed and Ozadali 2012). The quality preservation with respect to nutritional and sensory properties is higher when some of the non-thermal processing techniques are used. It is to be noted that most of the innovative technologies work reported in the literature is at research and development stages, with some technologies offering better commercialization potential than the others.

1.5 By-products from Processing Waste

Mango processing operations generate a substantial amount of waste while transforming raw fruits into finished products. For example, canning of mangoes is a major waste-generating operation, as shown in Table 1.5. Commercial processing of mango into juice, nectar, pulp, puree, fruit leather, and jam produces large quantities of waste (peels

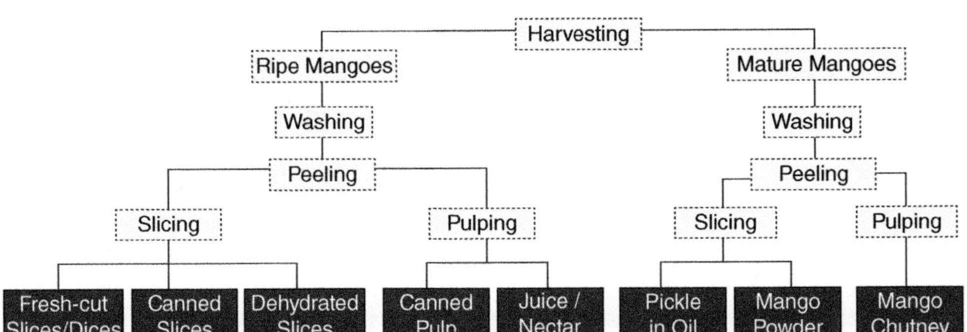

Figure 1.5 Commercial processing of different products from mangoes.

Table 1.5 Waste generation during typical mango canning operations.

Unit operations	Waste type
Washing	Wastewater, suspended solids, debris
Peeling	Solid pieces, wastewater
Pitting	Pits, Kernels
Cutting (slices, dices)	Solid pieces
Blanching	Wastewater, steam condensate
Filling and syruping	Syrup spillover
Thermal processing	Wastewater, steam condensate

and kernels). Fresh-cut processing, drying, and processing into other products also produces similar waste streams, with the exception of syrup in canning. In addition, a large number of non-marketable fruits are typically discarded, creating massive quantities of biowaste. Larrauri *et al.* (1996) reported that, depending on fruit and stone size, commercial processing results in 35–60% waste consisting of peel, kernel, and culled fruit. This waste contains significant amounts of nutrients and phytochemicals, which makes it suitable to be processed for value-added applications in functional foods and nutraceuticals.

Mango peel is rich in pectin, cellulose, hemicellulose, lipids, protein, polyphenols, and carotenoids, with excellent antioxidant and functional properties (Ajila *et al.* 2007). Mango peel flour has enormous potential as a functional ingredient in developing healthy food products such as noodles, bread, sponge cakes, biscuits, and other bakery products, besides using it in baby foods (Aziz *et al.* 2012). Mango contains various classes of polyphenols, carotenoids, and vitamins with different health-promoting properties, mainly antioxidant activity. Mango kernel is a rich source of gallic, ellagic, ferulic, and cinnamic acids, tannins, vanillin, coumarin, and mangiferrin, all having the potential to act as a source of natural antioxidants (Soong and Barlow 2006). Microwave assisted processes have been shown to produce extracts from mango seed with high antioxidant capacity (Dorta *et al.* 2012). Mango peel fibers with high hydration capacities have potential in dietary fiber-rich foods preparation (Koubala *et al.* 2013). Dried mango peel and kernel products can improve the nutritional, functional, and sensory properties, and oxidative stability of oil/oil-rich product (Abdalla *et al.* 2007); however, selection of a suitable drying method is important to ensure minimal quality losses.

Overall, mango processing wastes can be managed by adopting the following strategies:

- *Recovery:* waste utilization to produce by-products/co-products (e.g., non-fermented and fermented products);
- *Recycling:* the solid waste from processing contains nutrients that can be used for animal feed, as a source of energy in the form of biofuels, and for composting;
- *Disposal:* any leftover waste will need to be disposed of, which can be done without damaging the environment by incineration, land filling, etc.

1.6 Food Safety Considerations

In the last two decades, food safety has become the top-most concern of the fresh fruits industry, as well as different regulatory agencies (Kader and Siddiq 2012). The US Food and Drug Administration (FDA) published the *Guide to Minimize Microbial food Safety Hazards for Fresh Fruits and Vegetables* (Fan *et al.* 2009; Sapers 2005). This guide covers general principles, which prevent microbial contamination of fresh produce and recommends that efforts should be made to avoid food safety issues rather than focus on corrective actions alone. Paull and Lobo (2012) reported that the FDA has promulgated GMP regulations that apply to all food processing facilities, including fresh-cut operations and complements FDA's Current Good Manufacturing Practice regulations in *21 CFR 110*. Marketers and processors can benefit by following these guidelines, since mango fruit and fresh-cut mangoes are gaining popularity among consumers, especially in Europe and North America.

Sivakumar *et al.* (2011) suggested that food safety and quality maintenance of mangoes across supply chain channels depends on many factors, such as adequate orchard management practices, harvesting practices, packing operation, postharvest treatments, temperature management, transportation and storage conditions, and ripening at destination. Personnel working in the various fruit handling operations must be properly trained in good agricultural practices (GAPs). Moreover, development of the Hazard Analysis and Critical Control Points (HACCP) program is recommended with respect to mango handling and processing. Kader (2003) recommended maintaining cold chain throughout the food value chain.

1.7 Nutritional Profile

Mango, a flavorful and nutrient-rich fruit, is an excellent source of vitamins C and A, both important in human health. According to the NMB (2015), a one-cup serving of edible portion of mango has only 100 calories, and provides daily recommended allowance of 100% for vitamin C, 35% vitamin A, and 12% of daily fiber. Vitamin C, possessing antioxidant properties, promotes healthy immune function and collagen formation. Mango fruit has very low sodium (2 mg/100 g) and high potassium (156 mg/100 g) contents. Potassium, an important component of cell and body fluids, is helpful in controlling heart rate and blood pressure (Rudrappa 2015). Mango fruit is an excellent source of flavonoids, beta-carotene, and beta-cryptoxanthin. Consumption of natural fruits rich in carotenes is known to protect the body from lung and oral cavity cancers (Rudrappa 2015). Sivakumar *et al.* (2011) recommended including mango fruit and its processed products in the daily diet due to its health benefits, such as reduced risk of cardiac disease, anti-cancer, and anti-viral properties.

References

Abdalla, A.E.M., Darwish, S.M., Ayad, E.H.E. and El-Hamahmy, R.M. (2007) Egyptian mango by-product 2: Antioxidant and antimicrobial activities of extract and oil from mango seed kernel. *Food Chemistry*, **103**, 1141–1152.

Ahmed, J. and Ozadali, F. (2012) Novel processing technologies for fruits. In: *Tropical and Subtropical Fruits: Postharvest Physiology, Processing and Packaging* (ed, M. Siddiq), pp. 71–96. John Wiley & Sons, Ames, IA.

Ajila, C.M., Naidu, K.A., Bhat, S.G. and Rao, U.P. (2007) Bioactive compounds and antioxidant potential of mango peel extract. *Food Chemistry*, **105**, 982–988.

Aziz, N.A.A., Wong, L.M., Bhat, R. and Cheng, L.H. (2012) Evaluation of processed green and ripe mango peel and pulp flours (*Mangifera indica* var. Chokanan) in terms of chemical composition, antioxidant compounds and functional properties. *Journal of the Science of Food and Agriculture*, **92**, 557–563.

Baldwin, E.A., Burns, J.K., Kazokas, W., Brecht, J.K., Hagenmaier, R.D. *et al.* (1999) Effect of two edible coatings with different permeability characteristics on mango (*Mangifera indica* L.) ripening during storage. *Postharvest Biology and Technology*, **17**, 215–226.

Campbell, R.J. and Zill, G. (2009) Mango selection and breeding for alternative markets and uses. *Acta Horticulturae* 820:189–196.

CBI (2014) Product Factsheet: Fresh Mangoes in the European Market [Online]. Available: http://www.cbi.eu/market-information/fresh-fruit-vegetables/mangoes/europe [21 November 2014].

Dorta, E., Lobo, M.G. and Gonzalez, M. (2012) Reutilization of mango by-products: study of the effect of extraction solvent and temperature on their antioxidant properties. *Journal of Food Science*, **71**, 80–88.

Fan, X., Niemira, B.A., Doona, C.J., Feeherry, F.E. and Gravani RB (eds) (2009) *Microbial Safety of Fresh Produce*, pp. xv–xvi. Wiley-Blackwell, Ames, IA.

FAO (2015) Crops Production and Trade Statistics [Online]. Available: http://faostat3.fao.org/download/Q/QC/E [29 December 2015].

Kader, A.A. (2003) A perspective on postharvest horticulture (1978–2003). *HortScience*, **38**, 1004–1008.

Kader, A.A. and Siddiq, M. (2012) Introduction and overview. In: *Tropical and Subtropical Fruits: Postharvest Physiology, Processing and Packaging* (ed, M. Siddiq), pp. 3–16. John Wiley & Sons, Ames, IA.

Koubala, B.B., Kansci, G., Garnier, C., Thibault, J.F. and Ralet, M.C. (2013) Physicochemical properties of dietary fibers prepared from ambarella (*Spondias cytherea*) and mango (*Mangifera indica*) peels. *Food Bioprocess Technology*, **6**, 591–597.

Larrauri, J.A., Ruperez, P., Borroto, B. and Saura-Calixto, F. (1996) Mango peels as a new tropical fiber: Preparation and characterization. *LWT – Food Science and Technology*, **29**, 729–733.

Narayana, C.K., Rao, D.V.S. and Roy, S.K. (2012) Mango production, postharvest physiology and storage. In: *Tropical and Subtropical Fruits: Postharvest Physiology, Processing and Packaging* (ed, M. Siddiq), pp. 259–276. John Wiley & Sons, Ames, IA.

NMB [National Mango Board] (2015) Mango Nutrition [Online]. Available: http://www.mango.org/en/About-Mangos/Mango-Nutrition [14 August 2015].

Paull, R.E. and Lobo, M.G. (2012) Pineapple. In: *Tropical and Subtropical Fruits: Postharvest Physiology, Processing and Packaging* (ed, M. Siddiq), pp. 333–357. John Wiley & Sons, Ames, IA.

Pinheiro, J.L. and Lopez, J.L. (2012) A study of the Mango Market in the European Union [Online]. Available: http://www.bnb.gov.br/projwebren/exec/artigoRenPDF.aspx?cd_artigo_ren=1314 [12 December 2014].

Roy, S.K. and Joshi, G.D. (1989) An approach to integrated postharvest handling in mango. *Acta Horticulturae*, **231**, 649–661.

Rudrappa, U. (2015) Mango Fruit Nutrition Facts [Online]. Available: http://www.nutrition-and-you.com/mango-fruit.html [17 August 2015].

Sapers, G.M. (2005) Washing and sanitizing treatments for fruits and vegetables. In: *Microbiology of Fruits and Vegetables* (eds G.M. Sapers, J.R. Gorny and A.E. Yousef), pp. 375–400. CRC Press, Boca Raton, FL.

Siddiq, M., Akhtar, S. and Siddiq, R. (2012) Mango Processing, Products and Nutrition. In: *Tropical and Subtropical Fruits: Postharvest Physiology, Processing and Packaging* (ed, M. Siddiq), pp. 277–297. John Wiley & Sons, Ames, IA.

Sivakumar, D., Jiang, Y. and Yahia, E.M. (2011) Maintaining mango (*Mangifera indica* L.) fruit quality during the export chain. *Food Research International*, **44**: 1254–1263.

Soong, Y.Y., and Barlow, P.J. (2006) Quantification of gallic acid and ellagic acid from longan (*Dimocarpus longan* Lour.) seed and mango (*Mangifera indica* L.) kernel and their effects on antioxidant activity. *Food Chemistry*, **97**, 524–530.

USDA-APHIS [Animal and Plant Health Inspection Service] (2007a) Newsroom [Online]. Available: http://www.aphis.usda.gov/newsroom/hot_issues/indian_mango/indian_mango.shtml [15 August 2011].

USDA-APHIS [Animal and Plant Health Inspection Service] (2007b) Newsroom [Online]. Available: http://www.aphis.usda.gov/newsroom/hot_issues/thai_irradiated_fruit/index.shtml [15 August 2011].

USDA-APHIS [Animal and Plant Health Inspection Service] (2011) Newsroom [Online]. Available: http://www.aphis.usda.gov/import_export/plants/plant_imports/irradiation/PakistanMango/ [15 August 2011].

USDA-APHIS [Animal and Plant Health Inspection Service] (2014) Mangoes from Jamaica into the Continental United States [Online]. Available: https://www.regulations.gov/document?D=APHIS-2013-0018-0009 [26 July 2016].

USDA-ERS [Economic Research Service] (2014) Data Sets: Fruit and Tree Nuts Yearbook Data tables [Online]. Available: http://usda.mannlib.cornell.edu/MannUsda/viewDocumentInfo.do?documentID=1377 [26 July 2014].

USDA-FAS [Foreign Agricultural Service] (2014) Global Agricultural Trade System [Online]. Available: http://www.fas.usda.gov/gats/default.aspx [29 March 2015].

USDA-NASS [National Agricultural Statistics Service] (2014) Census of Agriculture. State Data Florida [Online]. Available: http://www.agcensus.usda.gov/Publications/2012/Full_Report/Volume_1,_Chapter_1_State_Level/Florida/st12_1_039_040.pdf [21 November 2014].

Ward, R.W. (2011) Evaluating the National Mango Board's Program Impact on US Demand for Mangoes [Online]. Available: http://www.mango.org/media/97579/mango_demand_report.pdf [15 June 2011].

2

Mango Production

Chantalak Tiyayon[1] *and Robert E. Paull*[2]

[1] *Department of Plant and Soil Sciences, Faculty of Agriculture, Chiang Mai University, Chiang Mai, Thailand*
[2] *Tropical Plant and Soil Sciences, College of Tropical Agriculture and Human Resources, University of Hawaii at Manoa, Honolulu, Hawaii, USA*

2.1 Introduction

Mango (*Mangifera indica* L.) is one of the most widely known and greatly appreciated tropical fruit. The leading producing countries are India, China, Thailand, Indonesia, and Pakistan. In 2012, the top five exporting countries were Mexico, India, Thailand, Brazil and Pakistan, which exported 66% of the 1.5 metric tons world exports (FAO 2015). Diverse production seasons (Figure 2.1), make fruit available year round in the world markets, with consumption patterns varying with culture. Aside from the fruit, young mango leaves are also used as vegetables, and the wood is used for carving.

Mango originated in the Indo-Burma region and has been cultivated in India for at least 4,000 years (De Candolle 1884 in Mukherjee and Litz 2009). The species has been spread to all tropical and subtropical countries, with cultivars selected based on local needs. The successful commercial production of mangoes requires an understanding of mango physiology and orchard care to achieve commercial yield and fruit quality. Every activity in the orchard, for example irrigation, pruning, spraying, fruit thinning, fruit bagging, fertilizing, and harvesting all affect yield and fruit quality. Mango production methods vary greatly worldwide, and even within the same country. Good agricultural practices (GAPs) (Poisot *et al.* 2004) have been introduced as a set of standards to ensure fruit quality, and for the well-being of growers, consumers, and the environment. This chapter provides an overview of research results and their application that have greatly contributed to improved production practices.

2.2 Cultivar Diversity and Potential

Over a thousand cultivar names are reported for India alone (Nakasone and Paull 1998), and hundreds more in other mango-producing countries. The number of cultivars may be fewer, as the same cultivar may have a different name in another area of cultivation. Mango fruit vary by shape, size, skin color (Figure 2.2), as well as the flavor and texture of the flesh.

Handbook of Mango Fruit: Production, Postharvest Science, Processing Technology and Nutrition, First Edition. Edited by Muhammad Siddiq, Jeffrey K. Brecht and Jiwan S. Sidhu.

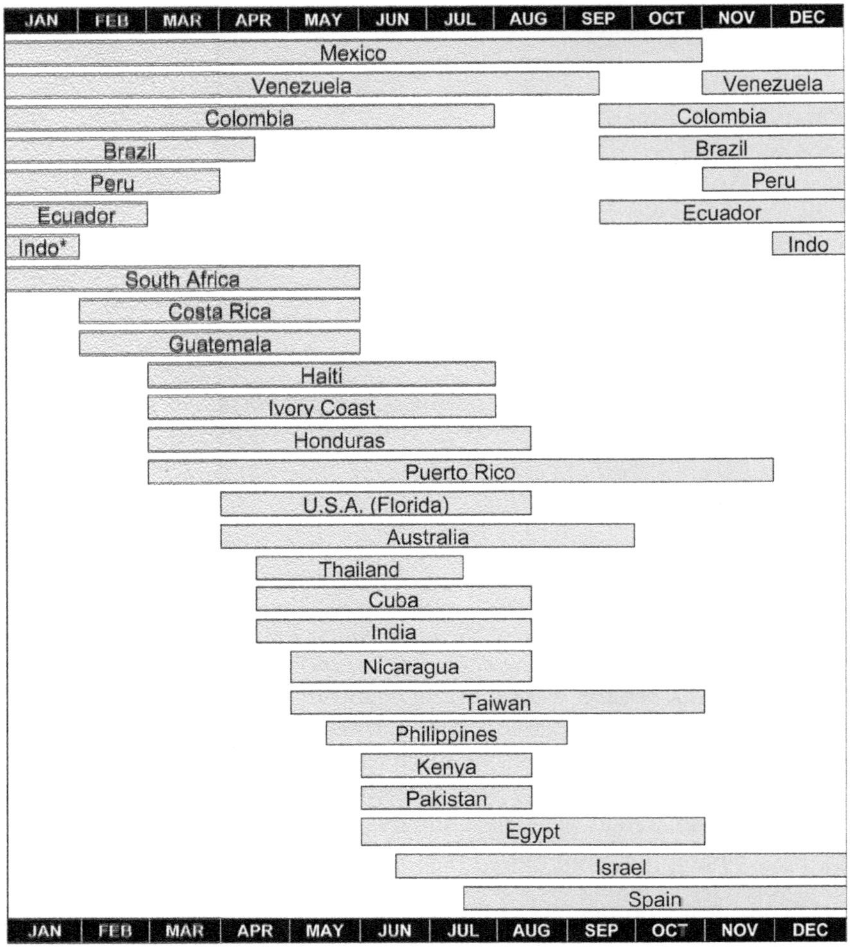

Figure 2.1 Mango production seasons in different countries. (*Indonesia) *Source:* Saúco (2004), PHEDB (2005).

The cultivars are divided into two broad groups by their place of origin: the Indian mango group and Indo-Chinese group, which originated in Southeast Asia (Table 2.1).

Mango fruit are generally consumed when ripe, though some cultivars are best consumed at the mature-green stage and are referred to as "starchy mangoes" or "crispy mangoes" (Yaacob and Subhadrabandhu 1995). Other cultivars are most often processed as chutney, dried fruit, juice, pickled mango, fruit leather, or other products. Some important commercial mango cultivars of the top five mango producing countries in the world are shown in Table 2.2. Other common commercial cultivars found in various countries are listed below:

- *Australia:* Calypso™, Haden, Irwin, Keitt, Kensington Pride, Kent, Palmer, R2E2
- *Bangladesh:* Ashwina, Kalia, Deori, Fazli, Langra, Surjapuri
- *Brazil:* Bourbon, Espada, Palmer, Rosa, Tommy Atkins

Figure 2.2 Morphological diversity of some mango cultivars (scale bar = 20 cm, ~8 inches).

- *Egypt:* Alphonso, Bullocks Heart, Langra, Mabrouka, Pirie, Taimour, Zebda
- *Kenya:* Boribo, Ngowe, Apple, Batawi
- *Malaysia:* Apple Mango, Arumanis, Malgoa, Tok Boon, Arumanis, Kuala Selangor
- *Mexico:* Ataulfo, Keitt, Kent, Manila, Palmer, Tommy Atkins
- *Nigeria:* Kerosene, Sherri
- *The Philippines:* Carabao, Pico

Table 2.1 Characteristics of mangoes in the Indian group and the Indo-Chinese group.

Characteristic	Indian group	Indo-Chinese group
Number of leaf veins	18–24	26–30
Fruit skin	Rough	Smooth
Fruit shape	Roundish	Long and flat
Young fruit color	Dark green	Green
Ripe fruit color	Red or yellow red	Golden yellow or greenish yellow
Fiber	Present or absent	Absent
Flavor and aroma	Sweet and strong aroma	Sweet and mild aroma
Embryo	Monoembryony	Polyembryony

Source: Singh (1960), Knight *et al.* (2009).

- *South Africa*: Keitt, Kent, Heidi, Sensation, Tommy Atkins, Zill
- *USA:* Haden, Irwin, Keitt, Kent, Tommy Atkins, Sensation

(Source: Crane *et al.* 2009, Kusumo *et al.* 1984, Mukherjee and Litz 2009, Narayana *et al.* 2012, Phahlane *et al.* 2013, Queensland Government 2012, Yaacob and Subhadrabandhu 1995, Ynus 1984).

Some cultivars (e.g., Haden, Irwin, Keitt, Kent, and Tommy Atkins from Florida) have been successfully moved to other areas with similar climates. However, some cultivars do not yield well with good fruit quality in every location. Developing local cultivars is

Table 2.2 Some important mango cultivars of leading mango producing countries and their fruit characteristics.

Country and cultivar	Embryo type[1]	Skin color[2]	Fiber[3] N	VS	S	M	Fruit Wt. (g)	Notes
India								
Alphonso	M	Y		✓			225–325	thin skin, characteristic aroma, irregular bearing
Chausa	M	G,Y		✓			245–350	delightful aroma, light bearer, tall and spreading tree
Bombay Green	M	G		✓			250	very good quality, medium bearer
China								
Tainung No.1	M	Y			✓		221	resistant to anthracnose, strong flavor, long storage
Jinhwang	M	YG, O	✓				1,200	aromatic, thin seed, very sweet
Thailand								
Namdokmai Sithong	P	Y	✓				350	very sweet with pleasant aroma, excellent quality
Khiaosawoey	P	G	✓				335–400	mature green consumption
Chok Anan	P	Y			✓		209–270	mild turpentine, easily set fruit, good for processing

Table 2.2 (Continued)

Country and cultivar	Embryo type[1]	Skin color[2]	Fiber[3] N	VS	S	M	Fruit Wt. (g)	Notes
Indonesia								
Arumanis	P	GY		✓			200–350	strong aroma, poor to fair quality, survive humid climate
Gedong	P	RY		✓			180–350	high sugar content, attractive fruit
Golek	P	GY				✓	200–365	mild aroma, nice flavor, poor to fair quality
Pakistan								
Anwar Ratol	M	Y	✓				108–130	high sugar content, nice flavor
Sindhri	M	Y	✓				340–406	strong aroma

Source: Kusumo *et al.* (1984), Cull and Lindsay (1995), Narain *et al.* (1998), Thailand DOA (2001), Hamdard (2004), Biosecurity Australia (2005), Crane *et al.* (2009), Hadi (2009), Knight *et al.* (2009), Gao *et al.* (2011), Ara *et al.* (2014).

[1]M, monoembryonic group; P, polyembryonic group.

[2]G, green; O, orange; R, red; Y, yellow; GY, greenish yellow; RY, reddish yellow; YG, yellowish green.

[3]N, nil; VS, very slight; S, slight; M, moderate.

a way to establish suitable mango cultivars for a specific location. This being the case, it is usually necessary to have a specific mango breeding program for a particular set of local conditions.

2.3 Breeding and Propagation

2.3.1 Breeding

Mango breeding is challenging due to the long juvenile phase, the high level of heterozygosity, the very low retention rate of crossed fruit, the fact that one fruit has only one seed, occurrence of polyembryony, and the need for a large area of orchard space to evaluate hybrids (Iyer and Schnell 2009). Even with these limitations, breeders have made great efforts towards the breeding and selection of new cultivars. One important objective of mango breeding is fruit quality, which includes fruit taste, texture, shape, and size, as well as the color of skin and flesh. All of these aspects of quality can be specific to different cultural preferences. Other desirable characteristics include yield, disease resistance, precocity, dwarfness, regular bearing, postharvest storage quality, absence of physiological disorders, and other particular growth characteristics such as salt and drought tolerances (Lavi *et al.* 2004; Usman *et al.* 2001).

Breeding mangoes is also often accomplished by chance seedlings from known mother plants. Controlled pollination is done at many levels, including hand pollination (Bally *et al.* 2009), with floral induction to synchronize flowering, and using caged top-worked trees (Pinto *et al.* 2004).

2.3.2 Propagation

Mangoes are propagated both sexually and asexually (Figure 2.3). Two types of seeds occur in mango: monoembryonic and polyembryonic. The embryo of monoembryonic

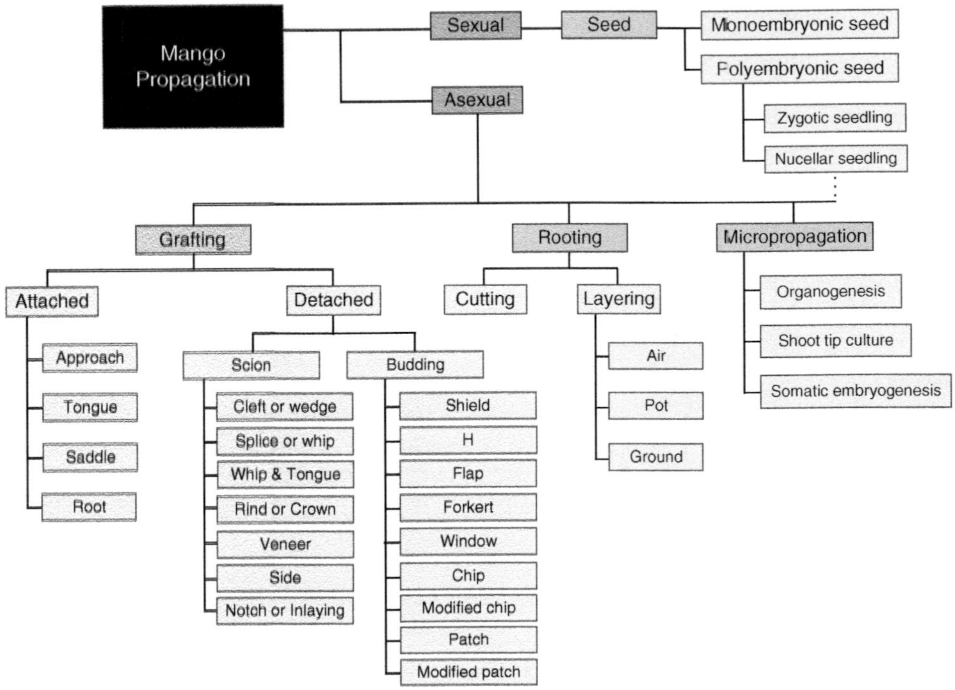

Figure 2.3 Methods of mango propagation. *Source:* Adapted from Singh (1960) and Ram and Litz (2009).

seeds develops from a zygote as a result of double fertilization. The polyembryonic seeds have one embryo that develops from fertilization, and other embryos that develop from the nucellus, the mother plant tissue. A polyembryonic seed can produce up to 12 seedlings (Meurant *et al.* 1999).

Planting mango from seeds is the easiest method of propagation; however, the seedlings of the monoembryonic group are never true to type due to their development from the zygote. Seeds also lose their viability quickly and should be planted within a few weeks of harvest (Nakasone and Paull 1998). Mango plants derived from seedlings require a relatively long juvenile period of 5–10 years before the first flowers appear, compared to grafted plants that usually bear fruit in their third or fourth year (Ram and Litz 2009). This long juvenile period is the major reason that vegetative propagation is widely used.

Vegetative propagation can be divided into three groups: grafting, rooting, and micropropagation (Figure 2.3). Various methods of grafting are used with the scion cultivars attached and detached from the mother plants (Ram and Litz 2009). In the attached methods, the scion is still attached to the mother plant until its tissue successfully fuses with the rootstock. In the detached methods, the scion is removed from the mother plant and attached to the rootstock (Figure 2.4). Rootstocks play a significant part in the success of grafting. A good rootstock should have a strong root system, compatible to the scion, and disease resistance is also an advantage. Scion growth and performance can be significantly affected by the rootstock (Reddy *et al.* 2003). The rootstock

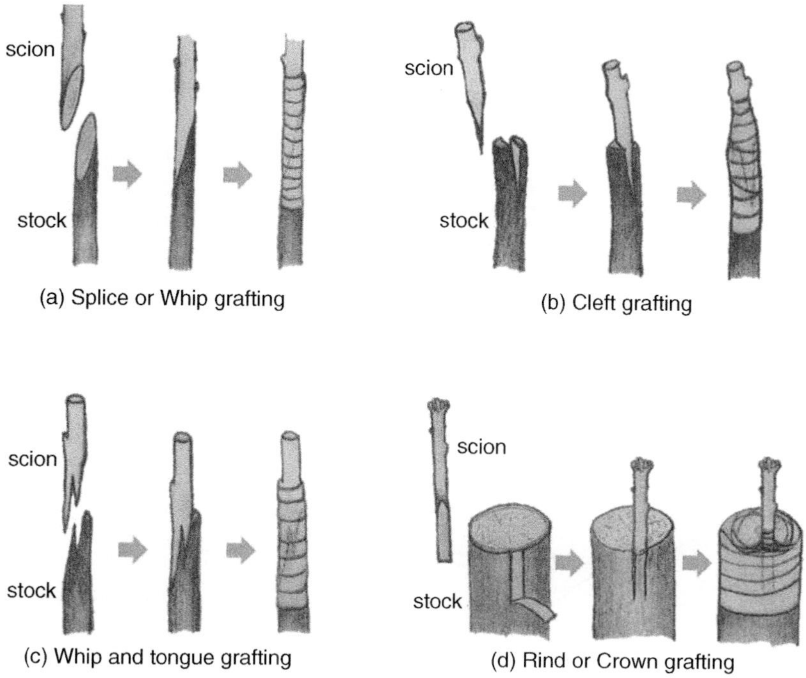

Figure 2.4 Some detached grafting methods often used in mango propagation.

seedlings are ready for use in propagation when they are approximately 1 year old (Singh 1960), about 50 cm tall, and 1 cm in diameter (Meurant *et al.* 1999). However, regularly irrigated and fertilized rootstocks can be used as soon as 6–8 months (Nakasone and Paull 1998). Rooting methods are mostly done by air layering. The details of vegetative propagation methods of mango are explained and illustrated thoroughly by Singh (1960) and Ram and Litz (2009).

Top-working is an application of grafting used to rejuvenate unproductive plants and for changing the cultivar of the plant already established. It is done by cutting the main limbs back to 60–90 cm from the main trunk. New sprouts will grow after about 1–2 months. The new scions are then grafted or budded onto this new growth (Ram and Litz 2009; Singh 1960). Top-working can be used to produce two cultivars on the same plant, for the benefit of natural pollination of cultivars with naturally low fruit set and for breeding purposes.

Micropropagation is a good choice to multiply the scions from a limited number of mother plants. However, there are problems with mango *in vitro* propagation, including exudation of phenol, medium discoloration, explant necrosis, latent systemic contamination, and tissue *in vitro* recalcitrance (Krishma and Singh 2007).

2.4 Field Practices

According to De Bie (2004), environmental factors account for 30% of the yield differences, management factors 49%, and year-to-year effects 21%. Therefore, all work done

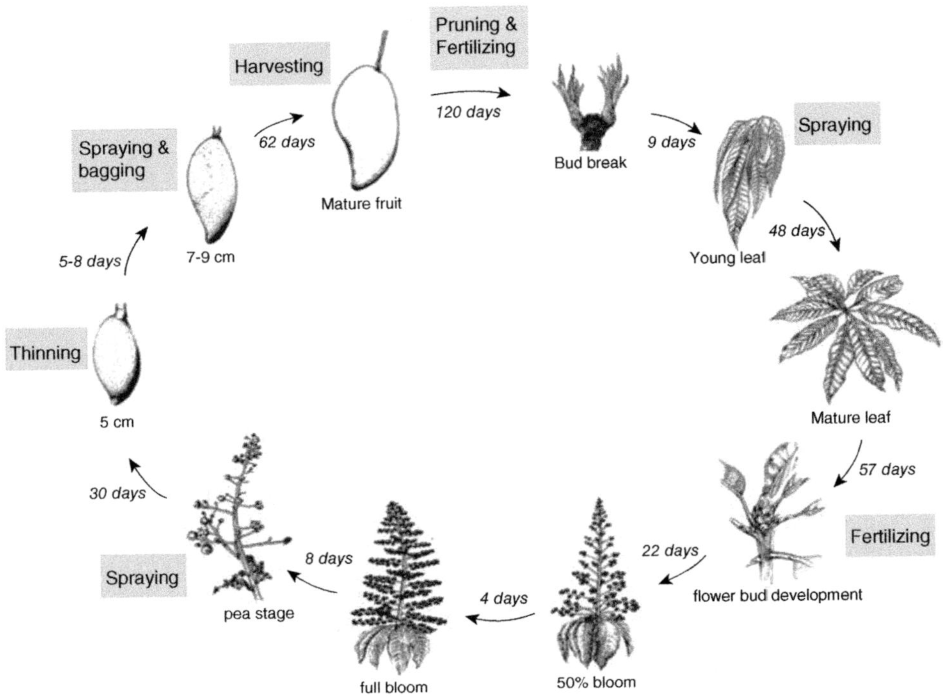

Figure 2.5 Growth cycle of Namdokmai Sithong mango in Chiang Mai Province, Thailand (gray highlights indicate orchard activities). Note that plants are left to rest 3–4 months after harvest and bud break will occur 10–15 days after pruning. *Source:* Adapted from Jaiphayak (2011).

in an orchard has an impact on yield and fruit quality. The cycle of management and stages of growth are shown in Figure 2.5.

2.4.1 Environmental Requirements

2.4.1.1 Elevation

Mangoes are found distributed in tropical Asia up to 27° latitude (Bompard 2009). Commercial cultivation is found also in subtropical areas at latitudes of 35–37° (Mukherjee and Litz 2009). Mango plants can grow and produce fruit from sea level to 1,200 m in the tropics, but most commercial varieties are not productive higher than 600 m above sea level (Bally 2006; Cheema *et al.* 1954).

2.4.1.2 Temperature

Mangoes grow over a wide temperature range of 10–42°C (Cull and Lindsay 1995). The optimum temperature range of mangoes is 24–27°C (Bally 2006). Temperatures below 0°C usually damage young trees and temperatures above 42°C inhibit growth in the reproductive stage and can damage very young plants (Narain *et al.* 1998).

2.4.1.3 Rainfall

An annual rainfall (or irrigation; see Section 2.4.3 below) of 750–2,500 mm is required for successful mango production (Singh 1978, in Narain *et al.* 1998), even though mangoes are drought tolerant. The periodic rainfall during the growing period is more important than the total amount (Nakasone and Paull 1998).

2.4.1.4 Soil

Mangoes prefer a deep, fertile and well-drained soil, although they can grow on a wide range of soil types (Narain *et al.* 1998). The optimum soil pH range is 5.5–7.5 (Bally 2006; Singh 1960).

2.4.1.5 Light

Full sunlight is required for photosynthesis and development of red fruit skin color cultivars (Schaffer *et al.* 2009). Shading can prevent or delay flower-bud formation and a higher percentage of perfect flowers occur on the side of the tree receiving direct sun. Pruning to facilitate light penetration into the canopy can increase fruit yield (Durand 1997).

2.4.2 Planting

The spacing used tends to vary from 3 m × 5 m to 16 m × 16 m (Crane *et al.* 2009). In recent years, canopy control has increased the ease of orchard management coupled to a significant reduction in plant height, often less than 3 m. This has resulted in the planting spacing of newer orchards being less than those previously used and all operations being easily carried out from the ground. The planting space also depends on orchard management. If heavy machinery is used (e.g., pruning machines to trim rows), then spacing between rows tends to be wider to facilitate mechanization.

2.4.3 Irrigation

Mangoes can grow without irrigation; however, irrigation systems greatly improve yield (Bally 2006). One common traditional irrigation method in some areas is using water hoses, which are labor-intensive and water-inefficient. Well established irrigation systems in orchards use mini-sprinklers or drip irrigation (Meurant *et al.* 1999). Schulze *et al.* (2013) recommended micro-sprinkler irrigation in export mango orchards. Micro-sprinklers are shown to be a more efficient use of water, therefore resulting in a higher gross return. In water-deficit areas, partial root-zone drying (PRD) irrigation increases water use efficiency (Spreer *et al.* 2011) and can induce early flowering in the tropics (Bally *et al.* 2000; Lu and Chako 2000).

2.4.4 Fertilization

Fertilization is not crucial for naturally grown mangoes, but it is important for commercial mango production to ensure high yield and good fruit quality. Usually the fertilizer is applied 2–3 times a year, depending on plant age. Normally, fertilizer is applied soon after pruning and again at the start of flowering. In addition, irrigation is needed after fertilizer application for proper uptake (Nakasone and Paull 1998). The amount of fertilizer given depends on soil type and fertility, plant age and cultivar, and the stage of growth. Suggested amounts of fertilizer for mature mango trees planted at 123 plants per ha is 100 (40–200) kg N, 25 (0–40) kg P, and 100 (60–100) kg K ha^{-1} $year^{-1}$ (Weir and Cresswell 1995). Dried manure can be applied yearly to improve organic matter and soil condition (A. Kasemsri, personal communication). The mineral nutrition requirements are well explained in Bally (2009).

2.4.5 Pruning

Pruning is an important practice to keep the plant architecture that allows growers to work within the tree canopy. Initial pruning is started in the nursery before transplanting to start the development of the desired branch scaffolds. After transplant and fruiting, pruning is normally done after harvest by removing unhealthy and/or diseased branches, as well as branches which grow inward or in a downward direction. Pruning depends on the age of the mango trees, height and width of the canopy the growers would like to maintain, and the desired levels of light penetration into the canopy (Figure 2.6).

Mango trees should be pruned with the appropriate tools. Small branches are normally cut with hand pruners. Large branches should be cut with a saw or chainsaw as close to the main stem as possible and cutting direction should be vertical to reduce the risk of disease introduction.

2.4.6 Pollination

Pollination processes are crucial for fruit set. Successful pollination depends on many factors including receptivity of stigma, pollen viability, pollen germination, adequate amount of pollen, and sufficient pollinator activity. The major pollinators of mangoes are insects, especially bees, flies, and to a lesser extent wasps and ants. Wind pollination in mango has been reported (Hartless 1914, in Davenport 2009), but is considered to be a minor contributor to pollination at large.

Inflorescences usually have a blooming period of approximately 2 weeks. Pollination has been reported to be most effective in the morning, with the peak time of flower opening between 9 and 10 a.m. The stigma has been reported to be receptive for 72 hours, but the most receptive period is the day the flowers open (Iyer and Schnell 2009). Pollen is generally compatible within and between varieties (Bally 2006); however, self-incompatibility in mangoes has also been reported (Mukherjee *et al.* 1968).

Figure 2.6 Semi-hard pruning of a 5-year-old grafted *Namdokmai Sithong* mango tree that had received minimal pruning in previous years (left: before, right: after).

2.4.7 Fruit Thinning

Fruit thinning is the removal of some fruit to maximize remaining fruit size and other fruit qualities, such as sugar content. Thinning reduces the risk of over usage of nutrients and helps to prevent branches from breaking. This practice is considered to decrease costs of production by reducing the cull rate after harvest, giving growers a higher harvest pack-out. Thinning also reduces the cost associated with fruit bagging and evens out labor needs between thinning and bagging to harvesting and packing.

Fruit are thinned about 30–35 days after fruit set for the Namdokmai mango (Radanachaless and Tiyayon 2010). The fruit that are malformed, diseased, or small are removed. Growers normally only retain one fruit per peduncle, with at least 8–10 subtending leaves. Sometimes more fruit per peduncle are kept if the adjacent shoots do not set fruit.

2.4.8 Fruit Bagging

Bagging protects mango fruit from fruit flies and other insects, sunburn, and scratching. It also helps to improve color in the yellow skin cultivars. Fruit bagging also improves many aspects of the internal quality of the fruit (Zhao *et al.* 2013).

There are many types of fruit coverings for specific purposes. The bags for yellow skin cultivars are made of two-layered paper, the inner layer a black carbon material to block sunlight and the outer layer a light brown wax paper to repel water. For the red skin cultivars, white paper bags and special one-sided clear plastic bags are used to allow the red color to develop. Using the right type of bag is important for quality of the fruit (Wu *et al.* 2013). Namdokmai mango fruit are bagged when they are 7–10 cm long or approximately 45 days after fruit set. Bagging when the fruit are too small can cause increased fruit drop (Radanachaless and Tiyayon 2010).

2.4.9 Harvesting

When the fruit are mature, they are picked by clipping with a pruner. Most growers pick the fruit with the bags on, and then remove the bags at the packinghouse. This practice is to avoid the sap contacting the skin and causing injury. Some cultivars, such as Kensington Pride, need to be washed with water soon after harvest to avoid sap burn on the skin. Picking poles with bamboo baskets, nylon net bags, or cotton bags are also used (Crane 2009), though these picking aids can lead to increased mechanical injury. The picking aids are generally unnecessary if the trees are pruned to maintain height to less than 2 to 3 meters. Picked fruit are kept in the shade. After the stem is cut to approximately 1 cm, the stem end of the fruit should be placed downward to allow the sap to drain. Fruit are placed in a basket with a foam net on to transport to the export company for further treatment or distribution. The fruit is kept at a temperature of no less than 13°C for transport and distribution to avoid chilling injury (Morton 1987).

2.4.10 Off-season Mango Production

Off-season mango production is possible using Paclobutrazol (PBZ), a growth retardant that inhibits gibberellic acid accumulation in the shoot. PBZ is also applied to on-season mangoes to ensure floral induction. Application is as a soil drench when the second or the third sets of leaves become mature and is followed by irrigation. The application rate is 1 g active ingredient per meter of canopy width. PBZ is not recommended for mango plants younger than 5 years old, since it may affect subsequent growth (Radanachaless and Tiyayon 2010).

2.5 Biotic and Abiotic Stresses

2.5.1 Biotic Stress

Biotic stress of mango results from plant pathogens, insects, and to a lesser extent, from weeds.

2.5.1.1 Plant Pathogens

Anthracnose is the most important disease of mangoes, especially in high humidity areas (Ploetz 2003). Anthracnose causes damage to leaves, inflorescences, and fruit. Other common mango diseases caused by pathogens are powdery mildew, algal leaf spot (red rust), bacterial black spot, blossom blight, mango scab, *Alternaria* rot, root rot and damping-off, stem-end rots, verticillium wilt, apical necrosis, phytophthora disease, and nematode damage (Ploetz 2003; Ploetz and Freeman 2009).

2.5.1.2 Insects

Fruit flies are an important pest of mangoes, which cause significant economic loss. Fruit bagging is an effective deterrent. Other common insect pests are thrips, the mango seed weevil, the mango hopper, midges, mites, scales, mealybugs, the mango shoot borer, the mango stem borer, and termites (Waite 2002).

2.5.1.3 Weeds

Weeds can interfere with young plant development, as they compete for nutrient absorption from the soil. Although weeds may not cause significant problems to mature mango trees, they can serve as insect pest refugia and indirectly increase the chance of insect infestation. Therefore, the orchard floor should be kept clean.

2.5.2 Abiotic Stress

2.5.2.1 Temperature

The correct temperature is crucial for flowering and fruit set, though lower temperatures can promote floral induction, particularly in sub-tropical areas (Núñez-Elisea and Davenport 1994). A temperature below 15°C greatly effects pollen germination. Ovule of young fruit may abort at below 12°C and above 44°C (Yahia *et al.* 2006, in Yahia 2011).

2.5.2.2 Drought

Mango can tolerate drought conditions for up to 8 months, but drought during fruit development causes fruit drop and low fruit yields (Bally 2006).

2.5.2.3 Floods

Flooding for 4–10 days normally kills mango plants; however, some plants can withstand flooding for up to 110 days (Larson *et al.* 1991). Surviving trees develop hypertropic lenticels, which are believed to remove toxic metabolites from anaerobic respiration (Larson *et al.* 1993).

2.5.2.4 Wind

Mangoes can withstand wind, but storm winds can damage mango trees by uprooting or breaking of branches, and hailstorms can destroy fruit, even after they have been bagged. Windbreaks are planted to reduce wind damage and improve pollination (Bally 2006).

2.5.2.5 Fire

Fire can cause serious damage to mango orchards. Firebreaks should be made around the orchard to prevent fire spreading from nearby areas.

2.5.2.6 Nutrient Disorders

Nutrient deficiency and/or toxicities lead to slower growth, lower yields, and poorer fruit quality (Weir and Cresswell 1995). In mangoes, appropriate nitrogen levels are important for flowering, yield, and fruit quality (Bally 2009), and calcium levels are shown to play an important role in fruit internal breakdown (Ploetz 2003).

2.6 Good Agricultural Practices (GAPs)

The GAP concept optimizes available technology to promote agricultural productivity and food safety, while achieving economic viability, environmental sustainability, and social responsibility (Poisot *et al.* 2004). In recent years, GAPs are required as a minimum practice in many crops. GAPs in different countries are utilized at many levels, from subnational, national, to international. Since 2007, GLOBAL GAP (previously EuroGAP or EurepGAP) was adopted as an international practice standard (The Agricultural Document Library 2011). One barrier for growers to get such certification is economic. Therefore, in some countries, government entities and companies provide support for these costs, for example Pakistan (ASF 2008), Peru (Lemeilleur 2013), and The Philippines (Protacio 2013). GLOBAL GAP group certification (called option 2) is also available for groups of growers who cannot afford to be certified, but wish to comply with the standards (GLOBAL GAP 2008). Growers should only consider seeking international certification if required for the marketing of their fruit.

2.7 Organic Mango Farming and Marketing

An underlying principle of organic production is to utilize biological processes as the basis for tree management. The principles of organic agriculture articulated by the International Federation of Organic Agriculture Movements (IFOAM) are health, ecology, fairness, and care (Luttikholt 2007). The growers' motivation and commitment is fundamental for organic farming success (Slaven 2015a). The organic regulations vary to some extent between countries and normally prohibit the use of synthetic fertilizers and pesticides.

Organic mango production is often not an option for a large commercial production, which depends on PBZ to promote flower induction. Growers would need to consider fertilization, pruning, and water stress to alter the flowering period.

Slaven (2015b) suggests that growers who aim to convert from conventional mango orchard to organic should first understand the organic standards and start with only a section rather than the whole orchard. The transition period for mango is normally at least 3 years. Growers in India mentioned higher plant immunity against disease, greater microbial activity, and their mangoes commanded nearly twice the normal price (Mehta 2014). An organic mango in the USA during the fall and winter season can also be nearly twice as expensive (Gallo 2015).

2.8 Biotechnological Innovations

Biotechnology innovations that are being applied to mangoes are related to molecular biology and *in vitro* propagation and storage. Litz and Gómez-Lim (2005) described many aspects of propagation *in vitro*, including somatic embryogenesis, organogenesis, and protoplast culture. *In vitro* storage conditions of embryogenic cultures and cryopreservation were also discussed.

Biotechnological innovations that involve molecular biology include gene cloning, genetic transformation, and the most studied being various types of molecular markers. Molecular markers are used for genetic identification and classification reviewed by Krishna and Singh (2007). Markers used to date are isozymes and DNA marker. The DNA markers evaluated include amplified fragment length polymorphisms (AFLP), restriction fragment length polymorphisms (RFLP), random amplified polymorphic DNA (RAPD), inter-simple sequence repeats (ISSR), simple sequence repeats (SSR), and variable number tandem repeats (VNTR). More recent molecular research expanded marker techniques to microsatellite (Ravishankar *et al.* 2011), start codon-targeted (SCoT) (Luo *et al.* 2011; Gajera *et al.* 2014), and expressed sequence tag-SSR (EST-SSR) (Dillon *et al.* 2014). In addition, Singh *et al.* (2014) reported that ongoing research on mango genome sequencing is already 95% accomplished for the variety Amrapali with assembly being the major challenge. The research team estimated the mango genome has a size of 450 Mbp and a predicted total of 63,130 genes.

2.9 Research Needs and Future Prospects

Mango research continues to develop a better understanding of factors controlling mango plant growth and development, and especially flowering control. Modern tools such as infrared thermography have assisted the development of plant water stress detection (Spreer *et al.* 2013), and may be associated with automated irrigation systems in the future. Various types of modeling will help in predicting stages of plant development and yield related to both biotic and abiotic factors. New predictive software has resulted in more user friendly programs with more accurate predictive power. This predictive ability coupled to improved orchard management skills will allow growers to reduce inputs, increase sustainability, predict harvest dates, improve fruit quality and orchard profitability. Molecular biotechnology will not only assist in accurate systematics of mango lines, but also will be utilized in breeding programs for screening progenies. The value of germplasm conservation of mango varieties can never be under appreciated. Organic mango production is another direction for sustainability. Finally, collaboration and information sharing among mango producers and researchers will accelerate the rate of success in improving mango orchard management worldwide.

References

Ara, R., Motalab, M., Uddin, M.N., Fakhruddin, A.N.M. and Saha, B.K. (2014) Nutritional evaluation of different mango varieties available in Bangladesh. *International Food Research Journal,* **21**, 2169–2174.

ASF [Agribusiness Support Fund] (2008) *ASF Pioneers GLOBALGAP Certification in Pakistan* [Online]. Available: http://www.asf.org.pk/doc/ASF-NEWSLETTER-DEC08.pdf [27 June 2015].

Bally, I.S.E. (2006) *Mangifera indica* (mango), ver. 3.1. In: *Species Profiles for Pacific Island Agroforestry* (ed. C.R. Elevitch), pp. 1–25. Permanent Agriculture Resources (PAR), Hōlualoa, Hawai'i. (http://www.traditionaltree.org).

Bally, I.S.E. (2009) Crop production: Mineral nutrition. In: *The Mango – Botany, Production and Uses* (ed. R.E. Litz), 2nd edition, pp. 404–431. CAB International, Wallingford, UK.

Bally, I.S.E., Harris, M. and Whiley, A.W. (2000) Effect of water stress on flowering and yield of Kensington Pride mango (*Mangifera indica* L.). *Acta Horticulturae,* **509**, 277–282.

Bally, I.S.E., Lu, P. and Johnson, P.R. (2009) Mango breeding. In: *Breeding Plantation Tree Crops: Tropical Species* (eds S.M. Jain and P.M. Priyadarshan), pp. 51–82. Springer Science and Business Media, LLC, New York.

Biosecurity Australia (2005) *Draft Extension of Policy for the Importation of Fresh Mangoes (Mangifera indica L.) from Taiwan.* Biosecurity Australia, Canberra, Australia.

Bompard, J.M. (2009) Taxonomy and systematics. In: *The Mango – Botany, Production and Uses* (ed. R.E. Litz), 2nd edition, pp. 19–41. CAB International, Wallingford, UK.

Cheema, G.S., Bhat, S.S. and Naik, K.C. (1954) *Commercial Fruits of India.* Macmillan, London.

Crane, J.H., Salazar-García, S., Lin, T.S., Pinto, A.C.Q. and Shü, Z.H. (2009) Crop Production: Management. In: *The Mango – Botany, Production and Uses* (ed, R.E. Litz), 2nd edition, pp. 432–483. CAB International, Wallingford, UK.

Cull, B. and Linsay, P. (1995) *Fruit Growing in Warm Climates for Commercial Growers and Home Gardens.* Reed Books, NSW, Australia.

Davenport, T.L. (2009) Reproductive Physiology. In: *The Mango – Botany, Production and Uses* (ed. R.E. Litz), 2nd edition, pp. 97–169. CAB International, Wallingford, UK.

De Bie, C.A.J.M. (2004) The yield gap of mango in Phrao, Thailand, as investigated through comparative performance evaluation. *Scientia Horticulturae,* **102**, 37–52.

De Candolle, A.P. (1884) *Origin of Cultivated Plants.* Hafner, London.

Dillon, N.L., Innes, D.J., Bally, I.S.E., Wright, C.L., Devitt, L.C. and Dietzgen, R.G. (2014) Expressed sequence tag-simple sequence repeat (EST-SSR) marker resources for diversity analysis of mango (*Mangifera indica* L.). *Diversity,* **6**, 72–87.

Durand, G. (1997) Effects of light availability on the architecture of canopy in mango (*Mangifera indica* L.) cv. Manzana trees. *Acta Horticulturae,* **455**, 217–227.

FAO [Food and Agriculture Organization of the United Nations] (2015) *Crop Production Data* [Online]. Available: http://faostat3.fao.org/browse/Q/QC/E [3 June 2015].

Gajera, H.P., Bambharolia, R.P., Domadiya, R.K., Patel, S.V. and Golakiya, B.A. (2014) Molecular characterization and genetic variability studies associated with fruit quality of indigenous mango (*Mangifera indica* L.) cultivars. *Plant Systematic and Evolution*, **300**(5), 1011–1020.

Gallo, E. (2015) Recent evolution of the world mango market. *Acta Horticulturae*, **1075**, 41–50.

Gao, A., Chen, Y., Crane, J.H., Zhu, M., Huang, J. and Luo, H. (2011) Status and analysis on mango production in China. *Advanced Biomedical Engineering*, **1–2**, 472–476.

GLOBAL GAP (2008) Bigger Market Chances for Small-Scale Farmers [Online]. Available: http://roap-ksp.ilo.org/services/meetings/meetings-and-workshops-2010/ilo-local-development-strategies-knowledge-sharing-and-learning-event/key-resources/field-trip/FTrip%20-%20Success%20Story%20GlobalGap.pdf [25 June 2015].

Hadi, A.A. (2009) Indonesian Potential Fruits [Online]. Available: http://gedong-gincu.blogspot.com/2009/07/gedong-gincu-mango.html [29 June 2015].

Hamdard, M.S., Rafique, R. and Farooq, U. (2004) Physico-chemical characteristics of various mango, *Mangifera indica* L. varieties. *Journal of Agricultural Research*, **42**(2), 191–199.

Hartless, A.C. (1914) Mango crops and some factors influencing them. *Agricultural Journal of India*, **9**, 141–159.

Iyer, C.P.A. and Schnell, J.J. (2009) Breeding and genetics. In: *The Mango – Botany, Production and Uses* (ed. R.E. Litz), 2nd edition, pp. 67–96. CAB International, Wallingford, UK.

Jaiphayak, S. (2011) Effects of pruning on flowering and fruit setting of Nam Dok Mai mango for delayed harvesting. MS Thesis, Chiang Mai University. (in Thai)

Knight, R.J., Campbell, R.J. and Maguire, I. (2009) Important mango cultivars and their descriptors. In: *The Mango – Botany, Production and Uses* (ed. R.E. Litz), 2nd edition, pp. 42–66. CAB International, Wallingford, UK.

Krishna, H. and Singh, S.K. (2007) Biotechnological advances in mango (*Mangifera indica* L.) and their future implication in crop improvement – a review. *Biotechnology Advances*, **25**, 223–243.

Kusumo, S., Lye, T.T., Vangnai, V., Yong, S.K. and Namuco, L.O. (1984) Mango: Commercial mango cultivars in ASEAN. In: *Mango: Fruit Development, Postharvest Physiology and Marketing in ASEAN* (eds D.B. Mendoza Jr. and R.B.H. Wills), pp. 12–20. ASEAN Food Handling Bureau, Kuala Lumpur, Malaysia.

Larson, K.D., Schaffer, B. and Davies, F.S. (1991) Flooding, leaf gas exchange and growth of mango trees in containers. *Journal of the American Society for Horticultural Science*, **116**, 156–160.

Larson, K.D., Schaffer B. and Davies, F.S. (1993) Floodwater oxygen content, lenticel hypertrophy, and ethylene evolution from mango (*Mangifera indica* L.) trees. *Journal of Experimental Botany*, **44**, 665–671.

Lavi, U., Kashkush, K., Sa'ada, D., Shats, H., Ravid, U. and Tomer, E. (2004) Mango breeding and the potential of modern biology. *Acta Horticulturae*, **645**, 51–59.

Lemeilleur, S. (2013) Smallholder compliance with private standard certification: the case of GlobalGAP Adoption by mango producers in Peru. *The International Food and Agribusiness Management Review*, **16**, 159–180.

Litz, R.E. and Gómez-Lim, M.A. (2005) *Mangifera indica* Mango. In: *Biotechnology of Fruit and Nut Crops* (ed. R.E. Litz), pp. 40–61. CAB International, Wallingford, UK.
Lu, P. and Chacko, E.K. (2000) Effect of water stress on mango flowering in low latitude tropics of Northern Australia. *Acta Horticulturae*, **509**, 283–290.
Luo, C., He, X., Chen, H. *et al.* (2011) Genetic diversity of mango cultivars estimated using SCoT and ISSR markers. *Biochemical Systematics and Ecology*, **39**, 676–684.
Luttikholt, L.W.M. (2007) Principles of organic agriculture as formulated by the International Federation of Organic Agriculture Movements. *Wageningen Journal of Life Sciences*, **54**(4), 347–360.
Mehta, Y. (2014) Going Organic is New Mantra of Mango Farmers. *The Times of India* [Online]. Available: http://timesofindia.indiatimes.com/city/surat/Going-organic-is-new-mantra-of-mango-farmers/articleshow/35101195.cms [22 June 2015].
Meurant, N., Holmes, R., MacLeod, N., Fullelove, G., Bally, I.S.E. and Kernot, I. (1999) *Agrilink-Mango Information Kit.* Queensland Department of Primary Industries, Brisbane, Australia.
Morton, J. (1987) Mango. In: *Fruits of Warm Climates* (ed. J.F. Morton), pp. 221–239. Miami, FL.
Mukherjee, S.K. and Litz, R.E. (2009). Introduction: Botany and importance. In: *The Mango – Botany, Production and Uses* (ed. R. E. Litz), 2nd edition, pp. 1–18. CAB International, Wallingford, UK.
Mukherjee, S.K., Singh, R.N., Majumder, P.K. and Sharma, D.K. (1968) Present position regarding breeding of mango (*Mangifera indica* L.) in India. *Euphytica*, **17**, 462–467.
Nakasone, H.Y. and Paull, R.E. (1998) *Tropical Fruits.* CAB International, Wallingford, UK.
Narain, N., Bora, P.S., Narain, R. and Shaw, P.E. (1998) Mango. In: *Tropical and Subtropical Fruits* (eds P.E. Shaw, H.T. Chan, Jr. and S. Nagy), pp. 1–77. Agscience, Inc., Florida.
Narayana, C.K., Rao, D.S. and Roy, S.K. (2012) Mango production, postharvest physiology and storage. In: *Tropical and Subtropical Fruits: Postharvest Physiology, Processing and Packaging* (ed. M. Siddiq), pp. 259–276. John Wiley and Sons, Inc. Ames, IA.
Núñez-Elisea, R. and Davenport, T.L. (1994) Flowering of mango trees in containers as influenced by seasonal temperature and water stress. *Scientia Horticulturae*, **58**, 57–66.
PHEDB [Pakistan Horticulture Export Development Board] (2005) Mango Marketing Strategies. [Online]. Available: http://phdec.org.pk/MktStrategies/Mango.pdf [21 June 2015].
Phahlane, H., Moobi, M. and Potelwa, Y. (2013) South African Fruit Trade Flow [Online]. Available: http://www.namc.co.za/upload/Fruit-Trade-Flow–June – 013-Issue-No-10.pdf [15 June 2015].
Pinto, A.C.Q., Andrade, S.R.M., Ramos, V.H.V. and Cordeiro, M.C.R. (2004) Intervarietal hybridization in mango (*Mangifera indica* L.): Techniques, main results and their limitations. *Acta Horticulturae*, **645**, 327–330.
Ploetz, R.C. (2003) Diseases of mango. In: *Diseases of Tropical Fruit Crops* (ed. R.C. Ploetz), pp. 327–364. CAB International, Wallingford, UK.
Ploetz, R.C. and Freeman, S. (2009) Foliar, floral and soilborne diseases. In: *The Mango – Botany, Production and Uses* (ed. R.E. Litz), 2nd edition, pp. 231–301. CAB International, Wallingford, UK.
Poisot, A., Speedy, A. and Kueneman, E. (2004) Good Agricultural Practices – a working concept. FAO, Rome, Italy [Online]. Available: http://www.fao.org/prods/gap/docs/pdf/5-gapworkingconceptpaperexternal.pdf [28 January 2016]paperexternal.pdf [22 June 2015].

Protacio, C.M. (2013) Towards a good agricultural practice (GAP)-compliant mango production system in the Philippines. *Acta Horticulturae*, **992**, 69–73.
Queensland Government (2012) Mango Varieties [Online]. Available: https://www.daf.qld.gov.au/plants/fruit-and-vegetables/fruit-and-nuts/mangoes/mango-varieties [27 June 2015].
Radanachaless, T. and Tiyayon, C. (2010) *Good Agricultural Practices for Mango 3: Pong Ta Long Export Mango Growers, Nakorn Rachasima Province*. Multiple Cropping Center, Faculty of Agriculture, Chiang Mai University. (in Thai)
Ravishankar, K.V., Mani, B.H., Anand, L. and Dinesh, M.R. (2011) Development of new microsatellite markers from mango (*Mangifera indica*) and cross-species amplification. *American Journal of Botany*, **98**, e96–e99.
Ram, S. and Litz, R.E. (2009) Crop production: Propagation. In: *The Mango – Botany, Production and Uses* (ed. R.E. Litz), 2nd edition, pp. 367–403. CAB International, Wallingford, UK.
Reddy, Y.T.N., Kurian, R.M., Ramachander, P.R., Singh, G. and Kohli, R.R. (2003) Long-term effects of rootstocks on growth and fruit yield patterns of Alphonso mango (*Mangifera indica* L.). *Scientia Horticulturae*, **97**, 95–108.
Saúco, V.G. (2004) Mango production and world market: Current situation and future prospects. *Acta Horticulturae*, **645**, 107–116.
Schaffer, B., Urban, L., Lu, P. and Whiley, A.W. (2009) Ecophysiology. In: *The Mango – Botany, Production and Uses* (ed. R.E. Litz), 2nd edition, pp. 170–209. CAB International, Wallingford, UK.
Schulze, K., Spreer, W., Keil, A., Ongprasert S. and Müller, J. (2013) Mango (*Mangifera indica* L. cv. Nam Dokmai) production in Northern Thailand – costs and returns under extreme weather conditions and different irrigation treatments. *Agricultural Water Management*, **126**, 46–55.
Singh, L.B. (1960) *The Mango: Botany, Cultivation, and Utilization*. The University Press, UK.
Singh, R.N. (1978) *Mango*. Indian Council of Agricultural Research, New Delhi.
Singh, N.K., Mahato, A.K., Sharma, N. *et al.* (2014) A Draft Genome of the King of Fruit, Mango (*Mangifera indica* L.) [Online]. Available: https://pag.confex.com/pag/xxii/webprogram/Paper11042.html [27 June 2015].
Slaven, T. (2015a) Organic Mangoes in Western Australia. [Online]. Available: https://www.agric.wa.gov.au/mangoes/organic-mangoes-western-australia [23 June 2015].
Slaven, T. (2015b) Converting to Organic Mango Production [Online]. Available: https://www.agric.wa.gov.au/mangoes/converting-organic-mango-production [23 June 2015].
Spreer, W., Schulze, K., Ongprasert, S., Wiriya-Alongkorn, W. and Müller, J. (2013) Mango and Longan Production in Northern Thailand: The role of water saving and water stress monitoring [Online]. Available: http://www.researchgate.net/profile/Wolfram_Spreer/publication/236632338_Mango_and_Longan_Production_in_Northern_Thailand_The_Role_of_Water_Saving_Irrigation_and_Water_Stress_Monitoring/links/00b7d51ae986080195000000.pdf [23 June 2015].
Spreer, W., Schulze, K., Srikasetsarakul, U., Ongprasert, S. and Müller, J. (2011) Introduction of Micro-Sprinkler Systems to Mango Production into The Uplands, Northern Thailand [Online]. Available: http://opus.uni-hohenheim.de/volltexte/2013/824/ [28 January 2016].
Thailand DOA (2001) *Plant Germplasm Database for Mango*. Plant Variety Protection Office, Department of Agriculture, Bangkok, Thailand. (in Thai)

The Agricultural Document Library (2011) Contacts Directory GLOBALG.A.P [Online]. Available: http://adlib.everysite.co.uk/adlib/defra/content.aspx?id=000IL3890W.16NTBY8YRUM1QR [25 June 2015].

Usman, M., Fatima, B. and Jaskani, M.J. (2001) Breeding mango. *International Journal of Agriculture and Biology*, **3**, 522–526.

Waite, G.K. (2002) Pests and Pollinators of Mango. In: *Tropical Fruit Pests and Pollinators: Biology, Economic Importance, Natural Enemies and Control* (eds J.E. Pena, J.L. Sharp and M. Wysoki), pp. 103–130. CAB International, Wallingford, UK.

Weir, R.G. and Cresswell, G.C. (1995) *Plant Nutrient Disorders 2: Tropical Fruit and Nut Crops*. Inkata Press, Melbourne, Australia.

Wu, H.X., Wang, S.B., Ma, X.W., Ma, W.H., Zhan, R.L. and Yao, Q.S. (2013) Effect of bagging on fruit quality in mango. *Acta Horticulturae*, **992**, 587–592.

Yaacob, O. and Subhadrabandhu, S. (1995) *The Production of Economic Fruits in South-East Asia*. Oxford University Press, Kuala Lumpur, Malaysia.

Yahia, E.M. (2011) Mango (*Mangifera indica* L.). In: *Postharvest Biology and Technology of Tropical and Subtropical Fruits: Cocona to Mango* (ed. E. Yahia), pp. 492–565. Woodhead Publishing Limited. Cambridge.

Yahia, E.M., de Ornelas-Paz, J. and Ariza, F.R. (2006) *The Mango*. Editorial Trillas, México City. (in Spanish)

Ynus, N. (1984) Mango Cultivation in Malaysia [Online]. Available: http://rfcarchives.org.au/Next/Fruits/Mango/MangoCultivation9-84.htm [29 July 2015].

Zhao, J., Wang, J., Zhang, X., Li, H. and Gao, Z. (2013) Effect of bagging on the composition of carbohydrate, organic acid and carotenoid contents in mango fruit. *Acta Horticulturae*, **992**, 537–542.

3

Biology, Postharvest Physiology, and Biochemistry of Mango

Maria Gloria Lobo[1] *and Jiwan S. Sidhu*[2]

[1] *Department of Tropical Fruits, Instituto Canario de Investigaciones Agrarias, Valle de Guerra, La Laguna, Tenerife, Canary Islands, Spain*
[2] *Department of Food Science and Nutrition, College of Life Sciences, Kuwait University, Safat, Kuwait*

3.1 Introduction

Mango (*Mangifera indica* L.) is an important member of the Anacardiaceae family, grown commercially in tropical and subtropical regions of the world. Mango fruit has attractive appearance, great taste, excellent nutritional composition (vitamins, fiber, minerals), and high concentration of bioactive compounds. Mango genotypes are divided into two distinct categories: monoembryonic mangoes, which are mostly subtropical (Indian types) and polyembryonic mangoes, which are mostly tropical (Southeast Asian types). A huge variation in fruit weight, shape, color, proximate composition, and bioactive compounds exists among cultivars, which are affected by the growing region, soil, climate, cultural practices, etc. During development, maturation and ripening fruit undergo several changes, such as cell wall modification and texture softening, degradation of chlorophylls, development of pigments, conversion of starch to sugars, and accumulation of aromas and flavors.

3.2 Botanical Description

Mango, known as the "king of the fruits" in many countries, is a juicy stone fruit botanically classified as follows:

- *Kingdom:* Plantae – Plants
- *Subkingdom:* Tracheobionta – Vascular plants
- *Superdivision:* Spermatophyta – Seed plants
- *Division:* Magnoliophyta – Flowering plants
- *Class:* Magnoliopsida – Dicotyledons
- *Subclass:* Rosidae
- *Order:* Sapindales
- *Family:* Anacardiaceae – Sumac family

Handbook of Mango Fruit: Production, Postharvest Science, Processing Technology and Nutrition, First Edition. Edited by Muhammad Siddiq, Jeffrey K. Brecht and Jiwan S. Sidhu.

- *Genus:* Mangifera L. – mango
- *Species: Mangifera indica* L. – mango

3.2.1 Mango Tree

Mango trees are long-lived, as some specimens still fruit after 300 years (Morton 1987). Mango tree can grow up to 35–40 m tall in the tropics and no more than 10 m in the subtropics (Figure 3.1) (Galán Saúco 2009), with a crown radius of 10 m. The trunk is monopodial, straight, and cylindrical (75–100 cm in diameter), whose gray-coffee bark has longitudinal cracks or cross-linked shallow grooves that sometimes contain droplets of resin. In deep soil, the root system is dense and vigorous with a taproot that descends to a depth of 6 m, with profuse, wide-spreading feeder roots; the tree also sends down many anchor roots, which penetrate the soil to a depth of 1.2 m and spread lateral as far as 7.5 m.

The leaves are evergreen, alternate, simple, 15–35 cm long and 6–16 cm broad, leathery, and oblong-lanceolate. The young leaves are orange-pink, rapidly changing to a dark, glossy red, then shiny dark green above, lighter beneath, with yellow or white venation as they mature.

Mature terminal branches bear profuse, showy, erect, pyramidal flower panicles 10–40 cm long with several hundred flowers (Figures 3.2A,B,C). Each flower is small (4 mm), greenish, white or pinkish, radially symmetrical and with 5 petals, streaked with red. There is usually only one fertile stamen per flower, the other four being sterile. The flower has a conspicuous 5-lobed disk between the petals and stamens. Both male and bisexual flowers are borne on the same tree. Most flowers function as males by providing pollen, while some are bisexual and set fruit. Pollination is by flies, wasps, and bees. A tree may eventually produce between 2,000 and 4,000 panicles with 200–10,000

Figure 3.1 A 20-year old mango orchard in the tropics, Kerala, India (left) and subtropics, Canary Islands, Spain (right). Photos from Wikipedia and M.G. Lobo (ICIA).

Figure 3.2 Flowering tree (A); Panicle and flowers (B,C); Tree with fruits (D); Fruits (E,F); Transversal and longitudinal cut of the fruit and seed (G). Photos from M.G. Lobo, D. Fernández Galván, P. Modesto and A. Rodríguez (ICIA).

flowers, depending on the cultivar and the location. Nevertheless, the flower to fruit yield is very low, reaching less than 1% (Mukherjee 1949).

3.2.2 Mango Fruit

Over 400 varieties of mangoes are known, many of which ripen in summer, while some give a double crop. The fruit is a drupe and takes three to six months to ripen (Figures 3.2D,E). The immature fruit has green skin that gradually turns to yellow, orange, purple, red, or combinations of these colors as the fruit matures with small yellowish, greenish, or reddish dots. The ripe fruit varies in size (6.25–25 cm in length), weight (0.17–2.26 kg), shape (nearly round, oval, ovoid-oblong, or somewhat kidney-shaped), and color (yellow, orange, red, purple, or green) (Figure 3.2F). Some have a "turpentine" odor and flavor, while others are richly and pleasantly fragrant. The flesh color ranges from pale yellow to deep orange. It is essentially peach-like, but much more fibrous (in some seedlings so excessive that it is referred to as "stringy") and is extremely juicy, with a flavor range from very sweet to semi-acidic to tart.

Mango is a stone fruit with a single flat, oval, oblong, or kidney-shaped pale yellowish pit that can be fibrous or hairy on the surface, and which does not separate easily from the pulp (Morton 1987). The seed contains one or more embryos, depending on the variety or type (Figure 3.2G).

3.3 Tree and Fruit Growth and Development

Graft propagation is the only commercial scale system used by nurseries worldwide. However, different methods have been developed that allow the rooting of mango by layering or stake with 100% success (Ram 1997; Reuveni and Castoriano 1997) and by micropropagation (Litz and Lavi 1997; Moyo and Van Staden 2013; Reuveni and Golubowick 1997).

Flowering and fruit set can occur even during the first year of growth. However, it is important to remove the flowers in the third and fourth year to allow the plant to have a good growth before enduring a full crop. Graft plants start bearing at the age of 3–4 years to give an optimum crop during 10th–15th year, which continues to increase up to the age of 40 years under good management. The terms alternate, biennial, intermittent, or irregular bearing for mangoes are used to describe the tendency of mango trees to bear heavy crops in one year and very little or no crop at all in the next year. This condition arises if the variety is inappropriate to the particular climate, which is not favorable to the appropriate vegetative reproductive balance for that variety to bear a profitable crop.

Maturation takes place 10–28 weeks after fruit set, depending on the variety, agro-ecological region, soil, etc.

The main preharvest factors that determine mango tree development and fruit quality are light, temperature, water and carbon availability, irrigation management, and thinning of fruit on the plant. These factors may also affect the action mode of mineral nutrients and thereby reflected in the fruit quality (Léchaudel and Joas 2007).

3.3.1 Factors Affecting Tree and Fruit Development

Mango can be grown under both tropical and subtropical climates from sea level to an altitude of 1,400 m. However, production decreases at higher altitudes.

3.3.1.1 Temperature

Temperature is the most influential climatic factor in the growth, development, and flowering of mango, being the optimum between 27 and 33 °C. Mango is susceptible to cold. Young trees may be killed by temperatures below 0.5 °C, while older trees will survive a few degrees of frost, but may be severely damaged. Under temperature conditions of below 10 °C and over 33 °C, pollen is affected, being one possible reason for the low fruit set, which shows in some of the commercial varieties of subtropical origin.

Certain cultivars are less tolerant to high temperatures and low humidity, and the fruit will show symptoms of sunburn (Sensation, Edward, Isis, Fascell, and Keitt). Cultivars with a high tolerance include Neldica, Tommy Atkins, Irwin, Lilley, Lippens, Chené, Kent, Ceriese, Kensington, Jubilee, Palmer, and Zillate.

The temperature differences between day and night are an important factor in the process of flowering induction. High temperatures overnight (28–32 °C) make the fruit sweet and mature well, while the hot days and cool nights (12–20 °C) help the fruit to develop a more attractive color. The main effects of temperature on the mango are presented in Figure 3.3.

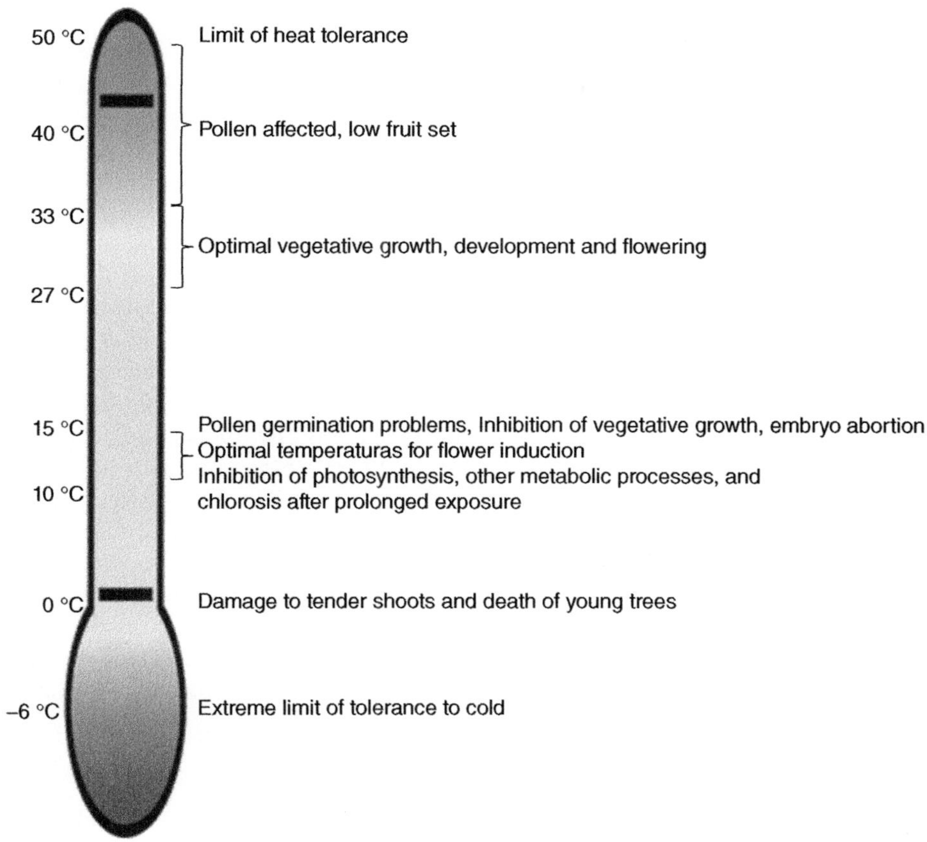

Figure 3.3 Effects of temperature on mango cultivation. *Source:* Adapted from Galán Saúco (2009).

3.3.1.2 Relative Humidity and Rainfall

The mango tolerates a wide variation in relative humidity. Whiley *et al.* (1986) pointing out that in Australia the relative humidity can vary from 40% in winter to 85% in the summer.

In general, the mango is successfully cultivated in areas where annual rainfall ranges from 500–2,500 mm. For a successful crop, most important is the annual distribution of rainfall rather than the amount. Most authors consider appropriate at least 700 mm and the rainy season has to alternate with the dry season as the latter must match the pre-flowering period (to induce flowering) because rain, heavy dews, or fog during the flowering, fruit set, early fruit growth, or harvest are harmful, as such conditions reduce pollination and fruiting and lead to loss of flowers and fruits due to attack by fungus, in particular anthracnose.

In subtropical areas, precipitation in the form of hail affects leaves and fruits, making them unacceptable for the market (Galán Saúco 2009).

3.3.1.3 Water

The soil has to be dry for 2 or 3 months before the flowering stage to promote good flower formation. Irrigation during the developmental stage of the fruit is essential to prevent fruit drop and to promote the development of young fruit. Additional irrigation from fruit set to ripening results in a considerable improvement in both fruit size and quality (National Department of Agriculture of South Africa 2000).

3.3.1.4 Soil

Mangoes grow on a wide range of soils, from light sands to heavy clays, preferring areas where soils are 1–2 m deep, well drained, and slightly acidic (pH 5.5–7.5). Heavy clays provide a slight dwarfing effect. Generally, on the heavier soils, nutritional management becomes more important, as this has a great bearing on fruit quality. Light sandy soils, while tending to have low nutritional status, lend themselves to easy management. Mangoes have only low to moderate tolerance to saline soils. The use of Paclobutrazol (PBZ) minimizes the effects of salt stress in mango by increasing the levels of photosynthetic pigments, water content, K^+ uptake and accumulation, and by reducing defoliation, the membrane injury index, and the uptake and accumulation of harmful Na^+ and Cl^- ions (Kishor *et al.* 2009). Research is currently being conducted to identify salt-tolerant rootstocks. Zuazo *et al.* (2004) evaluated the impact of salinity on the fruit yield of two rootstocks (Gomera-1 and Gomera-3). Results demonstrated that Gomera-1 was slightly more tolerant to salinity than Gomera-3. Pandey *et al.* (2014) suggested that the salt-tolerant behavior of the mango rootstocks Olour and Terpentine, is due to their ability to inhibit the uptake of Cl^- and Na^+ ions, combined with higher accumulations of proline, recommending these two rootstocks to be used to improve the yield and quality of commercial mango cultivars grown in salt-affected areas.

3.3.1.5 Mineral Nutrition

Fruit quality is the result of the action of several factors, in particular the individual and combined effect of mineral nutrients (Aular and Natale 2013).

Van Eeden (1992) found a direct relationship between Ca deficiency and higher incidence of the disorder characterized by a disorganization of cells and the breakdown of the cell wall. Evangelista *et al.* (2000, 2002), Rani and Brahmachari (2003), Hojo *et al.* (2009), and Wahdan *et al.* (2011) observed higher firmness in fruits when $CaCl_2$ sprays were used in pre-harvest, while other authors (Sampaio *et al.* 1999; Silva and Menezes 2001) did not found this clear positive effect. It is known that nutrient balance plays a key role in decreasing the incidence of internal collapse. Thus, de Assis *et al.* (2004) reported that high Ca and Mg and low ratios of N/Ca and K/Ca, both in pulp and in peel, were effective in preventing physiological disorders. Xiuchong *et al.* (2001) observed that the application of magnesium increased the number the fruits per tree and their weight, increasing significantly the yield. Excessive nitrogen fertilization also contributes to the incidence of lenticel damage during postharvest (Magwaza and Kruger 2005).

Deficiencies of manganese, iron, or a combination of both, seem to be related to the physiological disorder known as "mango decline" (Schaffer *et al.* 1988).

Poor yield and quality is related to boron deficiency. Raja *et al.* (2005) observed that the Alphonso mango was more responsive to foliar application of boron than soil application. Thus, foliar application of micronutrients such as iron ($FeSO_4$), boron (H_3BO_3), and zinc ($ZnSO_4$) improve fruit quality, increasing the total soluble solids content (Bhatt 2012; Moazzam *et al.* 2011).

Table 3.1 summarizes the main effects of mineral elements on the quality of mango fruit. Zn and B are involved in the formation and maintenance of mango fruit quality. Singh and Rajput (1977) observed increased fruit yield, fruit sugars, and ascorbic acid with the

Table 3.1 Main effects of mineral elements on the quality of mango fruit.

Quality attribute	Nitrogen	Calcium	Magnesium	Boron	Zinc
Fruit mass	--	(+) 13 (+) 12	(+) 14	(+) 16 (+) 18	(+) 19 (+) 20 (+) 21 (+) 22
Firmness	--	(0) 7 (+) 8 (0) 9 (+) 10 (+) 12 (+) 13	--	--	--
External appearance	--	(+) 6	--	--	--
Fruit color	(–) 1*	--	--	--	--
Internal appearance	--	--	--	--	--
Total Soluble solids (TSS)	(0) 2	(+) 10 (+) 12 (+) 12	--	(+) 17 (+) 18	(+) 19 (+) 21 (+) 22
Titratable Acidity (TA)	(0) 2	(+) 11 (+) 13	--	(+) 17 (+) 18	(+) 19 (+) 21 (+) 22
TSS/TA ratio	--	(+) 13	--	--	--
Internal disorders	(–) 3	(–) 5 (0) 6 (0) 7 (0) 9 (+) 11 (0) 12	(–) 15	--	--
Antracnose incidence	(+) 2	--	--	--	--
Lenticel damage	(+) 4	--	--	--	--

(+) Increase, (–) Decrease, (0) No effect, (--) Not reported

* *Source:* (1) Young *et al.* (1962), (2) Nguyen *et al.* (2004), (3) de Assis *et al.* (2004), (4) Magwaza and Kruger (2005), (5) Van Eeden (1992), (6) Freire-Júnior and Chitarra (1999), (7) Sampaio *et al.* (1999), (8) Evangelista *et al.* (2000, 2002), (9) Silva and Menezes (2001), (10) Rani and Brahmachari (2003), (11) de Assis *et al.* (2004), (12) Hojo *et al.* (2009), (13) Wahdan *et al.* (2011), (14) Xiuchong *et al.* (2001), (15) de Assis *et al.* (2004), (16) Raja and Kumar (2003), (17) Moazzam *et al.* (2011), (18) Bhatt *et al.* (2012), (19) Singh and Rajput (1977), (20) Raja and Kumar (2003), (21) Moazzam *et al.* (2011), (22) Bhatt *et al.* (2012)

treatment of the highest $ZnSO_4$ application rate. The foliar application of nutrients in the fruit before harvest ($CaCl_2$, borax, K_2SO_4, $Ca(NO_3)_2$, $ZnSO_4$, and $ZnCl_2$) indicated that trees sprayed with borax showed maximum yield, greatest weight and volume of the fruit, the highest values of total soluble solids, and reducing or non-reducing sugars (Bhatt *et al.* 2012).

3.3.1.6 Light

Day length does not seem to influence the floral initiation (Núñez-Elisea and Davenport 1995), but it has a great effect on the growth, development, and yield of the tree.

3.3.1.7 Wind

Wind, even moderate, causes scratch marks on fruits, making them unacceptable for marketing. Stronger winds result in fruit loss and lower yields. Certain cultivars, such as Zill, Haden, and Kent, are prone to a greater extent to fruit loss under windy conditions than others. To minimize damage by winds it is important to avoid windy areas, or establish windbreaks such as artificial structures or fast-growing trees on the upwind side of the prevailing winds. It is important to prune the non-bearing flower panicles as soon as it is evident that these will not bear any fruit, because when they become dry and hard, they cause scratch marks on the fruit.

3.3.1.8 Growth Regulators and other Treatments

Efficient techniques such as floral induction allow the establishment of commercialization strategies for favorable periods of market and economical success in the cultivation of mango fruits. Cardos *et al.* (2007) observed a higher percentage of flowering and fruit production for PBZ-treated mangoes in comparison to the control. The PBZ treatment also impacts flower initiation and fruit harvest, which means production out-of-season brings higher prices and more profitability.

Mouco *et al.* (2013) evaluated the effect of plant regulators applied to the leaves to minimize the risk of residues in the soil and plant, in the management of vegetative growth of Tommy Atkins mangoes. The results showed that spraying with prohexadione-Ca, trinexapac-ethyl, and chlormequat chloride in a dosage of 1.5 g/plant were as efficient as a dose of 4.0 g/plant of PBZ applied to the soil.

3.4 Fruit Ripening

3.4.1 Physiology and Biochemistry of Mango Ripening

The growth and development of mango fruit involves several physical, physiological, and chemical changes, starting with pollination and culminating at harvest. Mango is a climacteric fruit. All biochemical and physiological changes during fruit ripening are driven by the coordinated expression of fruit ripening-related genes. These genes encode enzymes that participate directly in biochemical and physiological processes.

3.4.2 Role of Ethylene and other Hormones

The physiological and biochemical changes that occur during mango ripening are initiated by autocatalytic production of ethylene and an increase in respiration (Matoo and Modi 1969).

Ethylene production is very low during fruit development and maturation and rises after a climacteric burst in respiration. A very low concentration of ethylene (0.01 ppm) can promote ripening, even if it is exogenous ethylene. Fruits treated with ethylene inhibitors, such as 1-methylcyclopropene (1-MCP), nitric oxide (NO), and aminoethoxyvinylglycine (AVG), delays ripening (Wang *et al.* 2009; Zaharah and Singh 2011), while the exogenous application of methionine or ethephon increases respiration rate and ethylene production (Lalel *et al.* 2003a; Matoo and Modi 1969).

Other hormones such as auxins, abscisic acid (ABA), jasmonic acid, brassinosteriods, and polyamines play a vital role in ripening. Auxins are involved in fruit development. During the pre-climacteric stage, endogenous indole-3 acetic acid (IAA) in the pulp fruit is higher than during climacteric and post-climacteric stages (Bouzayen *et al.* 2010). Zaharah *et al.* (2012) observed that mangoes treated with IAA were harder and firmer compared to those treated with ethylene and ABA, which plays an important regulatory role in plant growth and development including ripening (Parikh *et al.* 1990). The endogenous ABA, and not ethylene, is critical for the onset of ripening, while the role of ethylene is limited to later ripening stages (Singh *et al.* 2013). Jasmonic acid plays a role during fruit development and ripening. Exogenous treatment of methyl jasmonate vapors can trigger ethylene production, improving skin color, fatty acids concentration, and total aroma volatiles (Lalel *et al.* 2003b). During fruit ripening, total polyamines increase along with a climacteric rise in ethylene, suggesting that their biosynthesis are not competitive and free polyamines evolved as a response to increase biosynthesis of ethylene (Malik and Singh 2004). The postharvest application of putrescine, spermidine, and spermine retard fruit softening and color development during storage, maintain fruit firmness at the ripe stage without a significant reduction in ethylene production during ripening, and improve shelf life and fruit quality of mango (Malik *et al.* 2006).

3.4.3 Compositional Changes during Fruit Growth and Maturity

Ethylene production within the fruit activates many other enzymes, resulting in physiological changes such as the change of color, the development of aromas, and the softening of the fruit.

Titratable acidity, vitamin C, starch content, and reducing sugars are higher at the unripe state and gradually decrease during ripening, while phenol content, total soluble solid, total sugar, and non-reducing sugar contents gradually increase, as well as respiration rate, ethylene synthesis, and characteristic aromas. During maturation and ripening, organic acids (citric, succinic, malic, and tartaric acids) decrease. Chlorophyll degradation and carotenoids synthesis are other mayor biochemical changes that occur in the skin and the pulp (Hossain *et al.* 2014). Carotenoids and xanthophylls are the predominant pigments in those cultivars that turn yellow when ripe, while anthocyanins are present in those cultivars where the skin becomes reddish.

Polysaccharides such as starch, pectin, cellulose, and hemicelluloses decrease from unripe to ripe stage, with concomitant increase in soluble sugars (Yashoda *et al.* 2006).

The activities of amylase, α-mannosidase, α-glucosidase, and invertase increase sharply within the first few days and decrease significantly in the later stage of ripening. Meanwhile, polyphenol oxidase, β-galactosidase, and β-hexosaminidase predominantly increase significantly from unripe through late stages of ripening. Along with the ethylene production and respiration rate, the fruit ripening period is closely associated with an increased fruit softening (exo and endo polygalacturonases, pectatelyases, β-galactosidases, cellulases, galactanase, xylanase, etc.) and antioxidative enzymes, except for pectinesterase and superoxide dismutase (Razzaq *et al.* 2013). Hossain *et al.* (2014) suggested the suppression of β-galactosidase and β-hexosaminidase to enhance the shelf life of mango fruits.

Mango flavor is critical for consumer acceptability. More than 578 volatile compounds have been reported for mango, depending on the cultivar, fruit maturity, ripening methods, and storage temperature.

3.4.3.1 Proximate/Nutritional

The composition and nutritional profile of raw mangoes is shown in Table 3.2. Values are based on analyses of Tommy Atkins, Keitt, Kent, and/or Haden cultivars grown in the USA (USDA 2015). Varietal differences, climatic and solid conditions, agricultural practices, and postharvest handling may contribute to variations in the composition and nutritional profile.

One hundred grams of edible mango provides a very good source of vitamin A (15% of daily value (DV)), vitamin C (46% DV), vitamin B_6 (7% DV), vitamin E (6% DV), vitamin K (5% DV), dietary fiber (7% DV), copper (6% DV), and potassium (4% DV).

Moreover, it is very low in saturated fat (0% DV), cholesterol (0% DV), and sodium (0% DV). The glycemic load of mango is very low, 5 for 100 g serving, so blood glucose levels rise more slowly and insulin demand is lower than when a high glycemic load meal is ingested.

Table 3.2 Composition and nutritional profile of mango fruit (per 100 g).

Nutrient	Units	Raw
Proximate		
Water	g	83.46
Energy	kcal/KJ	60/250
Protein	g	0.82
Total lipid (fat)	g	0.38
Ash	g	0.36
Carbohydrate, by difference	g	14.98
Fiber, total dietary	g	1.6
Sugars, total	g	13.66
Sucrose	g	6.97
Glucose (dextrose)	g	2.01
Fructose	g	4.68
Minerals		
Calcium	mg	11
Iron	mg	0.16

Table 3.2 (Continued)

Nutrient	Units	Raw
Magnesium	mg	10
Phosphorus	mg	14
Potassium	mg	168
Sodium	mg	1
Zinc	mg	0.09
Copper	mg	0.111
Manganese	mg	0.063
Selenium	µg	0.6
Vitamins		
Vitamin C, total ascorbic acid	mg	36.4
Thiamin	mg	0.028
Riboflavin	mg	0.038
Niacin	mg	0.669
Pantothenic acid	mg	0.197
Vitamin B6 (Pyridoxine)	mg	0.119
Folate, total	µg	43
Choline, total	mg	7.6
Vitamin A, RAE	µg	54
β-Carotene	µg	640
α-Carotene	µg	9
β-Cryptoxanthin	µg	10
Vitamin A	IU	1082
Lycopene	µg	3
Lutein + zeaxanthin	µg	23
Vitamin E (α-tocopherol)	mg	0.90
Vitamin K (phylloquinone)	µg	4.2

Source: USDA (2015).

3.4.3.2 Phytochemicals and Antioxidants

Mango fruit, as well as other parts of the plant (kernel, flowers, leaves, gum, bark, and peel), are important sources of bioactive compounds with potential health promoting-activity, discussed in more detail in Chapter 12. Briefly, mango fruit is an excellent source of both pro-vitamin A carotenoids: carotenes such as α-carotene, β-carotene, and γ-carotene; and oxygenated carotenoids (xanthophylls), such as β-cryptoxanthin, lutein, zeaxanthin, violaxanthin, antheraxanthin, auroxanthin, and neoxanthin. Their antioxidant effect is related to its capacity to remove reactive oxygen species (ROS), as vitamin C does, protecting the human body against diseases associated with oxidative stress. Moreover, carotenoids affect gene expression regulation, which explains part of the associations between high carotenoids intake and lower risk of certain diseases (cardiovascular, some types of cancer, osteoporosis, infectious, cataracts, etc.). Many phenolic compounds have been described in the flesh, peel, and kernels of mango. The flavonols quercetin and kaempferol have biological activities, including anti-bacterial, anti-thrombotic, vasodilatory, anti-inflammatory, and anti-carcinogenic. The xanthone mangiferin is one of the major bioactive constituents in mango, with antioxidant, anti-lipid peroxidation, immunomodulation, cardiotonic, hypotensive, wound healing, anti-degenerative, anti-diabetic, and chemoprotective activities. Other polyphenols with antioxidant capacity are isomangiferin, tannins, and gallic acid derivatives (Ribeiro and Schieber 2010; Shah *et al.* 2010).

Mango contains terpenoids and fiber, healthy bioactive compounds that can protect against colon cancer (Corrales-Bernal *et al.* 2014), and significant amount of copper, a co-factor for many vital enzymes, including cytochrome c-oxidase and superoxide dismutase, which is required for the production of red blood cells.

Mango is a good source of tryptophan, the precursor for serotonin, the "happiness hormone," alongside glutamine, known to be an important protein that enhances concentration in humans as well as memory gain (Ubwa *et al.* 2014).

Some of the possible health benefits of consuming mango include a decreased risk of age-related macular degeneration, colon cancer, or cardiovascular disease, a protection from lung and oral cavity cancers, and an improvement in digestion and bone, skin, and hair health. In many countries, different parts of the mango tree and fruit are used for their anti-bacterial, antifungal, anthelmintic, anti-parasitic, anti-HIV, anti-spasmodic, anti-pyretic, anti-diarrhoeal, anti-allergic, immunomodulation, hypolipidemic, hepatoprotective, and gastroprotective properties.

3.4.4 Harvesting Indices

Mangoes normally reach maturity in 4–5 months from flowering, and harvesting indices depends on the variety. Generally, when the mango is fully-mature for picking, the stem will snap easily with a slight pull. If a strong pull is necessary, the fruit is still somewhat immature and should not be harvested. The development of a purplish-red or yellow blush at the base of the fruit is an additional indication of maturity in mangoes, which skin changes from green to red-purple or to yellow-orange. Other criteria used to determine harvesting include changes in the fruit shape and/or in the lenticels aspect as they lose their color and become dried near maturity, the appearance of a powder layer on the epidermis of the fruit, when two-thirds of the peduncle becomes dried and presents a brown color (peduncle lignification), and when the fruit shoulders are slightly beyond the insertion point of the peduncle, etc. Some harvesting indices are destructive, such as firmness, total soluble solids (°Brix), pulp color, dry matter, starch content, etc. (Galán Saúco 2009). The combination of absorption at 540 and 670 nm measured non-destructively in the intact fruit by time-resolved reflectance spectroscopy (TRS) can be used to classify mango fruit according to maturity and to predict the ripening of individual fruit (Zerbini *et al.* 2015). Susceptible mango varieties tend to break down further internally, the longer they are left on the tree (Yahia *et al.* 2006).

Harvesting must be done manually, carefully, using appropriate tools to avoid bruises and bumps, and in the cooler hours of the day, but not in the early morning or during heavy rains, because under these conditions the latex flow is greater and can produce sapburn in the fruits. It is done from the ground, using stairs or motorized platforms. The fruit is picked by hand using pruning clippers, or with a long-poled picking bag which should hold no more than four fruits (Figure 3.4). Field boxes of 18–25 kg capacity and baskets up to 200 kg are used. It is desirable to leave a 6–10 cm stem to avoid the spurt of milky/resinous sap that exudes if the stem is initially cut close and that causes sapburn to the skin. The cut stem must be placed face-down on the ground or box and sometimes fruits are put into tubs of water on trucks in order to wash off the sap that exudes from the stem end. Before packing, the stem is cut off 6 mm from the base of the fruit, and at the packinghouses desapping agents are used to reduce the effect of the sap. A recommended practice in Australia before cutting the fruit is to cover it with detergent or a wetting agent, which is then washed off in the packaging house together with the remains of latex. Barman *et al.* (2015) observed that fruits dipped for 5 min in sodium hydroxide (1%) showed significantly lower (7.6-fold) sapburn injury following alum (0.5%) treatment than untreated fruit.

Latex in the oily part contains antifungal resorcinols and in the aqueous layer chitinase activity. Nevertheless, latex disappears with ripening, so a decline in fruit resistance to fungal pathogens occurs. Karunanayake *et al.* (2015) found that the retention of latex at harvest enhances fruit resistance and reduces anthracnose and stem-end rot development during ripening. They achieved these results by harvesting and leaving the fruit with a 1 cm of pedicel.

At the packinghouse, the stem is cut and fruits are washed, graded, sized, and packed in bins for the hot water quarantine treatment if needed, or treated with waxes, calcium salts, 1-MCP, etc. Finally, fruits are usually packed manually in a single layer in corrugated boxes of 4–7 kg, depending on the cultivar and destination market. The filled cartons are stacked on pallets and fork-lifted into refrigerated trucks with temperatures set at no less than 13 °C. The increasing volumes of mangoes in international trade have shifted the mode of transport from air freight to marine containers. This fact has been

Figure 3.4 Mango harvest operations. Photos from J.M. Torres (COPLACA).

possible because of strict attention to fruit maturity, temperature control and use of modified or controlled atmospheres (Yahia 2011).

3.5 Fruit Quality Disorders

Both the quality and storage life of mango are affected by the development of many physiological disorders that occur in all growing regions of the world.

3.5.1 Spongy Tissue

Spongy tissue is a major problem in the Alphonso cultivar, caused by the shift of the seed to germination mode inside the fruit, sucking nutrients from the surrounding fleshy part of the fruit, making it appear white and spongy. Fruits from the outside look normal. Since the spot is devoid of nutrients, it starts to rot (appearance of black color). Inactivation of the ripening enzyme due to high temperature, convective heat, and postharvest exposure to sunlight are the causes.

3.5.1.1 Control Strategies

Harvesting the fruit when it is three-quarter mature versus fully mature will prevent any rotting. The use of mulching and postharvest exposure to low temperatures between 10 and 15 °C for 10–18 hours has been proved useful in reducing this malady.

Fruit disorders such as jelly seed, tip pulp, soft nose, insidious fruit rot, and stem-end cavity show symptoms similar to that of spongy tissue disorder, and most likely are also caused by the seed (Shivashankar 2014).

3.5.2 Jelly Seed

Jelly seed downgrades the fruit quality as it prematurely develops a watery and translucent over-ripe flesh around the seed, with off odors and flavors when the rest of the flesh is just beginning to soften. The flesh at the stem end is fibrous. Certain varieties, such as Tommy Atkins, Zill, and Sensation, are susceptible to jelly seed, while it is rare in Kensington Pride. This disorder can be confused with other physiological conditions such as soft nose and internal breakdown, a condition to which Keitt is particularly susceptible.

3.5.2.1 Control Strategies

Avoid growing varieties which are prone to jelly seed.

3.5.3 Soft Nose

Soft nose symptoms start to develop as the fruit approaches maturity on the tree, so consequently fruits harvested green are less affected than those harvested ripe. The distal end softens while the shoulders of the fruit remain firm. Flesh appears over-ripe and may discolor and become spongy. Soft nose is aggravated by heavy nitrogen fertilisation and calcium deficiency.

3.5.3.1 Control Strategies

To overcome this disorder, high nitrogen levels should not be allowed in sandy or acidic soil and proper calcium levels should also be maintained. Early picking is also suggested.

3.5.4 Internal Necrosis

Internal necrosis is an emerging physiological anomaly of developing mango fruits due to boron deficiency, which may also lead to fruit cracking. Boron is important in pollination, fruit set, and calcium uptake.

3.5.4.1 Control Strategies

Internal necrosis and fruit pitting disorders are caused by boron and/or calcium deficiency and are managed by supplying these trace elements.

Monitor the levels of boron in soil and leaves, and correct deficiencies as recommended. Foliar application is more effective than soil application in increasing yield and reducing internal necrosis and fruit cracking disorders (Saran and Ratan 2011).

3.5.5 Fruit Pitting

Fruit pitting leads to the development of some unattractive small sunken pits in the peel of the fruit. These pits appear on all sides of the fruit, with variable proportion and increase in size as the fruit increases in size. Development of such pits gives an unattractive and unhealthy look to the fruit, which affects consumer acceptability. Low calcium and boron are related to fruit pitting.

3.5.5.1 Control Strategies

Monitor the levels of calcium and boron and correct deficiencies (Sharma and Singh 2009).

3.5.6 Black Tip

Disorders such as black tip, taper tip, girdle necrosis, and stunted fruit are caused by pollutant gases such as carbon dioxide, fluoride, and sulfur dioxide that emanate from brick kiln fumes.

Black tip is a serious disorder, particularly in Dasheri cultivar. The damage to the fruit gets initiated right at the marble stage. The symptoms manifest as a small etiolated area at the distal end of the fruit with a characteristic yellowing of tissues which gradually spreads, turns nearly black, and covers the tip completely. At this stage, further growth and development of the fruit is retarded and the black ring at the tip extends towards the upper part of the fruit. This disorder has generally been detected in orchards located in the vicinity of brick kilns and it has been reported that their fumes damage the growing tip of fruits and give rise to the symptoms of black tip, making the fruit unmarketable and reducing the yield to a considerable extent. Apart from these factors, irrigation, condition of the tree, and management practices also play important roles in deciding the severity of the disorder.

3.5.6.1 Control Strategies

The practical control measure lies in keeping the brick kilns away from the mango orchards. The incidence of black tip can also be minimized by spraying Borax (1%) or other alkaline solutions such as caustic soda (0.8%) or washing soda (0.5%). The first spray of Borax should be done positively at pea stage, followed by two more sprays at 15 days interval.

3.5.7 Fruit Drop

Other disorders are related to low productivity and economic losses such as fruit drop, mango malformation, clustering in mango or *Jhumka*, fruit splitting, biennial bearing, and lenticel discoloration.

Fruit drop varies from variety and variety and is more or less a continuous process that can be classified into three phases: pinhead drop, post-setting drop, and May-month drop. The fruit drop in the first two phases is insignificant compared to the third phase, which significantly affects the final yield. Embryo abortion, climatic factors, disturbed water relation, lack of nutrition, attack of disease, and pest and hormonal imbalances are the major factors that lead to fruit drop.

3.5.7.1 Control Strategies

The extent of fruits drop can be reduced significantly by regular irrigation during the fruit development period, timely and effective control measures against major pests and diseases, and through the application of growth regulators such as naphthalene acetate (50 ppm) and 2,4-Dichloro-phenoxy acetic acid (20 ppm) at pea stage of fruit during "off years".

3.5.8 Mango Malformation

The complexity of mango malformation disorder is attributed to cultural, nutritional, and factors, such as mites and fungal, viral, and hormonal imbalance. The malformed panicles remain unproductive and are characterized by a compact mass of male flowers, greenish in color and stunted in growth. The main and secondary rachis is thick and short and bears flowers with relatively larger bracts, sepals, and petals as compared to normal flowers. The exact cause and control of the malady is yet to be established.

3.5.8.1 Control Strategies

Spraying of Planofix (200 ppm) during the first week of October followed by deblossoming at bud burst stage is recommended as a remedial measure against malformation.

3.5.9 Clustering in Mango (Jhumka)

A fruiting disorder, locally known as *Jhumka*, is characterized by the development of fruitlets in clusters at the tip of panicles. Such fruits cease to grow beyond pea or small marble stage and drop after a month of fruit set. Absence of sufficient population of pollinators in the orchards is the major reason, while other causes are old and overcrowding of trees, indiscriminate spraying against pests and diseases, use of synthetic pyrethroids, and bad weather during flowering.

3.5.9.1 Control Strategies

Introduce beehives into the orchards during the flowering season to increase the number of pollinators and restrict insecticidal sprays at full bloom to avoid killing of pollinators.

3.5.10 Biennial Bearing

The term biennial, alternate, or irregular bearing generally signifies the tendency of mango trees to bear a heavy crop in one year ("On year") and very little or no crop in the succeeding year ("Off year"). When a tree produces a heavy crop in one season, it becomes exhausted nutritionally and is unable to put forth new flush, thereby failing to yield in the following season. The problem has been attributed to genetical, physiological, environmental, and nutritional factors.

3.5.10.1 Control Strategies

For overcoming biennial bearing, de-blossoming is recommended to reduce the crop load in the "On year," so that it is balanced in the "Off year." Proper maintenance of the orchard by way of effective control of pests and diseases and regular cultural operations may also result in better performance of the tree every year. Application of PBZ or Cultar in September results in early flowering with higher fruit set and yield. It may be applied every year for regular fruiting, particularly in young trees. Ringing of branches is recommended as a means of inducing flowering in the "Off year." However, weak, stunted, unhealthy trees should not be ringed to force flowering.

3.5.11 Fruit Splitting

Fruits that grow rapidly will crack and split, thus becoming unmarketable. The split becomes brown and dried out, and sap oozes from the split. It is believed that significant fluctuations in soil moisture levels and atmospheric humidity during rapid growth of fruit are responsible for fruit splitting.

3.5.11.1 Control Strategies

Irrigation management and stable moisture levels are essential during fruit development. It is advisable to use soil moisture monitoring devices such as tensiometers.

3.5.12 Lenticel Discoloration

Lenticel discoloration is an inherent self-defense mechanism supported by continued metabolic activity that leads to a superficial, cosmetic defect. It is a superficial blemish that affects some cultivars by causing fruit to exhibit a speckled appearance making it less desirable, and it will be downgraded, despite the fact that the speckled appearance does not affect the internal quality of the fruit. Blemish development is limited to the lenticel perimeter and immediately adjacent area, and does not extend deeper than the outermost layers of the fruit rind. The development of a red halo is the first indication of lenticel discoloration. The redness sometimes intensifies and eventually results in a localized darkened appearance. Lenticels on unaffected fruit display a confluent coloration with the normal rind color of each cultivar (Du Plooy *et al.* 2006).

3.5.12.1 Control Strategies

Fruit bagging during growth controls lenticel discoloration (Mathooko *et al.* 2011).

3.5.13 Chilling Injury (CI)

Some disorders, such as chilling injury and CO_2 injury, are related to ambient or storage conditions. Chilling injury incidence and severity depend on cultivar, ripeness stage (riper mangoes are less susceptible), and temperature and duration of exposure. The symptoms of CI include grayish, scald-like discoloration of the skin, followed by pitting, uneven ripening, and poor flavor and color development and, in severe cases, flesh browning. Some varieties are more susceptible than others. Storage temperatures below about 10–13 °C but above freezing have been reported to injure mature green mangoes. Chilling injury occurs when the fruit is exposed to an injurious, low temperature for sufficient time to initiate irreversible injury.

3.5.13.1 Control Strategies

Avoid temperatures below 10–13 °C for long periods. Mangoes can be held for 2–3 weeks in controlled atmospheres (5–10% CO_2) at 8 °C for tree ripe fruit or 12 °C for mature green fruit without developing CI (Bender *et al.* 2000a,b). Hot water quarantine treatment and other time-temperature combinations reduced the susceptibility of Tommy Atkins and Keitt mangoes to CI (Brecht *et al.* 2000). Relative humidity should be kept between 90 and 95% to minimize water loss and shriveling.

3.5.14 CO_2 Injury

Exposure of mature green mango to oxygen levels below 2% and/or carbon dioxide levels above 10% for a few days induce skin discoloration, grayish or pale flesh color, uneven ripening, and off flavors due to fermentative metabolisms.

3.5.14.1 Control Strategies

Use the appropriate controlled atmosphere for each cultivar, because the susceptibility can be different for each one.

3.6 Conclusions

Mangoes are produced in tropical and subtropical regions, having a highly significant economic importance. The fruit is rich in nutraceuticals and due to its health benefits is recommended to be included in the daily diet. Preferable quality attributes include freedom from external damage such as bruises, sapburn, and also decay, uniform weight, color, aroma, firmness, shape, and size. Maintenance of mango fruit quality during the supply chain depends on many aspects, including adequate orchard management practices, harvesting practices, packing operations, postharvest treatments, temperature management, transportation and storage conditions, and ripening at destination. Postharvest losses are high during the supply chain due to harvesting fruit at improper maturity, mechanical damage during the whole chain, disorders development (sapburn, spongy tissue, jelly seed, soft-nose, lenticels discolouration, and chilling injury), and disease and pest damage. Adequate postharvest treatments and management practices are indispensable to create suitable conditions or environments to extend the storage life and retain the quality attributes, nutritional, and functional compounds.

References

Aular, J. and Natale, W. (2013) Nutrição mineral e qualidade do fruto de algumas frutíferas tropicais: goiabeira, mangueira, bananeira e mamoeiro. *Revista Brasileira de Fruticultura*, **35**(4), 1214–1231.

Barman, K., Asrey, R., Pal, R.K., Jha, S.K. and Sharma, S. (2015) Influence of different desapping agents on the incidence of sapburn, ripening behaviour and quality of mango. *Journal of Food Science and Technology*, **52**, 161–170.

Bender, R.J., Brecht, J.K., Baldwin, E.A. and Malundo, T.M.M. (2000a) Aroma volatiles of mature-green and tree-ripe Tommy Atkins mangoes after controlled atmosphere vs. air storage. *HortScience*, **35**, 684–686.

Bender, R.J., Brecht, J.K., Sargent, S.A. and Huber, D.J. (2000b) Low temperature controlled atmosphere storage for tree-ripe mangoes (*Mangifera indica* L.). *Acta Horticulturae*, **509**, 447–458.

Bhatt, A., Mishra, N., Mishra, D. and Singh, C. (2012) Foliar application of potassium, calcium, zinc and boron enhanced yield, quality and shelf life of mango. *HortFlora Research Spectrum*, **1**, 300–305.

Bouzayen, M., Latché, A., Nath, P. and Pech, J.C. (2010) Mechanism of fruit ripening. In: *Plant Developmental Biology-Biotechnological Perspectives* (eds E.C. Pua and M.R. Davey), vol. **1**, pp. 319–339, Springer, New York.

Brecht, J.K., Chen, W., Sargent, S.A. and Bartz, J.A. (2000) Hot water and ethylene treatments to reduce decay and chilling injury of Tommy Atkins and Keitt mangoes. Postharvest 2000: 4th International Conference Postharvest Science, Jerusalem, 26–31 March, 2000. Abstracts p. 18.

Cardos, M.G.S., Jose, A.R.S., Viana, A.E.S., Matsumoto, S.N. and Reboucas, T.N.H. (2007) Flowering and production of mango fruits (*Mangifera indica* L.) cv. Rosa promoted by different paclobutrazol doses. *Revista Brasileira de Fruticultura*, **29**, 209–212.

Corrales-Bernal, A., Amparo Urango, L., Rojano, B. and Maldonado M.E. (2014) *In vitro* and *in vivo* effects of mango pulp (*Mangifera indica* cv. Azucar) in colon carcinogenesis. *Archivos Latinoamericanos de Nutricion*, **64**, 16–23.

De Assis, J.S., Silva, D.J. and Moraes, P.L.D. (2004) Equilíbrio nutricional e distúrbios fisiológicos em manga Tommy Atkins. *Revista Brasileira de Fruticultura*, **26**, 326–329.

Du Plooy, G.W., Van Der Merwe, C.F. and Korsten, L. (2006) Lenticel discolouration in mango (*Mangifera indica* L.) fruit: a cytological study of mesophyll cells from affected tissue. *Journal of Horticultural Science and Biotechnology*, **81**, 869–873.

Evangelista, R.M., Chitarra, A.D. and Chitarra, M.I.F. (2000) Influência da aplicação pré-colheita de cálcio na textura e na atividade das enzimas poligalacturonase, pectinametilesterase e β-galactosidase de mangas Tommy Atkins armazenadas sob refrigeração. *Ciência e Agrotecnologia*, **24**, 174–181.

Evangelista, R.M., Chitarra, A.D. and Chitarra, M.I.F. (2002) Mudanças na ultraestrutura da parede celular de mangas Tommy Atkins tratadas com cloreto de cálcio na pré-colheita. *Revista Brasileira de Fruticultura*, **24**, 254–257.

Freire-Júnior, M. and Chitarra, A.B. (1999) Effect of calcium chloride application on mango fruit cv. Tommy Atkins hydrothermally treated. *Pesquisa Agropecuária Brasileira*, **34**, 761–769.

Galán Saúco, V. (ed.) (2009) *El Cultivo del Mango*, 2nd edn. Editorial Mundi Prensa, Madrid. 340 pp.

Hojo, R.H., Sao-Jose, A.R., Hojo, E.T.D., Alves, J.F.T., Reboucas, T.N.H. and Dias, N.O. (2009) Quality of "Tommy Atkins" mangoes in post-harvest with calcium chloride spray use in the preharvest period. *Revista Brasileira de Fruticultura*, **31**, 62–70.

Hossain, M.A., Rana, M.M., Kimura, Y. and Roslan, H.A. (2014) Changes in biochemical characteristics and activities of ripening associated enzymes in mango fruit during the storage at different temperatures. *BioMed Research International*. doi: 10.1155/2014/232969.

Karunanayake, K.O.L.C., Sinniah, G.D., Adikaram, N.K.B., Abayasekara, C.L. and Wijayasekara, D.S. (2015) Retention of latex at harvest, enhanced mango (*Mangifera indica* L.) fruit resistance and reduced anthracnose and stem-end rot. *Australasian Plant Pathology*, **44**, 113–119.

Kishor, A., Srivastav, M., Dubey, A.K., Singh, A.K., Sairam, R.K. et al. (2009) Paclobutrazol minimises the effects of salt stress in mango (*Mangifera indica* L.). *Journal of Horticultural Science and Biotechnology*, **84**, 459–465.

Lalel, H.D.J., Singh, Z. and Tan, S.C. (2003a) The role of ethylene in mango fruit aroma volatiles biosynthesis. *Journal of Horticultural Science and Biotechnology*, **78**, 485–496.

Lalel, H.D.J., Singh, Z. and Tan, S.C. (2003b) The role of methyljasmonate in mango ripening and biosynthesis of aroma volatile compounds. *Journal of Horticultural Science and Biotechnology*, **78**, 470–484.

Léchaudel, M. and Joas, J. (2007) An overview of preharvest factors influencing mango fruit growth, quality and postharvest behavior. *Brazilian Journal of Plant Physiology*, **19**(4), 287–298.

Litz, R.E. and Lavi, U. (1997) Biotechnology. In: *The Mango, Botany, Production and Uses* (ed. R.E. Litz), pp. 401–423. CAB International, Wallingford, UK.

Magwaza, L. and Kruger, F.J. (2005) Development of skin moisture and pulp nitrogen content parameters for the control of lenticel damage in Tommy Atkins and Keitt mangoes. *South African Mango Grower's Association*, **25**, 19–22.

Malik, A.U. and Singh, Z. (2004) Endogenous free polyamines of mango fruit in relation to growth, development and ripening. *Journal of the American Society for Horticultural Science*, **129**, 280–286.

Malik, A.U., Singh, Z. and Tan, S.C. (2006) Exogenous application of polyamines improves shelf life an fruit quality of mango. *Acta Horticulturae*, **619**, 291–297.

Mathooko, F.M., Kahangi, E.M., Runkuab, J.M., Onyangob, C.A. and Owinob, W.O. (2011) Preharvest mango (*Mangifera Indica* L. Apple) fruit bagging controls lenticel discolouration and improves postharvest quality. *Acta Horticulturae*, **906**, 55–62.

Matoo, A.K. and Modi, V.V. (1969) Ethylene and ripening of mangoes. *Plant Physiology*, **44**, 308–310.

Moazzam, A., Tahir, F.M., Shahzad, J. and Mahmood, N. (2011) Effect of foliar application of micronutrients on the quality of mango (*Mangifera indica* L.) cv. Dusehri fruit. *Mycopathologia*, **9**(1), 25–28.

Morton, J. (1987) Mango. In: *Fruits of Warm Climates* (ed. J.F. Morton), pp. 221–239. Miami, FL.

Mouco, M.A.D., Ono, E.O., Rodrigues, J.D. and Silva, G.J.N. (2013) Plant regulators on vegetative growth of Tommy Atkins mangoes. *Acta Horticulturae*, **992**, 187–192.

Moyo, M. and Van Staden, J. (2013) Micropropagation of Anacardiaceae species of economic importance: advances and future prospects. *In Vitro Cellular and Developmental Biology-Plant*, **49**, 85–96.

Mukherjee, S.K. (1949) The taxonomic value of the anatomical structures of the inflorescence axes of *Mangifera* L. *Journal of the Indian Botanical Society*, **28**, 162-171.

National Department of Agriculture of South Africa (2000) *Cultivation of Mangoes* [Online]. Available: http://www.nda.agric.za/docs/Infopaks/mango.htm [6 April 2015].

Nguyen, H., Hofman, P., Holmes, R., Bally, I., Stubbings, B. and McConchie, R. (2004) Effect of nitrogen on the skin colour and other quality attributes of ripe "Kensington Pride" mango (*Mangifera indica* L.) fruit. *The Journal of Horticultural Science & Biotechnology*, **79**, 204–210.

Núñez-Elisea, R. and Davenport, T.L. (1995) Effect of leaf age, duration of cool temperature treatment, and photoperiod on bud dormancy release and floral initiation in mango. *Scientia Horticulturae*, **62**, 63–73.

Pandey, P., Singh, A.K., Dubey, A.K. and Dahuja, A. (2014) Biochemical and salt ion uptake responses of seven mango (*Mangifera indica* L.) rootstocks to NaCl stress. *Journal of Horticultural Science and Biotechnology*, **89**, 367–372.

Parikh, H.R., Nair, G.M. and Modi V.V. (1990) Some structural changes during ripening of mangoes (*Mangifera indica* var. Alphonso) by abscisic acid treatment. *Annals of Botany*, **65**, 121–127.

Raja, E.M. and Anil Kumar, S.C. (2003) Boron deficiency: A serious yield-limiting factor in Alphonso mango of Konkan, Maharashtra. In: *Proceedings of 68th Annual Convention of the Indian Society of Soil Science*, Kanpur, India.

Raja, E.M., Anil Kumar, S.C. and Raju, S.Y. (2005) Boron deficiency in mango (*Mangifera indica* L.): A cause delineation study in acidic soils of Maharashtra, India. *Soil Science and Plant Nutrition*, **51**, 751–754.

Ram, S. (1997) Propagation. In: *The Mango, Botany, Production and Uses* (ed. R.E. Litz), pp. 363–400. CAB International, Wallingford, UK.

Rani, R. and Brahmachari, V.S. (2003) Effect of foliar application of chemicals on quality characters of mango during storage. *Progressive Horticulture*, **35**, 216–218.

Razzaq, K., Khan, A.S., Malik, A.U. and Shahid, M. (2013) Ripening period influences fruit softening and antioxidative system of Samar Bahisht Chaunsa mango. *Scientia Horticulturae*, **160**, 108–114.

Reuveni, O. and Castoriano, M. (1997) Beneficial effects of slow release fertilizers incorporated into the rooting medium of mango cuttings. *Acta Horticulturae*, **455**, 512–517.

Reuveni, O. and Golubowick, S. (1997) Trial of using *in vitro* techniques for vegetative propagation of mangoes. *Acta Horticulturae*, **455**, 496–504.

Ribeiro, S.M.R. and Schieber, A. (2010) Bioactive compounds in mango (*Mangifera indica*). In: *Bioactive Foods in Promoting Health: Fruits and Vegetables* (eds R.R. Watson and V.R. Preedy), Section C, pp: 507–524. Academic Press, Elsevier, UK.

Sampaio, V.R., Scarpare Filho, J.A. and Kluge, R.A. (1999) Distúrbios fisiológicos da manga: efeito da aplicação de Ca em pulverização foliar. *Scientia Agricola*, **56**, 459–463.

Saran, S.L. and Ratan, K. (2011) Boron deficiency disorders in mango (*Mangifera indica*): Field screening, nutrient composition and amelioration by boron application. *Indian Journal of Agricultural Sciences*, **81**, 506–510.

Schaffer, B., Larson, K.D., Snyder, G.H. and Sánchez, C.A. (1988) Identification of mineral deficiencies associated with mango decline by DRIS. *HortScience*, **23**, 617–619.

Shah, K.A., Patel, M.B., Patel, R.J. and Parmar, P.K. (2010) *Mangifera Indica* (Mango). *Pharmacognosy Review*, **4**, 42–48.
Sharma, R.R. and Singh, R. (2009) The fruit pitting disorder. A physiological anomaly in mango (*Mangifera indica* L.) due to deficiency of calcium and boron. *Scientia Horticulturae*, **119**, 388–391
Shivashankar, S. (2014) Physiological disorders of mango fruit. *Horticultural Reviews*, **42**, 313–348.
Singh, R.R. and Rajput, C.B.S. (1977) Effect of various concentrations of zinc on vegetative growth characters, flowering, fruiting and physicochemical composition of fruits in mango (*Mangifera indica* L.) cv. Chaunsa. Hary. *Journal of Horticultural Sciences*, **6**, 10–14.
Singh, Z., Singh, R.K., Sane, V.A. and Nath, P. (2013) Mango: Postharvest biology and biotechnology. *Critical Reviews in Plant Sciences*, **32**, 217–236.
Silva, A.V.C. and Menezes, J.B. (2001) Caracterização físico-química da manga Tommy Atkins submetida à aplicação de cloreto de cálcio em pré-colheita e armazenamento refrigerado. *Scientia Agricola*, **58**, 67–72.
Ubwa, S.T., Ishu, M.O., Offem, J.O., Tyohemba, R.L. and Igbum, G.O. (2014) Proximate composition and some physical attributes of three mango (*Mangifera indica* L.) fruit varieties. *International Journal of Agronomy and Agricultural Research*, **4**, 21–29.
USDA [United States Department of Agriculture] (2015) National nutrient database for standard reference release 27. http://ndb.nal.usda.gov/ndb/foods.
Van Eeden, S.J. (1992) Calcium infiltration as a possible postharvest treatment to increase storage potential of mango fruit. *South African Mango Growers' Association Yearbook*, **12**, 26–27.
Wahdan, M.T., Habib, S.E., Bassal, M.A. and Qaoud, E.M. (2011) Effect of calcium chloride and gibberellic acid on storability of "Succary Abiad" mango fruits under cold storage. *Journal of American Science*, **7**(5), 493–501.
Wang, B., Wang, J., Feng, X., Lin, L., Zhao, Y. and Jiang, W. (2009) Effects of 1-MCP and exogenous ethylene on fruit ripening and antioxidants in stored mango. *Journal of Plant Growth Regulation*, **57**, 185–192.
Whiley, A.W. (1986) Crop management: A review. *Proceedings of the First Australian Mango Research Workshop*, Melbourne, pp. 184–195.
Xiuchong, Z., Guojian, L., Jianwu, Y., Shaoying, A. and Lixian, Y. (2001) Balanced fertilization on mango in Southern China. *Better Crops International*, **15**, 16–20.
Yahia, E.M. (2011) Mango. In: *Postharvest Biology and Technology of Tropical and Subtropical Fruits, vol. 3: Cocona to Mango* (ed. E.M. Yahia), pp. 492–565. Woodhead Publishing Limited, Cambridge.
Yahia, E.M., Ornelas-Paz, J. de J. and Ariza, F.R. (2006) *El Mango*. Editorial Trillas, México City, 224 pp.
Yashoda, H.M., Prabha, T.N. and Tharanathan, R.N. (2006) Mango ripening: Changes in cell wall constituents in relation to textural softening. *Journal of Food Science and Agriculture*, **86**, 713–721.
Young, T.W., Koo, R.C.J. and Miner, J.T. (1962) Effects of nitrogen, potassium and calcium fertilization on Kent mangoes on deep, acid, sandy soil. *Proceedings of the Florida State Horticultural Society*, **75**, 364–371.
Zaharah, S.S. and Singh, Z. (2011) Mode of action of nitric oxide in inhibiting ethylene biosynthesis and fruit softening during ripening and cold storage of Kensington Pride mango. *Postharvest Biology and Technology*, **62**, 258–266.

Zaharah, S.S., Singh, Z., Symons, G.M. and Reid, J.B. (2012) Role of brassinosteroids, ethylene, abscisic acid, and indole-3-acetic acid in mango fruit ripening. *Journal of Plant Growth Regulation*, **31**, 363–372.

Zerbini, P.E., Vanoli, M., Rizzolo, A., Grassi, M., Pimentel, R.M.D. et al. (2015) Optical properties, ethylene production and softening in mango fruit. *Postharvest Biology and Technology*, **101**, 58–65.

Zuazo, V.H.D., Raya, A.M. and Ruiz, J.A. (2004) Impact of salinity on the fruit yield of mango (*Mangifera indica* L. cv. Osteen). *European Journal of Agronomy*, **21**, 323–334.

4

Pests of Mango

Daniel Carrillo[1], Andrea Birke[2], Larissa Guillen[2] and J.E. Peña[1]

[1] *Tropical Research and Education Center, University of Florida, Homestead, Florida, USA*
[2] *Instituto de Ecología A.C., Congregación El Haya, Veracruz, Mexico*

4.1 Introduction

Many species of insects and mites occur in mango. Some are pests, but most species cause little or no economic damage (Peña *et al.* 2008). Mango cultivation has increased exponentially in tropical America and other mango producing regions in response to increasing consumer demand. While increased demand has led to export opportunities for mango producing countries, the importing countries have become more restrictive with respect to fruit quality, demanding appropriate applications of phytosanitary treatments and reducing the levels of agrochemicals present in mangoes. Given these market and food safety dynamics, most mango growers try to implement best pest management practices for mango fruit. These include cultural control, development of resistant cultivars, biological control, and the use of registered pesticides. The most important mango pests are fruit flies, followed by scales, mealybugs, and thrips. This chapter reviews mites and the major insects in mango and the current technologies and practices for their management.

4.2 Mites

The most important tetranychids mites affecting mango are *Oligonychus yothersi* McGregor, *O. coffeae*, *O. biharensis*, *O. mangiferae*, *Vasates mangiferae*, *Paratetranychus* sp., *Panonychus citri* McGregor, and *Eotetranychus sexmaculatus* Riley (Yamaguchi *et al.* 2006; Zaher and Osman 1971; Zaman and Maiti 1994). Most of these mites attack and injure older mango leaves. As mites feed, dark patches appear on the leaf surface, until the entire leaf becomes deeply bronzed, necrotic, and with abscesses; thus growth is retarded.

The mango bud mite, *Aceria mangiferae* (Sayed) (Acari: Eriophyidae), attacks buds and inflorescences (Ochoa *et al.* 1994). According to Jeppson *et al.* (1975), this mite stunts and brooms twigs, causing bud proliferation and appears to be responsible for

Handbook of Mango Fruit: Production, Postharvest Science, Processing Technology and Nutrition,
First Edition. Edited by Muhammad Siddiq, Jeffrey K. Brecht and Jiwan S. Sidhu.

necrosis of bud tissue cells. However, *A. mangiferae* does not cause mango malformation, but may be a carrier of *Fusarium mangiferae*, which is recognized as the causal agent of mango malformation. Most of these mites can be controlled by natural enemies.

4.3 Insects

4.3.1 Hemipterans

The yellowish-green coreid bugs, *Amblypelta lutescens lutescens* (Distant) and *Amblypelta nitida* Stål, occur along the coast of Queensland, Australia, and attack most of the tropical and subtropical fruits, including mango (Cunningham 1984). They prefer to feed on young, green fruit, but *A. l. lutescens* also damages the terminals of a number of hosts. In tropical north Queensland, *A. l. lutescens* is the dominant species and feeds on young fruit, causing black lesions to develop and the fruit to fall. It also feeds on the terminals and leaf petioles, causing wilting and dieback (Cunningham 1984). Nymphs and adults inject salivary secretions into the plant tissue while feeding, which causes significant damage to fruit structures, thereby making it unfit for sale (Danne *et al.* 2013).

4.3.1.1 White and Black Flies

The two species of economic importance are the whitefly, *Aleurodicus dispersus* Russell and the blackfly, *Aleurocanthus woglumi* Ashby. The whiteflies suck cell sap from leaves and excrete honeydew upon which sooty mold develops. High infestations can almost blacken entire trees, reducing photosynthetic efficiency and causing defoliation (Angeles 1991; Peña 1993). A number of parasitoids, for example, *Encarsia opulenta* (Silvestri) and *Amitus hesperidus* (Silvestri) attack the immature stages and provide good control.

4.3.1.2 Leafhoppers

Peña *et al.* (2008) reported that approximately 18 species of leafhoppers have been reported as pests of mango. Of these, *Idioscopus clypealis* Leth., *I. niveosparsus* Leth., *I. magpurensis* Pruthi, and *Amritodus atkinsoni* Leth., are important (Fletcher and Dangerfield 2002; Viraktamath 1997). The females deposit their eggs in the panicles or midribs of tender leaves. The adults and nymphs preferentially feed on young leaves and flowers or shoots, and excrete honeydew upon which sooty mold develops. This interferes with photosynthesis, adversely affecting plant growth and yield (Godase *et al.* 2004). The affected inflorescences turn brown, become dehydrated, and fruit set does not occur.

Several natural enemies have been described from West and Southeast Asia. Mohyuddin and Mahmood (1993) reported the egg parasitoids, *Gonatocentrus* sp., *Miurfens* sp. nr. *mangiferae* Viggiani and Hayat, *Centrodora* sp. nr. *scolypopae* Valentine, *Aprostocetus* sp., and *Quadrastichus* sp., and the adult ectoparasitoid *Epipyrops fuliginosa* Tames in Pakistan. Peng and Christian (2005) reported that the weaver ant, *Oecophylla smaragdina* (Hymenoptera: Formicidae) is an efficient biocontrol agent of *I. nididulus* in Northern Australia. The entomopathogens, *Verticillium lecanii* (Zimmerman) Viegas, *Beauveria bassiana* Balsamo (Vuillemin), and *Isaria tax*, infect *I. clypealis* in India (Kumar *et al.* 1983), while the effectiveness of *Metarhizium anisopliae* var. *anisopliae* was tested under laboratory conditions against *A. atkinsoni* (Vyas *et al.* 1993).

4.3.2 Scales

4.3.2.1 Armored Scales

At least 26 species of diaspidids attack mangoes worldwide (Chua and Wood 1990). In India, *Aspidiotus destructor* Signoret causes serious damage, while *Parlatoria pergandii* Comstock and *Lepidosaphes gloverii* (Packard) typically damage 3-year-old plants (Chua and Wood 1990). Several diaspidids, for example, *Aulacaspis mangiferae* (*tubercularis*) Newstead, attack shoots and leaves, including oleander scale in Florida (Miller and Davidson 2005), and the mango scale in Ghana (Halteren 1970) cause similar damage. Scales are damaging not only because they feed on sap, but also because of the toxicity of their saliva (Singh 1991). The most effective biological control agents in Ecuador are *Coccidophilus* sp., and *Chrysopa* sp.

4.3.2.2 Soft Scales

The soft brown scale (*Coccus hesperidum* L.) and the mango shield scale (*Milviscutulus mangiferae* Green) occur in most tropical and subtropical countries. The damage is mainly caused from large amounts of honeydew excreted by the insects on the leaves and fruit, which cause thick layers of sooty mold. Leaf drop can occur, which may lead to crop reduction in the following season.

Several natural enemies of these pests are found around the world, including the parasitoids *Coccophagus pulvinariae* Compere, *Scutellista cyanea* Motschulsky, and *Aprostocetus* sp. *Tetrastychus* sp., as well as an unidentified predatory beetle (Peña *et al.* 2008).

4.3.3 Mealybugs

The margarodid mango mealybug, *Drosicha mangiferae* (Green), is a common pest of mango in India and Pakistan (Mohyuddin and Mahmood 1993). The pseudococcid fruit tree mealybug, *Rastrococcus invadens* Williams, is a serious pest of several crops, including mango in West Africa (Agounké *et al.* 1988). Tobih *et al.* (2002) reported that *R. invadens* reduced weight, size, and sugar content of infested mango fruit. Various control methods, including banding tree trunks with various materials to prevent *D. mangiferae* nymphs from climbing the tree, have been tried with little success (Lakra *et al.* 1980; Srivastava 1981). In Pakistan, the mango mealybug was controlled by hoeing or plowing the soil and conserving the predator, *Sumnius renardi* Weise, and by wrapping burlap around the trunks of the trees (Mohyuddin and Mahmood 1993). Ashfaq *et al.* (2005) reported that a funnel-type slippery trap made of a thick smooth transparent plastic ribbon was more effective against mango mealybug nymphs and to entrap females carrying eggs than the slippery bands.

In Africa, Neuenschwander *et al.* (1994) reported the establishment of the endophagous parasitoids, *Gyranusoidea tebygi* and *Anagyrus mangicola*, against *R. invadens* in Benin, Gabon, Sierra Leone, Nigeria, and Zaire; and Pitan *et al.* (2000) confirmed that *G. tebgi* reduced approximately 87% the mealybug infestation. Pitan *et al.* (2000) also reported the presence of the coccinellids, *Exochonus promputs* Weise, *Chilocorus nigritus* (F.), and *Nephus* sp., and the crysopids *Ceratochrysa autica* (Walker) and *Plesiochrysa* sp. in Nigeria.

4.3.4 Thrips

The western flower thrips, *Frankliniella occidentalis* (Pergande), damages flowers and fruits in Israel (Wysoki *et al.* 1993). In South Africa, Brink (1994) reported the presence of *F. occidentalis* and *Thrips tenellus* on mango inflorescences, but no evidence was found that these thrips affected fruit set. In Florida, the thrips complex consisting of *F. bispinosa* (Morgan) and *F. kelliae* (Sakimura) are the most frequently observed blossom pest on flowers and cause damage by ovipositing in the panicle and feeding on the floral nectaries and anthers, which may result in premature loss of pollen. Grove *et al.* (2001) reported that *Thrips acaciae* Thybom, *T. tenellus* Trybom, and *S. aurantii* were the most abundant species collected from mango flowers in South Africa, and the damage was on the newly set fruit. In Mexico, the thrips complex include species of the genera *Frankliniella* (*F. borinquen* Hood, *F. gossypiana* Hood, *F. williamsi* Hood complex, *F. cephalica* (Crawford), *F. gardenia* Moulton, *F. invasor* Sakamur, and *F. parvula* Hood), *Haplothrips, Arorathrips,* and *Scirtothrips* (Rocha *et al.* 2012). In Korea, the dominant thrips affecting mango are *Thrips palmi* and *S. dorsalis* (Choi *et al.* 2013). In Malaysia, *Thrips hawaiiensis* (Morgan) were the dominant thrips on mango panicles and weeds in insecticide-treated orchards, whereas *S. dorsalis* were the dominant thrips in untreated orchards (Aliakbarpour and Rawi 2012).

Several parasitoids and predators, for example, *Ceranisus menes* (Walker) in Israel (Rubin and Kuslitizky 1992), and the predators, *Orius* sp., *Anystis agilis* Banks, and *Hypoaspis aculifer* (Canestrini) (Loomans *et al.* 1995), are candidates for biological control of *F. occidentalis.*

The Mediterranean mango thrips, *Scirtothrips mangiferae* Priesner, is a severe pest of mango in Israel, causing the young leaves to curl along the midrib, distorting their shape, and leading to premature drop (Wysoki *et al.* 1993).

The red-banded thrips, *Selenothrips rubrocinctus* (Giard), is an important pest of cacao in the Caribbean Islands and attacks mango and avocado in Australia and USA (Florida and Hawaii). The adults feed on the underside of leaves, causing necrosis and subsequent leaf drop. According to Hill (1975), *S. rubrocinctus* is only a pest in mango nurseries, and rarely damages mature trees. Its biology was reviewed by Moznette (1922). Adult thrips are dark-bodied with a red band on the first abdominal segment. The immature stages are light orange with abdominal segments one and two and the anal segments bright red. The population of this species peaks during the dry season and declines during the rainy season.

The weaver ant, *Oecophylla smaragdina* (Fabricius) (Hymenoptera: Formicidae), is considered an effective biological control of *Selenothrips rubrocinctus* in the Northern Territories of Australia (Peng and Christian 2004).

4.3.5 Diptera

The mango gall midge or mango blister midge, *Erosomyia mangiferae* (Felt), is a major pest, destroying flowers and up to 70% of set fruit. Barnes (1948) recognized nine gall midges from mango; two of these, *Asynapta* sp. and *E. mangiferae*, are from the West Indies. Butani (1979) reported five cecidomyiid species on mango blossoms, including *E. indica* (Grover and Prasad). *Dasineura mangiferae* (Felt) was reported in Hawaii (Vannière *et al.* 2004). *Dasineura mangiferae* is known to cause up to 100% reduction in yield in Pakistan (Rehman *et al.* 2014). Male adults of *E. mangiferae* are 1.61 mm and females 1.32 mm long. Eggs are deposited in folds between sepals and petals of flower buds. The larva stage has four instars; young larvae are cream colored and late instar larvae are yellowish. Larval feeding prevents flower opening, thereby preventing any fruit development. Infested buds develop as long pointed galls, in which pupation occurs (Vannière *et al.* 2004).

Studies of population fluctuation of *Erosomyia* sp. have been conducted in India by Grover (1986a), who reported that emergence of adults was higher at 24 °C and 60–82% relative humidity (RH) compared to lower temperatures and RH. Abbas (1985) described systematic surveys to determine the percentage of infestation of *E. indica*, and showed that infestation follows a negative binomial infestation. The midge infests the newly emerged panicles by ovipositing at bud burst stage, and the first instar maggots bore into the growing panicle. The second generation then infests very young fruits, which drop before the marble stage. Sampling of mango midges needs to include affected tissue, different trapping devices, and pheromones, etc. *Dasineura mangiferae* was considered to be uniformly distributed on mango trees by Rehman *et al.* (2013). On citrus, use of colored sticky traps placed in the tree canopy provides a more efficient method of sampling the citrus midge, *Prodiplosis longifila* Gagné, than ground emergence traps and collection of larval samples (Peña and Duncan 1992). Rehman *et al.* (2014) determined that yellow sticky traps were more effective than green, blue, and transparent traps to monitor adults of *D. mangiferae.*

In a survey of parasitoids of cecidomyiid pests of mango in India, Grover (1986b) reported that *Platygaster* sp., *Systasis* sp., and *Eupelmus* sp. were associated with *Dasineura sp.*, and *Tetrastychus* sp. was associated with *Erosomyia indica*. An external parasitoid, the pteromalid, *Pirene* sp., attacked *Procystiphora mangiferae* (Felt). Predators of the cecidomyiids include *Formicai* sp., *Oecophylla* sp., and *Camponotus* sp.

Mango leaves are attacked by different Cecidomyiidae species, especially in Asia, but also in the Caribbean region. Two genera, *Procontarinia* Kieffer and Cecconi and *Erosomyia* Felt, are particularly associated with mango and all known species have been reared from mango (Harris and Schreiner 1992). Prasad (1971) described the biology of the main species attacking mango in India. A new species of gall midge, *P. schreineri* Harris, which attacks mango foliage in Guam, lays eggs on young mango leaves and larvae, and develop rapidly over about 5 days, inducing blister galls. According to Harris and Schreiner (1992), the main factors affecting populations of this midge are rainfall and location. More galls are present during rainy periods, possibly because RH improves larval and pupal survival. Daneel *et al.* (2000) suggested that products to control *P. matteiana* should be applied after harvest, coinciding with the first major flush and a second spray 6 weeks later. Rehman *et al.* (2014) reported that a neem seed kernel extract integrated with soil racking was the best option for the control of *D. mangiferae.*

4.3.6 Fruit Flies

Most mango-producing countries are in fruit fly infested tropical areas. Fruit flies affect the mango industry in several ways:

- by lowering of fruit production as larvae feed from mango pulp and make fruit unmarketable;

- causing an increase in production costs to protect fruits from fly oviposition; and
- restricting free trade and movement of fruits from countries and regions of fruit fly prevalence to fruit fly free areas (Aluja and Mangan 2008; Vannière *et al.* 2004).

For these reasons, tephritid fruit flies are considered to have the greatest impact on the international mango trade (Aluja and Mangan 2008; Badii *et al.* 2015).

Approximately 60 fruit fly species are reported to attack mango and a related species, *Mangifera foetida* Lour (Aluja *et al.* 1996; Clarke *et al.* 2005; Norrbom 2004; Vayssières *et al.* 2005; White and Elson-Harris 1992). Depending on the continent, fruit fly species attacking mango belong to different genera: *Anastrepha* in the Americas, *Bactrocera* in Africa, Asia, and Australia (only one invasive species, *Bactrocera carambolae* Drew and Hancock has been recorded in the Americas infesting mangoes, specifically in Surinam), and *Ceratitis* in Africa, although the most invasive fruit fly species, *Ceratitis capitata* (Wiedemann), has been recorded attacking mangoes in Africa and the Americas. Additionally, two species of *Dirioxa* have often been cited as mango pests, but they probably are opportunistic species that only attack previously damaged mango fruit and these records need to be verified (Lux *et al.* 2003, Norrbom 2004; SPC 2005; Vayssières and Kalabane 2000; White and Elson-Harris 1992).

The following species are referred to as the mango fruit fly: *Anastrepha obliqua* Macquart, *Bactrocera frauenfeldi* Schiner, *Bactrocera dorsalis* Hendel, and *Ceratitis cosyra* (Walker) (Aluja 1993; Lux *et al.* 2003; Migani *et al.* 2014; Steck 2003). *Bactrocera invadens* (Drew) also has been referred to as a mango fruit fly, but recent studies on mitochondrial DNA, wing shape, chemical ecology, chemotaxonomy, and mating compatibility have revealed that *B. invadens, B. dorsalis, Bactrocera papayae*, and *Bactrocera philippinensis* are the same biological species (Bo *et al.* 2014; Boykin *et al.* 2014; San Jose *et al.* 2013; Schutze *et al.* 2012; Tan *et al.* 2013).

4.3.6.1 Anastrepha

Anastrepha spp. are endemic to North and South America and their range extends from the southern USA to northern Argentina, including the Caribbean islands (Aluja 1994). Twelve *Anastrepha* species have been purportedly associated with mango (Norrbom 2004). Of these, *A. obliqua, Anastrepha ludens* (Loew), *Anastrepha suspensa* (Loew), and *Anastrepha fraterculus* (Wiedemann) stand out as economically important pests of mangoes (Aluja *et al.* 1996; Malavasi *et al.* 1980; Norrbom 2004; White and Elson-Harris 1992).

Most of the knowledge about *Anastrepha* fruit fly biology is based on studies carried out between 1900 and 1944 (Aluja 1994). The basic life cycle is very similar among the most pestiferous *Anastrepha* species. For example, egg incubation of *A. ludens* in mango requires 3.8 days, larval development requires 14.2 days, and pupal development requires 14.2 day at 27 ± 2°C (Leyva *et al.* 1991). Larvae pass through three instars before emerging from the fruit and burrowing into the ground to pupate (Aluja 1994). Clutch size varies between one egg/clutch in *A. obliqua* and more than 100 eggs in *Anastrepha grandis* Mcquart (Aluja 1994). Some adults can live for more than 150 days under laboratory conditions. Population fluctuation patterns can vary greatly among orchards within a fairly small geographic region, and Aluja *et al.* (2012, 2014) observed that these variations also are influenced by local and global climate variation and the presence of host tree reservoirs.

4.3.6.2 Bactrocera

Bactrocera spp. are pests of mango in Africa, Asia, and Australia (Clarke *et al.* 2005; Drew 1989; Leblanc *et al.* 2009; Tenakanai 1997), where they can cause losses of up to 70% in mango production, depending on the region (Stonehouse *et al.* 1998; Vayssières *et al.* 2009b).

The biology of fruit flies in the genus *Bactrocera* has been addressed by Aluja and Norrbom (2000), Bateman (1972), Christenson and Foote (1960), Clarke *et al.* (2005), Fletcher (1987), and White and Elson-Harris (1992). As in other pestiferous flies, females lay their eggs beneath the fruit skin or in the pulp, especially in ripening fruit, which is not the case for species belonging to other genera, such as *Anastrepha* spp. In the case of most female fruit flies, white banana-shaped eggs are usually deposited in clusters, hatching after 1–2 days (Fletcher (1987). In some species, a single female can lay over 1,000 eggs over her lifetime (White and Elson-Harris 1992). Studies of *B. dorsalis* in India (Singh 1991) indicated that the pupal period was longest (18 days) at 15 °C and shortest (6 days) at 35 °C. Warm, humid weather is favorable for most *Bactrocera* fruit flies, and the pests' population increases as mango ripening occurs.

In *B. invadens* (*B. dorsalis*), host acceptance is mainly influenced by egg load and not by mango ripeness. Although it has been reported that *B. invadens* mainly attacks ripe fruits, recent reports show that green mature fruit are also viable hosts (Diatta *et al.* 2013). This differs in other fruit fly species, which prefer ripe and overripe fruits (*C. cosyra*), or turning-yellow mangoes in the case of *A. obliqua* and *A. ludens* (Díaz-Fleischer and Aluja 2003; Migani *et al.* 2014).

4.3.6.3 Ceratitis

This genus is native to Africa, where eight *Ceratitis* spp. have been reported to attack mangoes. Depending on the locality, season, and variety, yield loss of mango due to *Ceratitis* spp. ranges between 30 and 70% (Badii *et al.* 2015; Lux *et al.* 2003). The polyphagous Mediterranean fruit fly, *C. capitata*, is the most recognized pest in mango-growing areas of Hawaii, Israel, Australia, Spain, Mexico, La Réunion, Brazil, and elsewhere in Central and South America (Barbosa 2005; Eskafi and Cunningham 1987; Galan-Sauco 1990; White and Elson-Harris 1992). *Ceratitis silvestrii* is considered to be of economic importance in Benin, where it attacks early mango cultivars (Vayssières *et al.* 2005, 2009b). *Ceratitis cosyra* is broadly distributed across Africa and causes enormous damage that can result in the total loss of the crop (Lux *et al.* 2003).

The biological cycle of *Ceratitis* spp. varies greatly according to the species, host, and environmental conditions, particularly temperature, altitude, and humidity (Duyck and Quilici 2002; Papadopoulos *et al.* 1996). The time when females seek suitable sites for oviposition and puncture mango fruit varies according to the species. The developmental period of immature stages is 3–7 weeks, and 5–10 generations per year can occur, depending on temperature and other factors intrinsic to the fly population (Delrio *et al.* 1986; Duyck and Quilici 2002; Duyck *et al.* 2006; Fletcher 1987; Hill 1975).

4.3.7 Fruit Fly Control

The global trends in control and eradication programs of fruit fly contribute to the development of economically-viable, practical, and environmentally benign tools that can be used alone, or in combination, in Integrated Pest Management (IPM) schemes for area-wide pest control programs (AW-IPM) (Aluja and Rull 2009; Badii *et al.* 2015; Dyck *et al.* 2005; Tan 2000). The major objective is to overcome the often ineffective and unsustainable control schemes resulting from uncoordinated actions by individual producers. It has been shown that orchard-by-orchard IPM practices are useful, but are insufficient for effective control since flies can move large distances (MacFarlane *et al.* 1987) and can be easily displaced from untreated host patches in adjacent areas to treated orchards (Aluja and Rull 2009; Lewis *et al.* 1997; Lloyd *et al.* 2010). Moreover, since fruit flies that attack mango, also attack other fruit crops in the same area, their management must be based not only on mango, but also on wild hosts and other commercially grown host plants (Aluja *et al.* 1996). Thus, to improve the efficiency of fruit fly management, host plant flowering, fruiting, mango variety, and abiotic factors such as daily rainfall, temperature, and humidity, among others, need to be considered, since these factors influence infestation patterns (Aluja *et al.* 1996, 2014; Vayssières *et al.* 2009a).

Recently, some programs applying the area-wide management concept to fruit fly control in mango and other fruit producer areas around of the world have been tested with promising results (Lloyd *et al.* 2010; Ndiaye *et al.* 2008; Patel *et al.* 2005). These programs have employed combinations of different techniques such as male annihilation, bait sprays, sanitation through the destruction of fallen fruit, Sterile Insect Technique (SIT), and cultural and biological control. Aluja (1993, 1996) and Aluja *et al.* (1996) have suggested that fruit fly management schemes could also include a border trapping system, enhancement of host-plant resistance through use of plant growth regulators, mass trapping for suppression purposes, and augmentative parasitoid releases (Dyck *et al.* 2005; Mangan and Moreno 2007; Montoya *et al.* 2000; Sivinski *et al.* 1996; Tan 2000). However, although these techniques or control measures are available, the accomplishment of these goals depends on the economic resources available to growers access to technology, cost, scale (single orchard, regional level, national level), globalization of markets, and local regulations that oversee the environment impact of pest control measures (Aluja 1993). The main control measures used in fruit fly management in mango orchards around the world are discussed below.

4.3.7.1 Monitoring

Monitoring fruit flies can help in achieving several goals:

- to determine which fruit fly pests are present in an area;
- to determine the distribution of pest species;
- to identify local host spots with high populations of the pest;
- to track changes in pest populations;
- to estimate the efficacy of control measures; and
- to facilitate early detection of new fruit fly pests in a particular area (Manrakhan 2006).

In general, infestation thresholds for adult fruit flies are quarantine-mediated (Beers *et al.* 1993). These thresholds vary from location to location but, depending on the fruit fly species, they are typically based on the capture of a single fruit fly of the target species. In other fruit crops, a threshold of 5 flies per trap is recommended, that can result in a reduction from 4 chemical sprays to 1.5 sprays per season (Beers *et al.* 1993).

Parapheromones are considered the most specific and efficient lures; their use for fruit fly control is referred to as the male annihilation technique (MAT) (Badii *et al.*

2015). The most frequently used parapheromones in fruit flies attacking mango are methyl eugenol (ME), analogs of raspberry ketone such as Cuelure and Melolure, and zingerone for *Bactrocera* spp. For *Ceratitis* spp., Trimedlure is frequently used, or analogues of this lure, such as Ceralure or terpinyl acetate. For *Anastrepha* spp., a parapheromone has not yet been identified and for this reason, food-based lures are employed (IAEA 2007).

ME has been used successfully for monitoring, control, and eradication of *B. dorsalis* in Hawaii (Steiner and Lee 1955), Rota Island (Steiner *et al.* 1965), Pakistan (Mohyuddin and Mahmood 1993), and several African countries (Badii *et al.* 2015; Lux *et al.* 2003), and also for *B. caryeae, B. correcta, B. tuberculata,* and *B. zonata* (Satarkar *et al.* 2009; White and Elson-Harris 1992). Cuelure and Melolure have been used for *B. frauenfeldi, B. kirki, B. melanota, B. occipitalis,* and *B. tryoni* (SPC 2005; White and Elson-Harris 1992), and Zingerone for *B. jarvasi* and *B. neohumeralis* (Fay 2012).

In the case of *Ceratitis* spp., Trimedlure is still considered an important parapheromone for the Mediterranean fruit fly. In contrast, terpinyl acetate is used for *C. cosyra,* which does not respond to Trimedlure, and for *C. rosa* both attractants can be utilized (Badii *et al.* 2015; Steck 2003).

Anastrepha spp. populations have been monitored using several food-based lures. since no parapheromones are known for this genus. The most frequently used lure to monitor fruit flies is a mixture of protein (torula yeast or hydrolyzed protein) and water. Occasionally, hydrolyzed cotton seed together with borax, molasses, or fermented fruit juices are also employed (Balock and López 1969; Jirón 1995). The Ceratrap® lure represents an efficient new option, which can be used in combination with any type of trap and is more effective for capturing flies than McPhail traps baited with hydrolyzed protein and water. Furthermore, it maintains its attraction activity for more than one month (Lasa and Cruz 2014). Also, dry synthetic food-based lures are available, such as BioLure® (Suterra LLC, Inc., Bend, Oregon, USA) (Heath *et al.* 1995; Robacker and Landolt 2002), and Nu-Lure® (Advanced Pheromone Technologies) (Robacker *et al.* 1997; Robacker 2001; Robacker and Warfield 1993). For *Anastrepha* spp., the two-component BioLure®, which contains ammonium acetate and putrescine, is effective, whereas for *C. capitata,* the three-component BioLure® that contains ammonium acetate, putrescine, and trimethylamine is used. Dry baits have been used for *A. ludens, A. obliqua, A. suspense,* and *C. capitata* with good results (Holler *et al.* 2009; Navarro-Llopis *et al.* 2008).

With respect to trap types, there are also several options (IAEA 2013). The most used trap until recently has been the McPhail trap; however, it has several drawbacks in that it is expensive, breaks easily, is cumbersome to clean or replace, and is inefficient. Consequently, this trap is being replaced by other McPhail-type traps comprising transparent plastic with a yellow colored lower half that are easier to service and clean. Aluja *et al.* (1989) found that only 31% of *Anastrepha* spp. flies landing on the McPhail trap were caught, with many flies entering the trap but then escaping. Although there are reports stating that high concentrations of McPhail traps can reduce the build-up of fly populations and protect mangoes from severe injury during certain periods of the year (Balock and López 1969), these results are not consistent (Díaz-Fleischer *et al.* 2009).

Due to its low efficiency and problems with its manipulation, the new McPhail-type traps such as MultiLure and Maxitrap are gaining importance. The new plastic traps are more efficient and practical and are compatible with liquid and dry baits/lures. Jackson and Lynfield traps baited with their respective dry lures have been used for *Ceratitis* spp. and *Bactrocera* spp. with good results. Lynfield traps are frequently preferred over McPhail and Jackson traps, because they are more practical, efficient, and cheaper (Cowley 1990; Dominiak and Nicol 2010). Other trap designs, such as "plastic bucket traps" baited with cotton wicks impregnated with a mixture of lure and Malathion, have been tested with good results for monitoring and for male annihilation schemes (Vargas *et al.* 2000).

For uncapitalized growers, the use of 500 mL transparent plastic bottles baited with liquid lures, with three 10 mm diameter lateral holes that allow the entry of flies, have been found to be as efficient as some commercial traps for the capture of *Anastrepha* spp. in Mexico (Lasa *et al.* 2014).

4.3.7.2 Chemical Control

Since 1950, poisoned-bait sprays have been recommended for fruit fly control in mango orchards. Usually, toxic bait sprays combine hydrolized protein with the insecticide malathion (Halawa *et al.* 2013; López *et al.* 1969; Mangan and Moreno 2007; Peck and McQuate 2000; Soto-Manatiú *et al.* 1987), although fruit flies are highly susceptible to many broad-spectrum insecticides including Dichlorvos, Dimethoate (organophosphate), Deltamethrin, and Alpha-cypermethrin.

The MAT is a measure in which many units of wooden blocks or cane disks (fiberboard made from sugarcane stalks with disks shape and yellow color) impregnated with a lure (methyl eugenol and/or, Cuelure) and Malathion are placed at sites with pest fruit fly infestations to massively attract males and to kill them (Cunningham and Steiner 1972; Ekesi and Billah 2006; Mwatawala *et al.* 2009). This technique has been successfully used for flies in the genus *Bactrocera* in different areas (Cunningham and Suda 1986; Vargas *et al.* 2000). However, restrictions on the use of Malathion and other insecticides imposed by importing countries, and its negative side effects, including insecticide resistance (Nadeem *et al.* 2014), secondary pest outbreaks (Knutson *et al.* 2011), and elimination of natural enemies (Ehler and Endicott 1984; Mohyuddin and Mahmood 1993), have motivated efforts to find more environmentally-friendly alternatives (Peck and McQuate 2000).

Phototoxic dyes such as Phloxine B, or GF120™ (Dow AgroScience), which is a mixture of a bacteria-derived insecticide (spinosad) and a food-based bait, have proven successful against various fruit fly species (Díaz-Fleischer *et al.* 1996; King and Hennessey 1996; McQuate *et al.* 2005; Peck and McQuate 2000). Since GF120™ contains a food bait, it is more specific and has proved to be one of the most effective alternative for fruit fly control against *C. capitata, C. cosyra, A. ludens, A. obliqua, A. suspensa, B. invadens* (*B. dorsalis*), and *B. cucurbitae* (Ekesi *et al.* 2014; King and Hennessey 1996; Vayssieres *et al.* 2009b). In Kenya, weekly bait sprays with Mazoferm E802-Spinosad have been tested to suppress *B. invadens* (*B. dorsalis*) populations in mango orchards with good results (Ekesi *et al.* 2014). Thomas and Mangan (2005) and Vargas *et al.* (2002) did not find negative effects in the fruit fly parasitoid *F. arisanus*, even after spinosad was sprayed in fruit-growing areas, which means that it is highly compatible with IPM programs (Vargas *et al.* 2001). Although GF120™ is effective, there are still hurdles to overcome, including its rapid decay (Cleveland *et al.* 2001), low persistence after rainfall (Prokopy *et al.* 2003), and the development of resistance (Hsu and Feng 2006) or the negative effects on some fruit fly parasitoids exposed to recent residues (Ruíz *et al.* 2008; Stark *et al.* 2004). However, spinosad remains an attractive option for the control of fruit fly populations. Another recent and more specific way to employ spinosad has been through the use of bait stations that comprises traps baited with a mixture of lure and insecticide (Mangan and Moreno 2007; Vayssieres *et al.* 2009b).

Mass trapping for suppression purposes is another system yielding positive results (Martínez-Ferrer *et al.* 2010; Navarro-Llopis *et al.* 2008) and that has the important added benefit of being environmentally-friendly and compatible with IPM programs. Its application cost and efficacy depends on the attractants, insecticides, and traps used. When depending on bait stations, its application cost can be high, but this limitation is being overcome with the development of new cost-effective lures and traps (Lasa *et al.* 2014; Vargas *et al.* 2001).

4.3.7.3 Biological Control

The most common natural enemies that attack fruit flies are insect parasitoids, followed by entomopathogens and predators.

4.3.7.3.1 Parasitoids

Insect parasitoids, usually wasps, reproduce by laying their eggs in fruit fly eggs, larvae, or pupae. Eggs hatch and the parasitoid larvae feed on the insect host until death. Classical biological control and repeated augmentative releases of mass-reared parasitoids have been used to suppress *Anastrepha, Bactrocera,* and *Ceratitis* populations (Montoya *et al.* 2000; Ndiaye *et al.* 2015; Sivinski *et al.* 1996, 1997). In Florida, Mexico, Costa Rica, Brazil, Colombia, and Peru, the parasitoids *Diachasmimorpha longicaudata* (Ashmead), *Fopius vandenboschi* (Fullaway), and *Aceratoneuromyia indica* (Silvestri) have been imported and released for the control of *A. suspensa, A. ludens, A. fraterculus,* and *A. obliqua* (Ovruski *et al.* 2000). Despite the widespread use of exotic parasitoids over the past 80–100 years, the current trend is to focus on the use of native species, which pose less of an environmental threat to local fauna (Aluja *et al.* 2009; García-Medel *et al.* 2007).

The use of parasitoids as biological agents of tephritid pests in mango orchards is hindered by the fact that fruits are very large and therefore provide larvae with a physical refuge from parasitism (López *et al.* 1999). As a consequence, Aluja (1993), Montoya *et al.* (2000), and Ndiaye *et al.* (2015) have recommended that in programs of augmentative biological control, parasitoids should be released outside the mango orchards to attack fly larvae in their much smaller native hosts and thereby significantly reduce the size of natural fly populations entering mango orchards. Recently, Aluja *et al.* (2014) proposed a natural scheme to reduce fruit fly pests in mango orchards through the conservation of fruit fly host trees that could act as parasitoid reservoirs, together with planting of such trees around the orchards. In this way, fruit trees of no economic importance attacked by other fruit fly species would be a source of native populations of parasitoids that could control or at least contribute to maintaining pest fruit fly populations at low levels around mango orchards. In line with this proposal, in Senegal Ndiaye *et al.* (2015) assessed the impact of releases of the Asian egg parasitoid *F. arisanus*, on fruit fly infestation in mango orchards and native vegetation surrounding mango orchards. They found that mango fruit was infested by *B. dorsalis* at a lower level than observed in control orchards as result of the action of the parasitoids. In their study, populations of *C. cosyrae* in wild fruits were parasitized by the native parasitoids *Fopius caudatus* (Szépligeti), *Fopius silvestrii* (Wharton), *Fopius desideratus* (Bridwell), *Diachasmimorpha fullawayi* (Silvestri), *Diachasmimorpha carinata* (Szépligeti), *Psyttalia cosyrae* (Wilkinson), and *Psyttalia concolor* (Szépligeti). Ndiaye *et al.* (2015) also suggested the use of parasitoid population conservation for the effective control of fruit flies.

In the case of *C. capitata*, several parasitoids, including *D. fullawayi*, *Diachasmimorpha kraussii* Fullaway, *Diachasmimorpha tryoni* (Cameron), *D. longicaudatus*, *Opius bellus* Gahan, and *F. arisanus* (Beardsley 1961; Wharton and Marsh 1978) have been recorded attacking this species. Bess *et al.* (1961) reported that the most important parasitoids collected from *C. capitata* in Hawaii, were *F. vandenboschi*, *F. arisanus*, and *D. longicaudata*. In Brazil, *Doryctobracon areolatus* (Szépligeti) (97%) and *D. longicaudata* (3%) parasitize fruit fly larvae in mango (Carvalho and de Queiroz 2002). In Kenya, Ghana, Tanzania, Uganda, and the Ivory Coast, the most important parasitoids obtained from *Ceratitis* spp. emerging from mangoes were *D. fullawayi*, *F. caudatus*, *P. cosyrae*, and *Tetrastichus giffardianus* Silvestri (Lux *et al.* 2003). In Mexico and other parts of Latin America, the most common native parasitoids attacking *Anastrepha* fruit flies that infest mangoes are *D. areolatus*, *Doryctobracon brasiliensis* (Szépligeti), *Doryctobracon crawfordi* (Viereck), *D. fluminensis* (Costa Lima), and *Utetes anastrephae* (Viereck) (López *et al.* 1999; Ovruski *et al.* 2000; Zucchi 2000), in addition to the exotic parasitoid *D. longicaudata*. In Pakistan, the parasitoids attacking *B. zonata* include *D. longicaudata*, *Dirhinus giffardii* Silvestri, and *Bracon* sp. Similarly, *D. longicaudata*, *D. giffardii*, and *Spalangia grotiusi* Girault were reported to attack *B. dorsalis*, albeit in small numbers (Syed *et al.* 1970a,b).

4.3.7.3.2 Insect Pathogens

The use of pathogens such as entomopathogenic fungi, bacteria, and nematodes, has been attempted against mango pests with varying degrees of success. Toxins from the bacterium *Bacillus thuringiensis* (Berliner), and spores from the fungi *Beauveria bassiana* (Bals.), *Isaria fumosorosea* (Wize), and *Metarhizium anisopliae* (Metschnikoff) Sorokin have been tested as well as the infective juvenile stages of the nematodes *Heterorhabditis bacteriophora* (Poinar and Hislop 1981), *Heterorhabditis heliothidis* Khan, Brooks and Hirschmann, *Steinernema feltiae* Filipjev, and *Steinernema carpocapsae* Filipjev (De la Rosa *et al.* 2002; Dimbi *et al.* 2013; Lezama-Gutierrez *et al.* 2000; Lux *et al.* 2003; Sánchez-Roblero *et al.* 2012; Toledo *et al.* 2009, 2014).

4.3.7.3.3 Predators

There are different kinds of tephritid predators belonging to the families Formicidae, Dermaptera, Staphylinidae, Carabidae, Coccinellidae, Chrysopidae, Pentatomidae, and Coreidae. Of these, only ants appear to be an efficient biocontrol agent (Fletcher 1987). Studies in Australia, Africa, and China have shown that the use of weaver ants (Hymenoptera: Formicidae) is a feasible alternative for fruit fly control in mango orchards and other tree crops (Peng and Christian 2006). Peng and Christian (2006) used the weaver ant, *Oecophylla smaragdina* [Fabricius]), to control the Jarvis fruit fly, *B. jarvisi*, in mango orchards in Australia. Van Mele *et al.* (2007) used an African weaver ant (*Oecophylla longinoda* [Latreille]) in Benin for control of larvae. In China, ants of the *Oecophylla* genus are considered to be a “living pesticide” and represent the earliest written record of biological control (Holldobler and Wilson 1990). The presence of *O. longinoda* ants on mango trees deters fruit fly oviposition, thus Van Mele *et al.* (2007) and other recent studies have suggested that olfactory and visual factors play a role in deterring these pests (Van Mele *et al.* 2007; Vayssieres *et al.* 2013). Fernandes *et al.* (2012) noticed that ants of the genus *Pheidole* are major predators of *Anastrepha* larvae. For additional details on the use of ants for biological control of pestiferous fruit flies, we recommend the recent review by Offenberg (2014).

4.3.7.4 Cultural Control

Fruit bagging is one of the best solutions to prevent fruit fly attack of mango and other tropical fruits (Aluja 1996; Peña *et al.* 1998). Success with mangoes can be high, but Bondad (1985) demonstrated that bagging materials are not always resistant to the effect of rain and wind. Therefore, while bagged mangoes tend to produce a greater amount of marketable fruit than those that are unbagged, more research is needed to determine the type of bags to use for different mango varieties and the best time to bag fruit (Chonhenchob *et al.* 2011; Love *et al.* 2003).

Jirón (1995) reported that *A. obliqua* populations could be reduced by increasing planting distances in order to reduce humidity and increase solar radiation within orchards. In India, cultural control practices include removal of fallen fruit and inter-tree plowing and raking, followed by insecticide cover sprays. Such practices can reduce fruit fly infestation by between 77 and 100% (Verghese *et al.* 2004).

As a part of an IPM program that was developed to control *B. invadens* (*B. dorsalis*) and *C. cosyra* in mango orchards in Senegal, Ndiaye *et al.* (2008) assessed sanitation actions such as weeding and destroying the fallen fruit. A number of fruit destruction techniques were tested, including the use of plastic bags, burying fruits in holes, burning fruits on the surface of the ground, and incineration in a modified barrel incinerator. They concluded that the use of a reinforced black plastic bag to destroy the collected fallen fruits was the best option.

4.3.7.5 Host Resistance

Several authors have reported that fruit flies do not attack all mango cultivars to the same extent (Aluja *et al.* 2014; Carvalho *et al.* 1996; Yee 1987). It is believed that mango resistance is related to fruit volatiles modulating cultivar preference; resin duct density and resin chemical content; and phenolic compound content (Jayanthi *et al.* 2014; Verghese *et al.* 2012). For example, Yee (1987) reported that the most susceptible cultivars to *B. dorsalis* attack in Hawaii are Hawaiian, Pirie, and Sandersha. Singh (1991) indicated that the frequency of *Bactrocera* injury to physiologically mature fruit of Dashehari ranged from 4–10%, while in fully ripe fruit the frequency of infestation ranged from 10–26%. The highest prevalence of damage was reported in fully ripe fruit of Mallika, followed by Totapuri.

The susceptibility of different mango cultivars to the attack of *A. obliqua* and *A. ludens* was measured by Carvalho *et al.* (1996) in Brazil and by Aluja *et al.* (2014) in Mexico. Certain varieties and cultivars such as Espada in Brazil showed no infestation by *A. obliqua*, whereas Carlota was highly infested, and Edward, Kent, Palmer, and Ataulfo cultivars in Mexico exhibited high tolerance to *A. obliqua* but less to *A. ludens*. In contrast, Vishi and Brooks were highly susceptible to *A. ludens*, while Tommy Atkins, Sensation, and Ataulfo niño (parthenocarpic fruit) were highly susceptible to *A. obliqua* infestation (Aluja *et al.* 2014). In the Brazilian study, the survival of adults of *A. obliqua* was lower when larvae fed in Espada compared to Carlota. Furthermore, Espada had an adverse effect on the longevity of *A. obliqua* females, possibly due to the presence of toxic substances (Carvalho and de Queiroz 2002) or the absence of essential nutrients. According to Joel (1980), mangoes contain resin ducts in the exocarp which confer protection against the vertical movement of the ovipositor and larval movement. Resin duct density in the pulp and sap content were negatively correlated with fruit infestation by *A. obliqua* and *A. ludens*. Verghese *et al.* (2012) suggested that phenolic compounds are a chemical barrier to *B. dorsalis* offspring development, and Jayanthi *et al.* (2012) proposed that volatiles play an important role in mango cultivar preference, which might explain why certain mango varieties are more preferred than others as hosts.

4.3.7.6 Sterile Insect Technique

The Sterile Insect Technique (SIT) consists of mass releases of sterilized males to introduce sterility into wild populations of pestiferous species (Knipling 1955). It is usually part of AW-IPM schemes forming part of eradication or suppression programs. SIT is expensive because it involves rearing large numbers of fruit flies, exposing them to gamma rays to induce sterility, and releasing these flies into the target population, inc incurring additional monitoring and quarantine costs and pre-treatment preparation (Mumford 2005). For this reason, most programs receive financial support from governments (national, regional, or local) (Dyck *et al.* 2005) and its use is only justified when the target pest is affecting a highly and productive industry or when the pest has recently invaded a fruit-growing area or country. This technique has been successfully used to control and suppress *C. capitata* in mango-producing countries such as Mexico, Guatemala, and the USA (Hawaii and California), Chile, Argentina, Costa Rica, South Africa, and Israel; and *A. ludens* and *A. obliqua* in Mexico, and *B. dorsalis* in Thailand (Dyck *et al.* 2005).

4.3.8 Coleoptera

4.3.8.1 Mango Seed Weevils

The mango seed weevil, *Sternochetus mangiferae* (Fabricius), and the mango pulp weevil, *S. frigidus* (Fabricius) (Coleoptera: Curculionidae), are important pests of mango. Quarantine restrictions prevent the export of fresh weevil-infested mangoes into uninfested areas. The flesh of ripe fruit is damaged when mango seed weevil adults emerge from the seeds, and weevil-damaged seeds may limit plant propagation in nurseries and orchards (Johnson 1989). Early fruit drop may be caused by severe weevil infestations (Subramanian 1925). The mango seed weevil occurs from India through Southeast Asia to Australia, on tropical Pacific Islands, in parts of Africa, in the Caribbean region, and in northern South America (Balock and Kozuma 1964; Johnson 1989; Shukla and Tandom 1985).

Follett and Gabbard (2000) reported that germination rates for infested seeds of polyembryonic "Common" are equal to those of uninfested seeds. Germination is significantly reduced for infested seeds of monoembryonic Haden compared with uninfested control seeds, although germination of infested seeds was more than 70%. Direct feeding damage to the pulp was found in only 0.11% of 3,602 mango fruits, which suggests that *S. mangiferae* is a less serious pest of mangoes than previously considered.

Field sanitation, such as the removal of all fallen fruit and seeds, is very labor intensive, and demands complete removal and disposal of fallen fruit.

Various insecticides have been evaluated for controlling adult weevils, particularly during oviposition (Shukla and Tandom 1985). The most effective control was provided by the organophosphate Fenthion, which reduced infestation to less than 17%. In another field test, the pyrethroid Deltamethrin and the carbamate Carbaryl were most effective, both resulting in more than 15% infestation rates. Spot application of Diazinon on tree trunks was recommended based on cost, efficiency, and least environmental damage. Verghese *et al.* (2004) reported that commercially-available Azadirachtin was not effective for management of *S. mangiferae* in India.

4.3.8.2 Ambrosia Beetles

Detailed coverage of a stem-boring species is beyond the scope of this chapter; however, *Hypocryphalus mangiferae* (Stebbing), *Apate monachus* Fabricius, and *Batocera rubus* L. are pests within this group.

The scolytids, *Hypocryphalus mangiferae* (Stebbing) and *Xylosandrus compactus* (Eickhoff), directly attack the main stem and branches (Wysoki *et al.* 1993). Fungus mycelium growth can extend terminally and basally from the beetle gallery in the mango tree and can kill the affected branches. The insects prefer trees that have been weakened by pathogens, wind, etc., but after a population has been established, the infestation spreads to healthy trees. *Hypocryphalus mangiferae* has been associated with mango wilt disease in Brazil and Oman (van Wyk *et al.* 2007). *H. mangiferae* is attracted to trees where the fungus is present (Berti Filho and Flechtmann 1986). Masood *et al.* (2010) determined that beetles were more attracted to trees that have been wounded than to healthy trees. Masood *et al.* (2009) reported that *H. mangiferae* females lay an average of 42 eggs, and the instar larvae completes their development in approximately 17 days. Pupation period lasts 4 days and the adult can live up to 23 days. Masood *et al.* (2011) consider that *H. mangiferae* is involved in the dissemination of the fungi that causes mango wilt. Masood *et al.* (2008) reported *H. mangiferae, Xyleborus* sp., and *Sinoxylon* sp., from mango trees affected with mango wilt in Pakistan.

Abbasi *et al.* (2008) reported that higher densities of *H. mangiferae* were trapped at 0.33 cm above ground and less at 2.7 m above ground. Besides, *H. mangiferae*, Souza *et al.* (2013) have reported the association of the cause of mango wilt with *Xyleborus affinis* in Brazil. Scolytid beetles are attracted to mango trees in response to visual stimuli, to host-specific chemicals, and to species-specific aggregation pheromones (Lindgreen *et al.* 1982). The evaluation of traps as tools for managing ambrosia beetles on mangoes in Florida is necessary in order to reduce their damage in newly-established groves. Saeed *et al.* (2010) used tree log traps and trapped beetles from at least 5 m from infested trees. Saeed *et al.* (2011) determined that the toxicity of Chlorpyrifos for *H. mangiferae* was significantly higher than the one caused by Deltamethrin and Bifenthrin.

4.4 Control Strategies

In general, most mango pests also occur on other fruit crop species. Fruit flies, scales, mites, thrips, lepidopteran flower feeders, weevils, and beetles are mostly generalists, and some of their management schemes need to be implemented with this in mind. In the case of fruit flies, Aluja (1996) suggests surveying vegetation adjacent to infested mango orchards, as populations are sustained and multiply in these locations and from whence adult flies move into commercial orchards to attack ripening fruit (Aluja *et al.* 1996). Management of key pests, such as fruit flies, seed weevils, etc., must be mandatory, in order to have an effect on a large region. The use of some measures, such as quarantine, etc., must involve neighboring producing countries in order to have a positive effect on sanitation. Management tactics that can be improved include the following.

4.4.1 Selective Pesticides

Pesticides that are used in IPM programs must have selective toxicity. The current trend is the development of chemicals that are highly effective for a limited group of insects. Cunningham (1984) suggested that oils could be utilized for control of scales in mango; however, most of the recommendations are based on highly toxic or illegal, non-registered persistent chemicals (de Bie 2004; Singh 1991; van Mele *et al.* 2001).

4.4.2 Biological Control

Biological control has great potential as a tactic for regulating pest populations in IPM programs in mango orchards; however, it will be difficult for biological control alone to reduce a pest from an economic to a completely non-economic status for pests attacking fruit. A combination of augmentative releases of parasitoids and the use of sterile insects, at least from a theoretical perspective, has been considered to be more effective for fruit flies than either method applied alone (Barclay 1987).

4.4.3 Host Plant Resistance

Tolerance of mango to pests is mentioned for *Noorda* sp. and *Idioscopus* sp. (Bagle and Prasad 1984), while mango resistance to *Sternochetum mangiferae* is mentioned by Hansen (1993). Carvalho *et al.* (1996) has also demonstrated the different degrees of susceptibility of mango cultivars to *A. obliqua*. Most of this research, however, needs to be assessed further.

4.4.4 Pheromones and Trapping Devices

Developments in the identification and synthesis of sex pheromones have resulted in their possible use for pest management in mango orchards (Chu *et al.* 1994; Khan *et al.* 2005). Food attractants, however, remain the most common monitoring tools. Trapping techniques can be utilized to reduce pesticide use by improving timing of sprays as a result of better monitoring of pest populations. It remains uncertain if trapping techniques can be used to predict infestations by fruit feeding pests and if they can be used for direct control (by mass trapping) over several years.

4.4.5 Cultural and Physical Control

Use of cultural and physical techniques, such as pruning and bagging, etc., depends on costs of control, availability of technical assistance, and market purposes. Greatly increased regulation of pesticides, heightened public awareness of environmental contamination, pesticide resistance problems in pests, and the high cost of chemical pest control have resulted in increasing reliance on integrated pest control as an important strategy in sustainable agriculture.

References

Abbas, S. (1985) Studies on the mango inflorescence midge, *Erosomyia indica* Grover. *Acta Horticulturae*, **231**, 593–596.

Abbasi, Q.D., Jan, N.D., Mahar, A.N., Khuhro, R.D., Nizamani, S.M. and Panhwar, A. (2008) Monitoring of ambrosia bark beetle through installation of sticky color traps at different heights in mango trees. *International Journal of Fruit Science*, **7**, 65–79.

Agounké, D., Agricola, U. and Bokonon-Ganta, H. (1988) *Rastrococcus invadens* Williams (Hemiptera: Pseudococcidae), a serious exotic pest of fruit trees and other plants in West Africa. *Bulletin of Entomology Research*, **78**, 695–702.

Aliakbarpour, H. and Rawi, C.S.M. (2012) Seasonal abundance of *Thrips hawaiiensis* (Morgan) and *Scirtothrips dorsalis* (Hood) (Thysanoptera: Thripidae) in mango orchards in Malaysia. *Pertanika Journal of Tropical Agricultural Science*, **35**, 637–645.

Aluja, M. (1993) *Manejo Integrado de la Mosca de la Fruta*. Editorial Trillas, Mexico DF. [in Spanish]

Aluja, M. (1994) Bionomics and management of *Anastrepha*. *Annual Review of Entomology*, **39**, 155–178.

Aluja, M. (1996) Future trends in fruit fly management. In: *Fruit Fly Pests: A World Assessment of their Biology and Management* (eds B. McPheron and G. Steck), pp. 309–320. St Lucie Press, Delray Beach, FL.

Aluja, M. and Mangan, R.L. (2008) Fruit fly (Diptera: Tephritidae) host status determination: Critical conceptual, methodological, and regulatory considerations. *Annual Review of Entomology*, **53**: 473–502.

Aluja, M. and Norrbom, A.L. (eds) (2000) Fruit flies (Tephritidae). In: *Phylogeny and Evolution of Behavior*. CRC Press, Boca Raton, FL. 984 p.

Aluja, M. and Rull, J. (2009) Managing pestiferous fruit flies (Diptera: Tephritidae) through environmental manipulation. In: *Biorational Tree Fruit Pest Management* (eds M. Aluja, T. Leskey and C. Vincent), pp. 171–213. CAB International, Wallingford, UK.

Aluja, M., Cabrera, M., Guillén, J., Celedonio, H. and Ayora, F. (1989) Behavior of *Anastrepha ludens, A. obliqua* and *A. serpentina* (Diptera: Tephritidae) on a wild mango tree (*Mangifera indica*) harbouring three McPhail traps. *Insect Science and its Application*, **10**, 309–318.

Aluja, M., Celedonio-Hurtado, H., Liedo, P., Cabrera, M., Castillo, F. *et al.* (1996) Seasonal population fluctuations and ecological implications for management of *Anastrepha* fruit flies (Diptera: Tachinidae) in commercial mango orchards in southern Mexico. *Journal of Economic Entomology*, **89**, 654–667.

Aluja, M., Sivinski, J., Ovruski, S., Guillén, L., López, M. *et al.* (2009) Colonization and domestication of seven species of native New World Hymenopterous larval-prepupal and pupal fruit fly (Diptera: Tephritidae) parasitoids. *Biocontrol Science and Technology*, **19**, 49–79.

Aluja, M., Ordano, M., Guillén, L. and Rull, J. (2012) Understanding long-term fruit fly (Diptera: Tephritidae) population dynamics: implications for area-wide management. *Journal of Economic Entomology*, **105**, 823–836.

Aluja, M., Arredondo, J., Díaz-Fleischer, F., Birke, A., Rull, J. *et al.* (2014) Susceptibility of 15 mango (Sapindales: Anacardiaceae) cultivars to the attack by *Anastrepha ludens* and *Anastrepha obliqua* (Diptera: Tephritidae) and the role of under-developed fruit as pest reservoirs: Management implications. *Journal of Economic Entomology*, **107**, 375–388.

Angeles, D.E. (1991) Mangifera atissima In: *Edible Fruits and Nuts, Plant resources of Southeast Asia 2 (PROSEA)* (eds R.E. Coronel, E.W.M. Verheij), pp. 206–207. Prudoc, Wageningen.

Ashfaq, M., Khan, R.A., Khan, M.A., Rasheed, F. and Hafeez, S. (2005) Complete control of mango mealybug using funnel type slippery trap. *Pakistan Entomologist*, **27**, 45–48.

Badii, K.B., Billah, M.K., Afreh-Nuamah, K. and Obeng-Ofori, D. (2015) Species composition and host range of fruit-infesting flies (Diptera: Tephritidae) in northern Ghana. *International Journal of Tropical Insect Science*, **35**, 137–151.

Bagle, B. and Prasad, V. (1984) Varietal incidence and control of stone weevil *Sternochetus mangiferae* Fabricius. *Indian Journal of Entomology*, **46**, 389–392.

Balock, J.W. and Kozuma, T. (1964) Notes on the biology and economic importance of the mango weevil, *Sternochetus mangiferae* (Fabricius), in Hawaii (Coleoptera: Curculionidae). *Proceedings of the Hawaiian Entomological Society*. **8**, 353–364.

Balock, J.W. and López, D.F. (1969) Trapping for control of the Mexican fruit fly in mango and citrus groves. *Journal of Economic Entomology*, **62**, 53–56.

Barbosa, F.R. (2005) Arthropodes-pragas associados a cultura da mangueira no Brasil e seu controle. In: *Pragas da Mangueira, Embrapa Semi-arido* (eds E. Menezes, and F.R. Barbosa), pp. 1–50. Ministerio da Agricultura, Pecuaria e Abastecimiento, Petrolina.

Barclay, H.J. (1987) Models for pest control: complementary effects of periodic releases of sterile pests and parasitoids. *Theoretical Population Biology*, **32**, 76–89.

Barnes, H.F. (ed.) (1948) Gall midges of economic importance, vol. 3: *Gall Midges of Fruit*. Crosby Lockwood & Sons, London, 184 p.

Bateman, M.A. (1972) The ecology of fruit flies. *Annual Review of Entomology*, **17**, 493–518.

Beardsley, J.W. (1961) A new review of the Hawaiian Braconidae (Hymenoptera). *Proceedings of the Hawaiian Entomology Society*, **27**, 333–366.

Beers, E., Hull, L. and Jones, V. (1993) Sampling pest and beneficial arthropods of apples. In: *Handbook of Sampling Methods for Arthropods in Agriculture* (eds L. Pedigo and D. Buntin), pp. 383–416. CRC Press, Boca Raton, FL.

Berti Filho, E. and Flechtmann, C.A. (1986) A model of ethanol trap to collect scolytidae and platypodidae (Insecta, Coleoptera). *Instituto de Pesquisas e Estudios Florestais*, **34**, 53–56.

Bess, H.A., van den Bosch, R. and Haramoto, F.H. (1961) Fruit fly parasites and their activities in Hawaii. *Proceedings of the Hawaiian Entomology Society*, **27**, 367–378.

Bo, W., Ahmad, S., Dammalage, T., Tomas, U.S., Wornoayporn, V., Haq, I.U. *et al.* (2014) Mating compatibility between *Bactrocera invadens* and *Bactrocera dorsalis* (Diptera: Tephritidae). *Journal of Economic Entomology*, **107**, 623–629.

Bondad, N.D. (1985) Mango bagging. *Animal Husbandry and Agricultural Journal*, June, 11–14.

Boykin, L.M., Schutze, M.K., Krosch, M.N., Chomič, A., Chapman, T.A. *et al.* (2014) Multi-gene phylogenetic analysis of south-east Asian pest members of the Bactrocera dorsalis species complex (Diptera: Tephritidae) does not support current taxonomy. *Journal of Applied Entomology*, **138**, 235–253.

Brink, T. (1994) The occurrence of thrips on mango inflorescences. *Yearbook*, 14, 8–81. South African Mango Growers' Association, Tzaneen.
Butani, D.K. (1979) *Insects and Fruits.* Periodical Expert Book Agency, Delhi, 415 p.
Carvalho, P.J and De Queiroz, A.C. (2002) *A Cultura da Mangueira.* CIP-Brasil, Brazil.
Carvalho, R, da S., Nascimento, A., Morgante, J. and Fonseca, N. (1996) Susceptibility of different mango varieties to the attack of the fruit fly, *Anastrepha obliqua.* In: *Fruit Fly Pests: A World Assessment of Their Biology and Management* (eds B. McPheron and G. Steck), pp. 325–331. St Lucie Press, Delray Beach, FL.
Choi, K.S., Yang, J.Y., Park, Y.M., Kim, S., Choi, H. *et al.* (2013) Pests' lists and their damages on mango, dragon fruit and atemoya in jejum Korea. *Korean Journal of Applied Entomology*, **52**, 45–51.
Chonhenchob, V., Kamhangwong, D., Kruenate, J., Khongrat, K., Tangchantra, N. *et al.* (2011) Preharvest bagging with wavelength-selective materials enhances development and quality of mango (*Mangifera indica* L.) cv. Nam Dok Mai # 4. *Journal of the Science of Food and Agriculture*, **91**, 664–671.
Christenson, L.D. and Foote, R.H. (1960) Biology of Fruit Flies. *Annual Review of Entomology*, **5**, 171–192.
Chu, Y., Lee, K.T. and Tseng, Y.H. (1994) Occurrence of melon and oriental fruit fly in Republic of Naru. *Plant Protection Bulletin*, **36**, 131–140.
Chua, T.H. and Wood, B.J. (1990) Other tropical fruit trees and shrubs. In: *Armored Scale Insects, their Biology, Natural Enemies and Control, 4B* (ed. D. Rosen), pp. 543–552. Elsevier, New York.
Clarke, A.R., Armstrong, K.F., Carmichael, A.E., Milne, J.R., Raghu, S. *et al.* (2005) Invasive phytophagous pests arising through a recent tropical evolutionary radiation: The *Bactrocera dorsalis* complex of fruit flies. *Annual Review of Entomology*, **50**, 293–319.
Cleveland, C.B., Mayes, M.A. and Cryer, S.A. (2001) An ecological risk assessment for spinosad use on cotton. *Pest Management Science*, **58**, 70–84.
Cowley, J.M. (1990) A new system of fruit fly surveillance trapping in New Zealand. *New Zealand Entomologist*, **13**, 81–84.
Cunningham, I.C. (1984) Mango insect pests. In: *Proceedings of Australian Mango Research Workshop*, pp. 211–224. CSIRO, Melbourne.
Cunningham, R.T. and Steiner, L.F. (1972) Field trial of cue-lure+ naled on saturated fiberboard blocks for control of the melon fly by the male-annihilation technique. *Journal of Economic Entomology*, **65**, 505–507.
Cunningham, R.T. and Suda, D.Y. (1986) Male annihilation through mass-trapping of male flies with methyl eugenol to reduce infestation of oriental fruit fly (Diptera: Tephritidae) larvae in papaya. *Journal of Economic Entomology*, **79**, 1580–1582.
Daneel, M.S., de Jager, K., Steyn, W. and Husselman, J. (2000) Efficacy of different insecticides against gall fly on mangoes. *South African Mango Growers' Association Yearbook*, **19**, 20.
Danne, A.W., Llewellyn, R., Huwer, R. and Furlong, M. (2013) Fruitspotting bugs, *Amblypelta nitida* Stål and *A. lutescens* Distant (Hemiptera: Coreidae): A review of the potential for integrated management practices. *Austral Entomology*, **53**, 112–123.
de Bie, C.A. (2004) The yield map of mango in Phrao, Thailand, as investigated through comparative performance evaluation. *Scientia Horticulturae*, **102**, 37–52.
de la Rosa, W., López, F.L. and Liedo, P. (2002) *Beauveria bassiana* as a pathogen of the Mexican fruit fly (Diptera: Tephritidae) under laboratory conditions. *Journal of Economic Entomology*, **95**, 36–43.

Delrio, G., Conti, B. and Crovetti, A. (1986) Effect of abiotic factors on *Ceratitis capitata* (Wied.) (Diptera: Tephritidae), vol. I: Egg development under constant temperatures. In: *Fruit Flies of Economic Importance* (ed. R. Cavalloro), pp. 133–139. Balkema, Rotterdam.

Diatta, P., Rey, J.Y., Vayssieres, J.F., Diarra, K., Coly, E.V. *et al.* (2013) Fruit phenology of citruses, mangoes and papayas influences egg-laying preferences of *Bactrocera invadens* (Diptera: Tephritidae). *Fruits*, **68**, 507–516.

Díaz-Fleischer, F. and Aluja, M. (2003) Clutch size in frugivorous insects as a function of host firmness: the case of the tephritid fly *Anastrepha ludens*. *Ecological Entomology*, **28**, 268–277.

Diaz-Fleischer, F., Toledo, J., Enkerlin, W. and Fernandez, J. (1996) Cyromazine: effects on three species of *Anastrepha*. In: *Fruit Fly Pests: A World Assessment of their Biology and Management* (eds B. McPheron and G. Steck), pp. 336–337. St Lucie Press, Delray Beach, FL.

Díaz-Fleischer, F., Arredondo, J., Flores, S., Montoya, P. and Aluja, M. (2009) There is no magic fruit fly trap: Multiple biological factors influence the response of adult *Anastrepha ludens* and *Anastrepha obliqua* (Diptera: Tephritidae) individuals to Multilure traps baited with BioLure or NuLure. *Journal of Economic Entomology*, **102**, 86–94.

Dimbi, S., Maniania, N.K. and Ekesi, S. (2013) Horizontal transmission of *Metarhizium anisopliae* in fruit flies and effect of fungal infection on egg laying and fertility. *Insects*, **4**, 206–216.

Dominiak, B.C. and Nicol, H.I. (2010) Field performance of Lynfield and McPhail traps for monitoring male and female sterile *Bactrocera tryoni* (Froggatt) and wild *Dacus newmani* (Perkins). *Pest Management Science*, **66**, 741–744.

Drew, R.A.I. (1989) The tropical fruit flies (Diptera: Tephritidae: Dacinae) of the Australasian and Oceanian regions. *Memoirs of the Queensland Museum*, **26**, 521 p.

Duyck, P-F. and Quilici, S. (2002) Survival and development of different life stages of three *Ceratitis* spp. (Diptera: Tephritidae) reared at five constant temperatures. *Bulletin of Entomological Research*, **92**, 461–469.

Duyck, P-F., David, P. and Quilici, S. (2006) Climatic niche partitioning following successive invasions by fruit flies in La Réunion. *Journal of Animal Ecology*, **75**, 518–526.

Dyck, V.A., Hendrichs, J. and Robinson, A.S. (eds) (2005) *Sterile Insect Technique. Principles and Practice in Area-Wide Integrated Pest Management*. Springer, Dordrecht, 787 p.

Ehler, I.E. and Endicott, P.C. (1984) Effect of malathion-bait sprays on biological control of insect pests of olive, citrus and walnut. *Hilgardia*, **52**, 1–47.

Ekesi, S. and Billah, M.K. (eds) (2006) *A field guide to the management of economically important tephritid fruit flies in Africa*. ICIPE, Nairobi, Kenya. 206 p.

Eskafi, F.M. and Cunningham, R.T. (1987) Host plant of fruit flies (Diptera: Tephritidae) of economic importance in Guatemala. *Florida Entomologist*, **70**, 166–123.

Ekesi, S., Mohamed, S. and Tanga, C.M. (2014) Comparison of food-based attractants for *Bactrocera invadens* (Diptera: Tephritidae) and evaluation of Mazoferm-Spinosad bait spray for field suppression in mango. *Journal of Economic Entomology*, **107**, 299–309.

Fay, H.A.C. (2012) A highly effective and selective male lure for *Bactrocera jarvisi* (Tryon) (Diptera: Tephritidae). *Austral Entomology*, **51**, 189–197.

Fernandes, W.D., Sant'Ana, M.V., Raizer, J. and Lange, D. (2012) Predation of fruit fly larvae Anastrepha (Diptera: Tephritidae) by ants in grove. *Psyche: A Journal of Entomology*, 2012, Article ID 108389.

Fletcher, B.S. (1987) The biology of dacine fruit flies. *Annual Review of Entomology*, **32**, 115–144.

Fletcher, M.J. and Dangerfield, P.C. (2002) *Idioscopus clypealis* (Lethierry), a second new leafhopper pest of mango in Australia (Hemiptera: Cicadellidae: Idiocerinae). *Australian Journal of Entomology*, **41**, 35–38.

Follett, P. and Gabbard, Z. (2000) Effect of mango weevil (Coleoptera: Curculionidae) damage on mango seed viability in Hawaii. *Journal of Economic Entomology*, **93**, 1237–1240.

Galan-Sauco, V. (ed.) (1990) *Los frutales tropicales en los subtropicos*. pp. 83–84. MundiPrensa, Madrid.

García-Medel, D., Sivinski, J., Díaz-Fleischer, F., Ramirez-Romero, R. and Aluja, M. (2007) Foraging behavior by six fruit fly parasitoids (Hymenoptera: Braconidae) released as single- or multiple-species cohorts in field cages: influence of fruit location and host density. *Biological Control*, **43**, 12–22.

Godase, S.K., Bhole, S.R., Shivpuje, P.R. and Patil, B.P. (2004) Assessment of yield loss in mango (*Mangifera indica*) due to mango hopper *(Idioscopus niveoparvasus*) (Homoptera: Cicadelidae). *Indian Journal of Agricultural Sciences*, **74**, 370–372.

Grover, P. (1986a) Population fluctuation of *Erosomya indica* and *Dasineura amaramanjarae* and co-related extent of damage. *Cecidologia Internationale* **7**, 43–57.

Grover, P. (1986b) Integrated control of midge pests. *Cecidologia Internationale*, **7**, 1–28.

Grove, T., Guillomee, J.H. and Pringle, K.L. (2001) *Thrips* (Thysanoptera) species associated with mango trees in South Africa. *African Entomology*, **9**, 153–162.

Halawa, S.M., El-Hosary, R.A., Mosallam, A.M.Z., El-Khayat, E.F. and Ismail, M.M. (2013) Toxicological, biological and biochemical effects of certain insecticides on *Bactrocera zonata* (Saunders) (Diptera, Tephritidae). *American Eurasian Journal of Toxicol Sciences*, **5**, 55–65.

Halteren, P. van. (1970) Notes on the biology of the scale insect *Aulacaspis mangiferae* Newst. (Diaspididae, Hemiptera) on mango. *Ghana Journal of Agricultural Science*, **3**, 83–85.

Hansen, J.D. (1993) Dynamics and control of the mango seed weevil. *Acta Horticulturae*, **341**, 415–420.

Harris, K.M. and Schreiner, I.H. (1992) A new species of the gall midge (Diptera: Cecidomyiidae) attacking mango foliage in Guam, with observations on its pest status and biology. *Bulletin of Entomological Research*, **82**, 41–48.

Heath, R.R., Epsky, N.D., Guzman, A., Dueben, B.D., Manukian, A. and Meyer, W.L. (1995) Development of a dry plastic insect trap with food-based synthetic attractant for the Mediterranean and Mexican fruit flies (Diptera: Tephritidae). *Journal of Economic Entomology*, **88**, 1307–1315.

Hill, D. (1975) *Agricultural Insect Pests of the Tropics and Their Control*. Cambridge University Press, New York, 516 p.

Hölldobler, B. and Wilson, E.O. (1990) *The Ants*. Harvard University Press, Cambridge MA, 746 p.

Holler, T.C., Peebles, M., Young, A., Whiteman, L., Olson, S. and Sivinski, J. (2009) Efficacy of the suterra biolure individual female fruit fly attractant packages vs. the unipak version. *Florida Entomologist*, **92**, 667–669.

Hsu, J.-C. and Feng, H.-T. (2006) Development of resistance to Sinosad in Oriental fruit fly (Diptera: Tephritidae) in laboratory selection and cross-resistance. *Journal of Economic Entomology*, **99**, 931–936.

IAEA [International Atomic Energy Agency] (2007) Development of improved attractants and their integration into fruit fly SIT, IAEA-TECDOC-1574. International Atomic Energy Agency, Vienna.

IAEA [International Atomic Energy Agency] (2013) Trapping Manual for Area-wide Fruit Fly Programs [Online]. Available: http://www-pub.iaea.org/MTCD/Publications/PDF/TG-FFP_web.pdf [21 April 2016]

Jayanthi, P.D.K., Woodcock, C.M., Caulfield, J., Birkett, M.A. and Bruce, T.J. (2012) Isolation and identification of host cues from mango (*Mangifera indica*) that attract gravid female oriental fruit fly, *Bactrocera dorsalis*. *Journal of Chemical Ecology*, **38**: 361–369.

Jayanthi, P.D.K., Kempraj. V., Aurade, R.M., Venkataramanappa, R.K., Nandagopa, I.B. *et al.* (2014) Specific volatile compounds from mango elicit oviposition in gravid *Bactrocera dorsalis* females. *Journal of Chemical Ecology*, **40**, 259–266.

Jeppson, L., Keiffer, H. and Baker, E. (1975) *Mites Injurious to Economic Plants*. University of California Press, Berkeley, CA. 528 p.

Jirón, L.F. (1995) Opciones al uso de insecticidas en mango. In: *Opciones al uso Unilateral de Plaguicidas en Costa Rica* (eds J. Garcia, G. Fuentes and J. Monge), pp. 129–155. Editorial Euned, San Jose 2, Costa Rica.

Joel, D.M. (1980) Resin ducts in the mango fruit: a defense system. *Journal of Experimental Botany*, **30**, 1707–1718.

Johnson, P.J. (1989) *Mango Seed Weevil (Sternochetus mangiferae F.): Identification, biology and management*, University of Idaho, Moscow, ID, pp. 3–7.

Khan, M.A., Ashfaq, M., Akram, W. and Lee, J.-J. (2005) Management of fruit flies (Diptera: Tephritidae) of the most perishable fruits. *Entomological Research*, **35**, 79–84.

King, J.R. and Hennessey, M.K. (1996) Spinosad bait for the Caribbean fruit fly (Diptera: Tephritidae). *Florida Entomologist*, **79**, 526–531.

Knipling, E.F. (1955) Possibilities of insect control or eradication through the use of sexually sterile males. *Journal of Economic Entomology*, **48**, 459–462.

Knutson, A.E., Butler, J., Bernal, J., Bográn, C. and Campos, M. (2011) Impact of area-wide Malathion on predatory arthropods and secondary pests in cotton during boll weevil eradication in Texas. *Crop Protection*, **30**, 456–467.

Kumar, D., Roy, C.S., Khan, Z. R., Yazdani, S. S., Hameed, S.F. and Mahmood, M. (1983) An entomogenous fungi *Isaria tax* parasitizing mango *Idioscopus clypealis*. *Science and Culture*, **49**, 253–254.

Lakra, R. K., Kharub, W. and Singh, Z. (1980) Pest management systems for the mango mealybug *Drosicha mangifera* Green – A polyphagous pest of fruit trees in Haryana. *Indian Journal of Entomology* **42**, 153–165.

Lasa, R. and Cruz, A. (2014) Efficacy of commercial traps and the lure CeraTrap® against *Anastrepha obliqua* (Diptera: Tephritidae). *Florida Entomologist*, **97**, 1369–1377.

Lasa, R., Toxtega, Y., Herrera, F., Cruz, A., Navarrete, M.A. and Antonio, S. (2014) Inexpensive traps for use in mass trapping *Anastrepha ludens* (Diptera: Tephritidae). *Florida Entomologist*, **97**, 1123–1130.

Leblanc, L., Rubinoff, D. and Vargas, R. (2009) Attraction of non-target species to fruit fly (Diptera: tephritidae) male lures and decaying fruit flies in traps in Hawaii. *Environmental Entomology*, **38**, 144–1461.

Lewis, W.J., van Lenteren, J.C., Phatak, S.C. and Tumlinson, J.H. (1997) A total system approach to sustainable pest management. *Proceedings of the National Academy of Sciences, USA*, **94**, 12243–12248.

Leyva, J.L., Browning, H. and Gilstrap, F. (1991) Development of *Anastrepha ludens* (Diptera: Tephritidae) in several host fruits. *Environmental Entomology*, **20**, 1160–1165.
Lezama-Gutierrez, R., Trujillo, A., Molina-Ochoa, J., Rebolledo-Dominguez, O., Pescador, A. *al et.* (2000) Virulence of *Metarrizium anisopliae* (Deuteromycotina: Hypomycetes) on *Anastrepha ludens* (Diptera: Tephritidae): Laboratory and field trials. *Journal of Economic Entomology*, **93**, 1080–1084.
Lindgreen, S., Borden, J., Gray, D., Lee, P., Palmer, D. and Chong, L. (1982) Evaluation of two trap log techniques for ambrosia beetles (Coleoptera: Scolytidae) in timber processing areas. *Journal of Economic Entomology*, **75**, 577–586.
Lloyd, A.C., Hamacek, E.L., Kopittke, R.A., Peek, T., Wyatt, P.M. *et al.* (2010). Area-wide management of fruit flies (Diptera: Tephritidae) in the Central Burnett district of Queensland, Australia. *Crop Protection*, **29**, 462–469.
Loomans, A.J.M., van Lenteren, J.C., Tommasini, M.G., Maini, S. and Riudavets, J. (1995) *Biological Control of Thrips Pests. Wageningen Agricultural University Papers, No. 95*. Veenman Drukkers, Wageningen.
López, F., Chambers, D., Sanchez, M. and Kamasaki, H. (1969) Control of the Mexican fruit fly by bait sprays concentrated at discrete locations. *Journal of Economic Entomology*, **62**, 1255–1257.
López, M., Aluja. M. and Sivinski, J. (1999) Hymenopterous larval-pupal and pupal parasitoids of *Anastrepha* flies (Diptera: Tephritidae) in Mexico. *Biological Control*, **15**, 119–129.
Love, K., Bowen, R. and Paull, R. (2003) Increasing marketable production of exotic tropical fruit with protective covering. *Western Region, Sustainable Agriculture Research and Education, Project Report,* Farmer/Rancher, Captain Cook, Hawaii.
Lux, S.A., Ekesi, S., Dimbi, S., Mohamed, S. and Billah, M. (2003) Mango-infesting fruit flies in Africa: Perspectives and limitations of biological approaches to their management. In: *Biological Control in IPM Systems in Africa* (eds P. Neuenschwander, C. Borgemeisterm and J. Langewald), pp. 277–293. CAB International, Wallingford, UK.
MacFarlane, J.R., East, R.W., Drew, R.A.I. and Betlinski, G.A. (1987) Dispersal of irradiated Queensland fruit flies, *Dacus tryoni* (Froggatt) (Diptera: Tephritidae), in southeastern Australia. *Australian Journal Zoology*, **35**, 275–281.
Malavasi, A., Morgante, J.S. and Zucchi, R.A. (1980) The biology of fruit flies (Diptera, Tephritidae), vol. 1: List of hosts and occurrence. *Revista Brasileira de Biologia*, **40**, 9–16. (in Portuguese)
Mangan, R.L. and Moreno, D.S. (2007) Development of bait stations for fruit fly population suppression. *Journal of Economic Entomology*, **100**, 440–450.
Manrakhan, A. (2006) Fruit fly monitoring – Purpose, tools and methodology. In: *Field Guide to the Management of Economically Important Tephritid Fruit Flies in Africa* (eds S. Ekesi and M.K. Billah), pp. 14–19. ICIPE Science Press, Nairobi.
Martínez-Ferrer, M.T., Campos, J.M. and Fibla, J.M. (2010). Field efficacy of *Ceratitis capitata* (Diptera: Tephritidae) mass trapping technique in clementine groves in Spain. *Journal of Applied Entomology*, **136**, 181–190.
Masood, A., Saeed, S. and Sajjad, A. (2008) Characterization and damage patterns of different bark beetle species associated with mango sudden death syndrome in Punjab, Pakistan. *Pakistan Journal of Zoology*, **30**, 163–168.
Masood, A., Saeed, S., Sajjad, A. and Ali, M. (2009) Life cycle and biology of the mango bark beetle, *Hypocryphalus mangiferae* (Stebbing), a possible vector of mango sudden death disease in Pakistan. *Pakistan Journal of Zoology*, **41**, 281–288.

Masood, A., Saeed, S., Erbilgin, N. and Jung Kwon, Y. (2010) Role of stressed mango host conditions in attraction of and colonization by the mango bark beetle *Hypocryphalus mangiferae* Stebbing (Coleoptera: Curculionidae: Scolytinae) and in the symptom development of quick decline of mango trees in Pakistan. *Entomological Research*, **40**, 316–327.

Masood, A., Saeed, S., da Silveira, S.F., Nges Akim, C., Hussain, N. and Farooq, M. (2011) Quick decline of mango in Pakistan: Survey and pathogenicity of fungi isolated from mango trees and bark beetle. *Pakistan Journal of Botany*, **43**, 1793–1798.

McQuate, G.T., Peck, S.L., Barr, P.G. and Sylva, C.D. (2005) Comparative evaluation of spinosad and phloxine B as toxicants in protein baits for suppression of three fruit fly (Diptera: Tephritidae) species. *Journal of Economic Entomology*, **98**, 1170–1178.

Migani, V., Ekesi, S. and Hoffmeister, T.S. (2014) Physiology vs. environment: What drives oviposition decisions in mango fruit flies (*Bactrocera invadens* and *Ceratitis cosyra*)? *Journal of Applied Entomology*, **138**, 395–402.

Miller, D.R. and Davidson, J. (2005) *Armored Scale Insect Pests of Trees and Shrubs (Hemiptera: Diaspididae)*. Cornell University Press, Ithaca, NY, 422 p.

Mohyuddin, A.I. and Mahmood, R. (1993) Integrated control of mango pests in Pakistan. *Acta Horticulturae*, **341**, 467–483.

Montoya, P., Liedo, P., Benrey, B., Cancino, J., Barrera, J.F. *et al.* (2000) Biological control of *Anastrepha* spp. (Diptera: Tephritidae) in mango orchards through augmentative releases of *Diachasmimorpha longicaudata* (Ashmead) (Hymenoptera: Braconidae). *Biological Control*, **18**, 216–224.

Moznette, G. (1922) Insects injurious to the mango in Florida and how to combat them. *Farmers Bulletin*, **1257**.

Mumford, J.D. (2005). Application of benefit/cost analysis to insect pest control using the sterile insect technique. In: *Sterile Insect Technique. Principles and Practice in Area-wide Integrated Pest Management* (eds V.A. Dyck, J. Hendrichs and A.S. Robinson), pp. 481–498. Springer, Dordrecht.

Mwatawala, M., De Meyer, M., Makundi, R. and Maerere, A.P. (2009) Host range and distribution of fruit-infesting pestiferous fruit flies (Diptera: Tephritidae) in selected areas of Central Tanzania. *Bulletin of Entomological Research*, **99**, 629–641.

Nadeem, M.K., Ahmed, S., Nadeem, S., Ishfaq, M. and Fiaz, M. (2014) Assessment of insecticides resistance in field population of *Bactrocera zonata* (Saunders) (Diptera: Tephritidae). *Journal of Animal and Plant Sciences*, **24**, 172–178.

Navarro-Llopis, V., Alfaro, F., Dominguez, J., Sanchis, J. and Primo, J. (2008) Evaluation of traps and lures for mass trapping of Mediterranean fruit fly in citrus groves. *Journal of Economic Entomology*, **101**, 126–131.

Ndiaye, M., Dieng, E.O. and Delhove, G. (2008) Population dynamics and on-farm fruit fly integrated pest management in mango orchards in the natural area of Niayes in Senegal. *Pest Management in Horticultural Ecosystems*, **14**, 1–8.

Ndiaye, O., Ndiaye, S., Djiba, S., Tidiane Ba, C., Vaughan, L. *et al.* (2015) Preliminary surveys after release of the fruit fly parasitoid *Fopius arisanus* Sonan (Hymenoptera Braconidae) in mango production systems in Casamance (Senegal). *Fruits*, **70**: 91–99.

Neuenschwander, P., Boavida, C., Bokonoganta, A., Gado, A. and Herren, H. (1994) Establishment and spread of *Gyranusoidea tebygi* Noyes and Noyes (Hymenoptera: Encyrtidae), two biological control agents released against the mango mealybug i

Williams (Homoptera: Pseudococcidae) in Africa. *Biocontrol Science and Technology*, **4**, 61–69.

Norrbom, A.L. (2004) Host plant database for Anastrepha and Toxotrypana (Diptera: Tephritidae: Toxotrypanini). Diptera Data Dissemination Disk (CD-ROM), North American Dipterist's Society, Washington, DC.

Ochoa, R., Aguilar, H. and Vargas, C. (eds) (1994) *Phytophagous Mites of Central America: An illustrated guide.* Technical Service 6, Tropical Agricultural Research and Higher Education Center, Cartago, Costa Rica.

Offenberg, J. (2014) Pest repelling properties of ant pheromones. *IOBC-WPRS Bulletin*, **99**, 173–176.

Ovruski, S., Aluja, M., Sivinski, J. and Wharton, R. (2000) Hymenopteran parasitoids on fruit infesting Tephritidae (Diptera) in Latin America and the southern United Status: Diversity, distribution, taxonomic status and their use in fruit fly biological control. *Integrated Pest Management Reviews*, **5**, 81–107.

Papadopoulos, N.K., Carey, J.R. and Katsoyannos, B.I. (1996) Overwintering of the Mediterranean fruit fly (Diptera: Tephritidae) in Northern Greece. *Annals of the Entomological Society of America*, **89**, 526–534.

Patel, R.K., Verghese, A., Patel, V.M., Joshi, B.K., Stonehouse, J.M. and Mumford, J.D. (2005) Bait, lure and cultural IPM of fruit flies in mangoes in Gujarat. *Pest Management in Horticultural Ecosystems*, **11**, 155–158.

Peck, S.L. and McQuate, G.T. (2000) Field tests of environmentally friendly Malathion replacements to suppress wild Mediterranean fruit fly (Diptera: Tephritidae) populations. *Journal of Economic Entomology*, **93**, 280–289.

Peng, R.K. and Christian, K. (2004) The weaver ant, *Oecophylla smaragdina* (Hymenoptera: Formicidae), an effective biological control agent of the red banded thrips *Selenothrips rubrocinctus* (Thysanoptera: Thripidae) in mango crops in the Northern Territory of Australia. *International Journal of Pest Management*, **50**, 107–114.

Peña, J.E. (1993) Pests of mango in Florida. *Acta Horticulturae*, **341**, 395–406.

Peña, J.E. and Duncan, R. (1992) Sampling methods for *Prodiplosis longifila* (Diptera: Cecidomyiidae) in limes. *Environmental Entomology*, **2**, 996–1001.

Peña, J. E., Mohyuddin, A. I. and Wysoki, M. (1998) A review of the pest management situation in mango agro-ecosystems. *Phytoparasitica*, **26**, 129–148.

Peña, J.E., Aluja, M. and Wysoki, M. (2008) Pests. In: *The Mango – Botany, Production and Uses* (ed. R.E. Litz), pp. 317–366. CAB International, Wallingford, UK.

Peng, R.K. and Christian, K. (2005) The control efficacy of the weaver ant, *Oecophylla smaragdina* (Hymenoptera: Formicidae), on the mango leafhopper, *Idioscopus nitidulus* (Hemiptera: Cicadellidae) in mango orchards in the Northern Territory. *International Journal of Pest Management*, **51**, 297–304.

Peng, R.K. and Christian, K. (2006) Effective control of Jarvis's fruit fly *Bactrocera jarvisi* (Diptera: Tephritidae), by weaver ant, *Oecophyla smaragdina* (Hymenoptera: Formicidae), in mango orchards in the Northern Territory of Australia. *International Journal of Pest Management*, **52**, 275–282.

Pitan, O.R., Akinlosotu, T.A. and Odebiyi, J.A. (2000) Impact of *Gyranosuoidea tebygi* Noyes (Hymenoptera: Encyrtidae) on the mango mealybug *Rastrococcus invadens* Williams (Homoptera: Pseudococcidae) in Nigeria. *Biocontrol Science and Technology*, **10**, 245–254.

Poinar, G.O., Jr. and Hislop, R.G. (1981) Mortality of Mediterranean fruit fly adults (*Ceratitis capitata*) from parasitic nematodes (*Neoaplectana* and *Heterorhabditis* spp.). *IRCS Medical Science – Biochemistry*, **9**, 641.

Prasad, S.N. (1971) *The Mango Midge Pests.* Cecidological Society, Allahabad. 172 p.

Prokopy, R.J., Miller, N.W., Piñero, J.C., Barry, J.D., Tran, L.C. *et al.* (2003) Effectiveness of GF-120 fruit fly bait spray applied to border area plants for control of melon flies (Diptera: Tephritidae). *Journal of Economic Entomology*, **96**, 1485–1493.

Rehman, M.H., Mahmood, R. and Razaq, M. (2013) Phenology, population dynamics and within tree distribution of *Dasyneura mangiferae* Grover, 1965 (Diptera: cecidomyiidae) in Punjab, Pakistan. *Pakistan Journal of Zoology*, **45**, 1563–1572.

Rehman, M.H., Mahmood, R. and Razaq, M. (2014) Occurrence, monitoring techniques and management of *Dasyneura mangiferae* Grover (Diptera: Cecidomyiidae) in Punjab, Pakistan. *Pakistan Journal of Zoology*, **46**, 45–62.

Robacker, D.C. (2001) Roles of putrescine and 1-pyrroline in attractiveness of technical-grade putrescine to the Mexican fruit fly (Diptera: Tephritidae). *Florida Entomologist*, **84**, 679–685.

Robacker, D.C. and Warfield, W.C. (1993) Attraction of both sexes of Mexican fruit fly, *Anastrepha ludens*, to a mixture of ammonia, methylamine, and putrescine. *Journal of Chemical Ecology*, **19**, 2999–3016.

Robacker, D.C., DeMilo, A.B. and Voaden D.J. (1997) Mexican fruit fly attractants: Effects of 1-pyrroline and other amines on attractiveness of a mixture of ammonia, methylamine, and putrescine. *Journal of Chemical Ecology*, **23**, 1263–1280.

Robacker, D.C. and Landolt, P.J. (2002) Importance and use of attractants. In: *Invasive Arthropods in Agriculture: Problems and solutions.* (eds G.J. Hallman and C.P. Schwalbe), pp. 169–205. Science Publishers, Inc., Enfield, NH.

Rocha, F., Infante, F., Quilantan, J., Golderazena, A. and Funderburk, J.E. (2012) Ataulfo mango flowers contain a diversity of thrips (Thysanoptera). *Florida Entomologist*, **95**, 1711–178.

Rubin, A. and Kuslitizky, W. (1992) First record of *Ceranisus menes* (Hymenoptera: Eulophidae) in Israel. *Phytoparasitica*, **20**, 123–124.

Ruiz, L., Flores, S., Cancino, J., Arredondo, J., Valle, J. *et al.* (2008) Lethal and sublethal effects of spinosad-based GF-120 bait on the tephritid parasitoid *Diachasmimorpha longicaudata* (Hymenoptera: Braconidae). *Biological Control*, **44**, 296–304.

Saeed, S., Masood, A., Sajjad, A. and Muhammad Zahid, D. (2010) Monitoring the dispersal potential of bark beetle *Hypocryphalus mangiferae* Stebbing (Scolytidae: Coleoptera) in mango orchards. *Pakistan Journal of Zoology*, **42**, 473–479.

Saeed, S., Masood, A., Sayyed, A. and Kwon, Y.J. (2011) Comparative efficacy of different pesticides against mango bark beetle *Hypocryphalus mangiferae* Stebbing (Coleoptera: Scolytidae). *Entomological Research*, **41**, 142–150.

Sánchez-Roblero, D., Huerta-Palacios, G., Valle, J., Gómez, J. and Toledo, J. (2012) Effect of *Beauveria bassiana* on the ovarian development and reproductive potential of *Anastrepha ludens* (Diptera: Tephritidae). *Biocontrol Science and Technology*, **22**, 1075–1091.

San Jose, M., Leblanc, L., Geib, S.M. and Rubinoff, D. (2013) An evaluation of the species status of *Bactrocera invadens* and the systematics of the *Bactrocera dorsalis* (Diptera: Tephritidae) complex. *Annals of the Entomological Society of America*, **106**, 684–694.

Satarkar, V.R., Krishnamurthy, S.V., Faleiro, J.R. and Verghese, A. (2009) Spatial distribution of major *Bactrocera* fruit flies attracted to methyl eugenol in different ecological zones of Goa, India. *International Journal of Tropical Insect Science*, **29**, 195–201.

Schutze, M.K., Krosch, M.N., Armstrong, K.F., Chapman, T.A., Englezou, A. *et al.* (2012) Population structure of *Bactrocera dorsalis* s.s., *B. papayae* and *B. philippinensis* (Diptera: Tephritidae) in Southeast Asia: Evidence for a single species hypothesis using mitochondrial DNA and wing-shape data. *BMC Evolutionary Biology*, **12**, 130.

Shukla, R.P. and Tandom, P.L. (1985) Bio-ecology and management of the mango weevil, *Sternochetus mangiferae* (Fabricius) (Coleoptera: Curculionidae). *International Journal of Tropical Agriculture*, **3**, 293–303.

Singh, G. (1991) Loss assessment, ecology and management of mango fruit fly, *Dacus* sp. *Acta Horticulturae*, **291**, 425–436.

Sivinski, J., Aluja, M. and López, M. (1997) Spatial and temporal distributions of parasitoids of Mexican *Anastrepha* species (Diptera: Tephritidae) within the canopies of fruit trees. *Annals of the Entomological Society of America*, **90**, 604–618.

Sivinski, J.M., Calkins, C.O., Baranowski, R.M., Harris, D., Brambila, J. *et al.* (1996) Suppression of Caribbean fruit fly (*Anastrepha suspensa* (Loew) Diptera: Tephritidae) population through releases of the parasitoid *Diachasmimorpha longicaudata* (Ashmead) (Hymenoptera: Braconidae). *Biological Control*, **6**, 177–185.

Soto-Manitiu, J., Jirón, L. and Hernández, R. (1987) Chemical control and ecological observations of fruit flies of the genus *Anastrepha* Schiner (Diptera: Tephritidae) on mango. *Turrialba*, **37**, 245–251.

Souza, A.G., Maffia, L., Murta, H.M., Alves, Y.H., Pereira, R. and Picanco, M.C. (2013) First report on the association between *Ceratocystis fimbriata*, an agent of mango wilt, *Xyleborus affinis* and the sawdust produced during beetle colonization in Brazil. *Plant Disease*, **97**: 1116.

SPC [Secretariat of the Pacific Community] (2005) *Fruit Flies in Palau*. Pest Advisory Leaflet No. 44, Palau, Pacific Islands.

Srivastava, R.P. (1981) Comparative efficacy of various insecticidal dusts against mango mealybug eggs. *Indian Journal of Entomology*, **43**, 225–229.

Stark, J.D., Vargas, R. and Miller, N. (2004) Toxicity of Spinosad in protein bait to three economically important tephritid fruit fly species (Diptera: Tephritidae) and their parasitoids (Hymenoptera: Braconidae). *Journal of Economic Entomology*, **97**, 911–915.

Steck, G.J. (2003) Mango Fruit Fly, Marula Fruit Fly, *Ceratitis cosyra* (Walker) (Insecta: Diptera: Tephritidae) [Online]. Available: http://edis.ifas.ufl.edu/IN563 [21 July 2015].

Steiner, L.F. and Lee, R.K. (1955) Large area tests of male annihilation method for oriental fruit fly control. *Journal of Economic Entomology*, **48**, 311–317.

Steiner, L.F., Mitchell, W.C., Harris, E., Kozuma, T. and Fujimoto, M. (1965) Oriental fruit fly eradication by male annihilation. *Journal of Economic Entomology*, **58**, 961–964.

Stonehouse, J.M., Mumford, J.D. and Mustafa, G. (1998) Economic losses to tephritid fruit flies (Diptera: Tephritidae) in Pakistan. *Crop Protection*, **17**, 159–164.

Subramanian, C.K. (1925) A note on the life history of *Cryptorhynchus mangiferae* Fab. *Madras Agriculture Department Yearbook*, pp. 29–36.

Syed, R.A., Ghani, M.A. and Murtaza, M. (1970a) Studies on trypetids and their natural enemies in West Pakistan. vol. III: *Dacus* (*Strumeta*) *zonatus* (Saunders). *Technical Bulletin Commonwealth Institute of Biological Control*, **13**, 1–16.

Syed, R.A., Ghani, M.A. and Murtaza, M. (1970b) Studies on the Trypetids and their natural enemies in West Pakistan. vol. IV: Further observations on *Dacus* (*Strumeta*) *dorsalis* Hendel. *Technical Bulletin Commonwealth Institute of Biological Control*, **13**, 17–30.

Tan, K.H. (ed.) (2000) *Area-Wide Control of Fruit Flies and Other Insect Pests.* Penerbit University Sains Malaysia, Pulau Pinang. CAB International, Wallingtord, UK. 780 p.

Tan, K.H., Wee, S.L., Ono, H. and Nishida, R. (2013) Comparison of methyl eugenol metabolites, mitochondrial COI, and rDNA sequences of *Bactrocera philippinensis* (Diptera: Tephritidae) with those of three other major pest species within the dorsalis complex. *Applied Entomology and Zoology*, **48**, 275–282.

Tenakanai, D. (1997) Fruit fly fauna in Papua New Guinea. In: *Management of fruit flies in the Pacific* (eds A.J. Allwood and R.A.I. Drew), pp. 87–94. ACIAR Proceedings No. 76, Canberra, Australia.

Thomas, D.B. and Mangan, R.L. (2005) Non-target impact of spinosad GF-120 bait sprays for control of the Mexican fruit fly (Diptera: Tephritidae) in Texas citrus. *Journal of Economic Entomology*, **98**, 1950–1956.

Tobih, F.O., Omoloye, A., Ivbijaro, M.F. and Enobakhare, D.A. (2002) Effects of field infestation by *Rastrococcus invadens* Williams (Hemiptera: Pseudococcidae) on the morphology and nutritional status of mango fruits, *Mangifera indica* L. *Crop Protection*, **21**: 757–761.

Toledo, J., Williams, T., Pérez, C., Liedo, P., Valle, J. and Ibarra, J.E. (2009) Abiotic factors affecting the infectivity of *Steinernema carpocapsae* (Rhabditida: Steinernematidae) on larvae of *Anastrepha obliqua* (Diptera: Tephritidae). *Biocontrol Science and Technology*, **19**, 887–898.

Toledo, J., Sánchez, J.E., Williams, T., Gómez, A., Montoya, P. and Ibarra, J.E. (2014) Effect of soil moisture on the persistence and efficacy of *Heterorhabditis bacteriophora* (Rhabditida: Heterorhabditidae) against *Anastrepha ludens* (Diptera: Tephritidae) larvae. *Florida Entomologist*, **97**, 528–533.

van Mele, P., Cuc, T.T. and van Huis, A. (2001) Farmer's knowledge, perceptions and practices in mango pest management in the Mekong Delta, Vietnam. *International Journal of Pest Management*, **47**, 7–16.

van Mele, P., Vayssières, J.F., van Tellingen, E. and Vrolijks, J. (2007) Effects of an African weaver ant, *Oecophylla longinoda*, in controlling mango fruit flies (Diptera: Tephritidae) in Benin. *Journal of Economic Entomology*, **100**, 695–701.

Vannière, H., Didier, C., Rey, J.Y., Diallo, T.M., Kéita, S. and Sangaré, M. (2004) La mangue en Afrique de l'Quest francophone: les systèmes de production et les itinéraires techniques. *Fruits*, **59**, 383–398.

Van Wyk, M., Al Adawi, A., Khan, I., Deadman, M., Al Jahwari, A. *et al.* (2007) *Ceratocystis manginecans* sp. nov., causal agent of a destructive mango wilt disease in Oman and Pakistan. *Fungal Diversity*, **27**, 213–230.

Vargas, R.I., Stark, J.D., Kido, M.H., Ketter, H.M. and Whitehand, L.C. (2000) Eugenol and Cue-Lure traps for suppression of male oriental fruit flies and melon flies (Diptera: Tephritidae) in Hawaii: Effects of lure mixtures and weathering. *Journal of Economic Entomology*, **93**, 81–87.

Vargas, R.I., Peck, S.L., McQuate, G.T., Jackson, C.G., Stark, J.D. and Armstrong, J.W. (2001) Potential for area wide integrated management of Mediterranean fruit fly

(Diptera: Tephritidae) with a braconid parasitoid and a novel bait spray. *Journal of Economic Entomology*, **94**, 817–825.

Vargas, R.I., Miller, N.W. and Prokopy, R.J. (2002) Attraction and feeding responses of Mediterranean fruit fly and a natural enemy to protein baits laced with two novel toxins, phloxine B and spinosad. *Entomologia Experimentalis et Applicata*, **102**, 273–282.

Vayssières, J.F. and Kalabane, S. (2000) Inventory and fluctuations of the catches of Diptera Tephritidae associated with mangoes in Coastal Guinea. *Fruits*, **55**, 259–270.

Vayssières, J.F., Goergen, G., Lokossou, O., Dossa, P. and Akponon, C. (2005) A new *Bactrocera* species in Benin among mango fruit fly (Diptera: Tephritidae) species. *Fruits*, **60**, 371–377.

Vayssières, J.F., Korie, S. and Ayegnon, D. (2009a) Correlation of fruit fly (Diptera Tephritidae) infestation of major mango cultivars in Borgou (Benin) with abiotic and biotic factors and assessment of damage. *Crop Protection*, **28**, 477–488.

Vayssières, J.F., Sinzogan, A., Korie, S., Ouagoussounon, I. and Thomas-Odjo, A. (2009b) Effectiveness of spinosad bait sprays (GF-120) in controlling mango-infesting fruit flies (Diptera: Tephritidae) in Benin. *Journal of Economic Entomology*, **102**, 515–521.

Vayssières J.F., Sinzogan, A., Adandonon, A., Van Mele, P. and Korie, S. (2013) Ovipositional behaviour of two mango fruit fly species (Diptera Tephritidae) in relation to Oecophylla cues (Hymenoptera Formicidae) as compared to natural conditions without ant cues. *International Journal of Biological and Chemical Sciences*, **7**, 447–456.

Verghese, A., Nagaraju, D.K., Jayanthi, P.D.K., Vasudev, V. and Madhura, H.S. (2004) Is azadirachtin useful in the management of the mango stone weevil (*Sternochetus mangiferae* Fabricius)? *Journal of Food Agriculture and Environment*, **2**, 213–216.

Verghese, A., Soumya, C.B., Shivashankar, S., Manivannan, S. and Krishnamurthy, S.V. (2012) Phenolics as chemical barriers to female fruit fly, *Bactrocera dorsalis* (Hendel) in mango. *Current Science*, **103**, 563–566.

Viraktamath, C.A. (1997) A revision of the Idiocerine leafhopper genus *Amritodus* (Hemiptera: Cicadellidae) breeding on mango. *Entomon-Trivandrum*, **22**, 111–117.

Vyas, R.V., Patel, J.J., Godhani, P.H. and Yadav, D.N. (1993) Evaluation of green muscardine fungus (*Metarrhizium anisopliae* var. *anisopliae*) for control of mango hopper (*Amritodus atkinsoni*). *Indian Journal of Agricultural Sciences*, **63**, 602–603.

Wharton, R.A. and Marsh, P.M. (1978) New World Opiinae (Hymenoptera: Braconidae) parasitic on Tephritidae (Diptera). *Journal of the Washington Academy of Science*, **68**, 147–165.

White, I.M. and Elson-Harris, M. (1992) *Fruit Flies of Economic Significance*. CAB International, Wallingford, UK. 600 p.

Wysoki, M., Ben-Dov, Y., Swirski, E. and Izhar, Y. (1993) The arthropod pests of mango in Israel. *Acta Horticulturae*, **341**, 452–466.

Yamaguchi, T., Iwamoto. J., Goto, H., Nojima, H., Omatu, N. *et al.* (2006) Insect pests of the mango plant *Mangifera indica* on the Amani islands, Japan. *Kyushu Plant Protection Research*, **52**: 60–65.

Yee W. (1987) *The Mango in Hawaii. Cooperative Extension Service, University of Hawaii Circular*, **388**, 19–22.

Zaher, M., and Osman, A. (1971) Population studies on miotes associated with mango trees in Egypt (Acarina). *Bulletin Societe Entomologique d'Egypte*, **54**, 141–148.
Zaman, Z. and Maiti, B. (1994) Insects and mites infesting mango in Bengal. *Journal of Environmental Entomology*, **12**, 734–736.
Zucchi, R.A. (2000) Espécies de *Anastrepha*, sinonímias, plantas hospedeiras e parasitóides. In: *Moscas-Das-Frutas de Importância Econômica no Brasil: Conhecimento Básico e Aplicado* (eds A. Malavasi and R.A. Zucchi). Holos Editora, Ribeirao Preto. 327 p.

5

Mango Pathology and Diseases

Andressa de Souza-Pollo[1] *and Antonio de Goes*[2]

[1] *Laboratory of Molecular Epidemiology, Sao Paulo State University (UNESP), Faculty of Agricultural and Veterinarian Sciences, Jaboticabal, São Paulo, Brazil*
[2] *Department of Plant Pathology, Faculty of Agricultural and Veterinarian Sciences (FCAV), Sao Paulo State University (UNESP), Jaboticabal, São Paulo, Brazil*

5.1 Introduction

Mango is widely cultivated in the tropics, as well as in subtropical areas. Because of the high moisture content and nutrient-rich profile, mango fruit is highly susceptible to different pathogens including fungi and bacteria. Table 5.1 gives a summary of mango fruit, flowers, leaves, and stem diseases and the corresponding pathogens. A review of the major diseases affecting mango tree and fruit in both pre- and post-harvest stages is provided in this chapter.

5.2 Anthracnose

Mango is among the plants commonly affected by *Colletotrichum* genera, which ranks as one of the most studied phytopathogenic fungi. Anthracnose on mango is a major disease, limiting fruit production in both pre- and in post-harvest stages and is prevalent worldwide. The incidence and severity of the disease are higher in countries where high humidity prevails during the cropping season (Akem 2006; Arauz 2000). For example, in Ethiopia, its incidence ranged from 41% to 74% in leaves and fruits in the field, and from 64% to 82% at the market (Tucho *et al.* 2014).

Anthracnose symptoms occur in leaves, panicles, and fruits. In leaves, the symptoms appear as irregularly black necrotic spots that coalesce and form large necrotic areas frequently in leaves margin. In the panicle, blossom blight can affect the inflorescence stalk and individual flowers. Once in fruit, symptoms are present in both pre- and post-harvest as large black lesions that are usually confined to the peel, though the fungus can invade the pulp in severe cases (Figure 5.1) (Arauz 2000). Mango anthracnose is mainly caused by *C. gloeosporioides*, while *C. acutatum* has also been reported as a causal agent of this disease (Freeman *et al.* 1998; Rivera-Vargas *et al.* 2006; Souza *et al.* 2013). In addition, five other *Colletotrichum* species associated with anthracnose symptoms are *C. asianum*, *C. fructicola*, *C. tropicale*, *C. karstii*, and *C. dianesei* (Lima *et al.*

Handbook of Mango Fruit: Production, Postharvest Science, Processing Technology and Nutrition,
First Edition. Edited by Muhammad Siddiq, Jeffrey K. Brecht and Jiwan S. Sidhu.

Table 5.1 Mango diseases of fruit, flower, leaves, and stem.

Disease	Pathogen
Fruit Diseases:	
Anthracnose	*Colletotrichum gloeosporioides*
Stem-end decay	*Lasiodiplodia theobromae, Phomopsis mangiferae*
Bacterial black spot	*Xanthomonas campestris* pv. *mangiferae indica*
Rhizopus soft rot	*Rhizopus stolonifera, R. arrhizus*
Soft brown rot	*Hendersonia crebemma*
Flower and Leaf Diseases:	
Anthracnose	C. *gleosporioides*
Powdery mildew	*Oidium mangiferae*
Mango malformation	*Fusarium monilifonne* var. *subglutinans*
Bacterial black spot	*Xanthomonas campestris* pv. *mangiferae indica*
Scab	*Elsinoe indica*
Blossom blight	*Botrytis cinerea*
Phytophthora disease	*Phytophthora palmivora*
Stem Diseases:	
Anthracnose	*Colletotrichum gloeosporioides*
Bacterial black spot	*Xanthomonas campestris* pv. *mangiferae indica*
Mango malformation	*Fusarium monilifonne* var. *subglutans*
Verticillium wilt	*Verticillium albo-atrum*
Dieback	*Lasiodiplodia theobromae* (Botryosphaeria)
Sclerotium rot	*Sclerotium rolfsii* Sacc.

Source: Nishiyama (2009), Prusky *et al.* (1997).

2013). These new *Colletotrichum* species were found to be pathogenic for all mango cultivars, with the exception of *C. karstii*, which did not cause infection in the Palmer and Keith cultivars (Lima *et al.* 2015).

Colletotrichum conidia are abundantly produced in the mango canopy, on leaves, twigs, panicles lesions, and mummified fruit. Although conidia are considered the main inoculum source, ascospores can be produced in dry leaves on the ground and contribute to the spread of the disease (Arauz 2000). As *C. gloeosporioides* affects a wide range of crops, the fungus presents lack of host specificity and a high genetic diversity, which makes anthracnose control rather difficult (Souza *et al.* 2013).

Anthracnose control consists mainly of fungicide application during fruit development. Prophylactic treatments with fungicides before the beginning of the rainy season and previously manifested infections are recommended. The active ingredients reported as more effective in the disease control are thiophanate methyl and azoxystrobin (Diedhiou *et al.* 2014). Carbendazim was largely applied on mango, however, *C. gloeosporioides* has acquired resistance to this active ingredient (Kongtragoul *et al.* 2011). Postharvest

Figure 5.1 Anthracnose disease: (A) Blossom blight in inflorescence stalk; (B) small fruits infected and mummified fruits attached to the tree; (C) spots on green fruits spreading in the way of water conidia suspension; (D) large black lesions on mango fruit postharvest; (E) spots on leaves in the field; and (F) lesions on inoculated leaves evolving to necrosis (Souza *et al.* 2013). Source of A, B, and E: Nivaldo Lima.

application of non-fungicide products has been suggested, such as β-aminobutyric acid (Zhang *et al.* 2013), propionic acid, and bee-carnauba wax (Thinh and Kunasakdakul 2013), essential oils (Abd-Alla and Haggag 2013), and nitric oxide (Hua *et al.* 2014).

5.3 Powdery Mildew

Powdery mildew is distributed widely throughout the world and was one of the first diseases reported in mango. The disease affects almost every cultivar, irrespective of geographical location (Nasir *et al.* 2014). Crop losses can reach up to 90% owing to its effect on fruit set and development (Nelson 2008).

(A) (B)

Figure 5.2 Powdery mildew symptoms on: (A) inflorescences and (B) leaves of mango.
Source: Nelson (2008).

The causal agent of this disease is an obligated plant pathogen, formerly known as *Oidium mangiferae* Berthet; its name was changed to *Pseudoidium anacardii* (F. Noack) U. Braun and R.T.A. Cook in 2012. The fungus initially attacks young tissues, leaves, inflorescence, and fruits, which become coated with a whitish powdery mass. Young leaves fall when covered on the underside and mature leaves develop purplish brown spots. As a result, inflorescence become unfertilized and fall prematurely. Infected fruits are often malformed and off-colored (Figure 5.2) (Nasir *et al.* 2014; Nelson 2008). The disease surges from inocula prevailing in pre-infected trees in the same orchard or through conidia dispersed in the air from other sources. Mycelium or haustoria overwinter in the buds or in malformed green panicles under dense foliage (Nasir *et al.* 2014). The mildew is favored by high humidity, cloudy weather, and low night temperature between panicle development and fruit set (Gupta 1979).

Disease control is based mainly on the fungicides application. However, cultural management by the removal of infected leaves and malformed panicles reduce the primary inoculum, contributes to the reduction of disease incidence and, consequently, increases fungicidal control. Generally, 2–3 applications of fungicides at 15–20 day intervals are effective in disease control. Inorganic cooper as a protector followed by systemic fungicides applications are recommended. Among the systemic fungicides, Imidazole has excellent curative and preventive properties, benzimidazols control powdery mildew of mango from florescence to opening of flowers, and strobilurins and triazoles are also very important fungicides for the disease control. There is no mango variety that is immune, although Espada, Imperial, and Coquinho among others may have some resistant behavior, while Haden, Glenn, Carrie, Keitt, and Tommy Atkins among others are moderately susceptible (Nasir *et al.* 2014).

5.4 Mango Malformation Disease

Mango malformation disease (MMD) is one of the most serious diseases worldwide. It was first reported in India in 1891 and now has spread elsewhere in Asia (Israel,

Figure 5.3 Symptoms of malformed inflorescence of mango.

Malaysia, and Pakistan), Africa (Egypt, South Africa, Sudan, Swaziland, and Uganda), Europe (Spain), and the Americas (Brazil, El Salvador, Mexico, Nicaragua, the USA, and Venezuela) (Crespo *et al.* 2014; Marasas *et al.* 2006). Floral malformation can cause loss of the entire crop or 50–80% yield reduction (Kumar *et al.* 2011).

Vegetative malformation symptoms are more pronounced in seedlings that present loss of apical dominance, swollen and miss-shapen axillary buds, shortened internodes, and narrow leaves. Blossom malformation appears with the emergence of inflorescences on mature trees. The symptoms result in enlargement and crowded inflorescence, hypertrophied axes of the panicle, no fruit production, or premature abortion (Figure 5.3) (Kumar *et al.* 2011; Marasas *et al.* 2006). Several *Fusarium* species belonging to the *Gibberella fujikuroi* complex are associated with MMD, such as *F. mangiferae*, *F. sterilihyphosum*, and *F. tupiense* from Brazil and South Africa (Lima *et al.* 2012; Marasas *et al.* 2006), *F. proliferatum* from China and Malaysia (Nor *et al.* 2013; Zhan *et al.* 2010), *F. pseudocircinatum* from Africa, and *F. mexicanum* from Mexico (Otero-Colina *et al.* 2010).

Sexual compatibility is commonly reported in the *G. fujikuroi* species complex (Crespo *et al.* 2014; Lima *et al.* 2009; Marasas *et al.* 2006). Vegetative compatibility groups (VCG) contribute to genetic diversity and can be clearly identified by molecular markers such as AFLP (Lima *et al.* 2009). VCG tests showed that the Spanish population of *F. mangiferae* is different from those in Egypt, the USA, and Israel (Crespo *et al.* 2014). The same discrimination was obtained by AFLP analysis, which was also able to differentiate Fusaria within the Asian isolates clade comprised of *F. concentricum*, *F. fractiflexum*, *F. proliferatum*, and *F. sacchari* (Newman *et al.* 2012).

The epidemiology of the disease is by Fusaria conidial infection. Conidia from diseased panicles and malformed vegetative tissue disseminate passively in the air until their contact with dormant buds on the same tree and on others surroundings. Thus, the disease control consists of orchard management by protecting apical buds from airborne infections and maintaining strict sanitation in the orchard by immediate removal of malformed tissues (Gamliel-Atinsky *et al.* 2009). Also, field experiments conducted over a number of seasons indicated that combined sanitation with timely applications of prochloraz, in comparison with other fungicides, significantly reduced severity and incidence of MMD. These treatments used in combination can take MMD to negligible levels over time (Freeman *et al.* 2014).

5.5 Mango Bacterial Canker

Mango bacterial canker (MBC), or bacterial black spot, is a major concern for the export market where high fruit quality is required. The disease has been identified in many mango producing countries, including Australia, Brazil, Egypt, Ghana, India, Iran, Japan, South Africa, Sudan, Taiwan, Philippines, Queensland, Mali, and the Ivory Coast, among others (Gagnevin and Pruvost 2001; CABI/EPPO, 2012; Pruvost *et al.* 2012, 2014). Infections on highly susceptible cultivars can result in drastic yield losses from 50–80% (Ah-You *et al.* 2007; Gagnevin and Pruvost 2001).

The MBC symptoms on leaves appear initially as small water-soaked spots, which enlarge and become angular and black. These lesions are usually limited in the vicinity of the vein and surrounded by chlorotic haloes. Severe leaf infection may result in abscission. After months of infection, leaf lesions dry and become light brown or ash gray. On twigs, dark and cracked lesions called canker appear along the axis and result in weakening of branches that become subjected to damage by wind (Gagnevin and Pruvost 2001). In Brazil, yellow to brown lesions are reported, which are longitudinally cracked, and evolve as a generalized dieback of the youngest vegetative flush (Ah-You *et al.* 2007). Fruit symptoms appear as small water-soaked spots on the lenticels and may cause premature fruit drop in severe infections (Figure 5.4). Bud necrosis and inflorescence cankers are rarely observed. The root system is not affected by MBC infection, since the disease is not systemic (Gagnevin and Pruvost 2001).

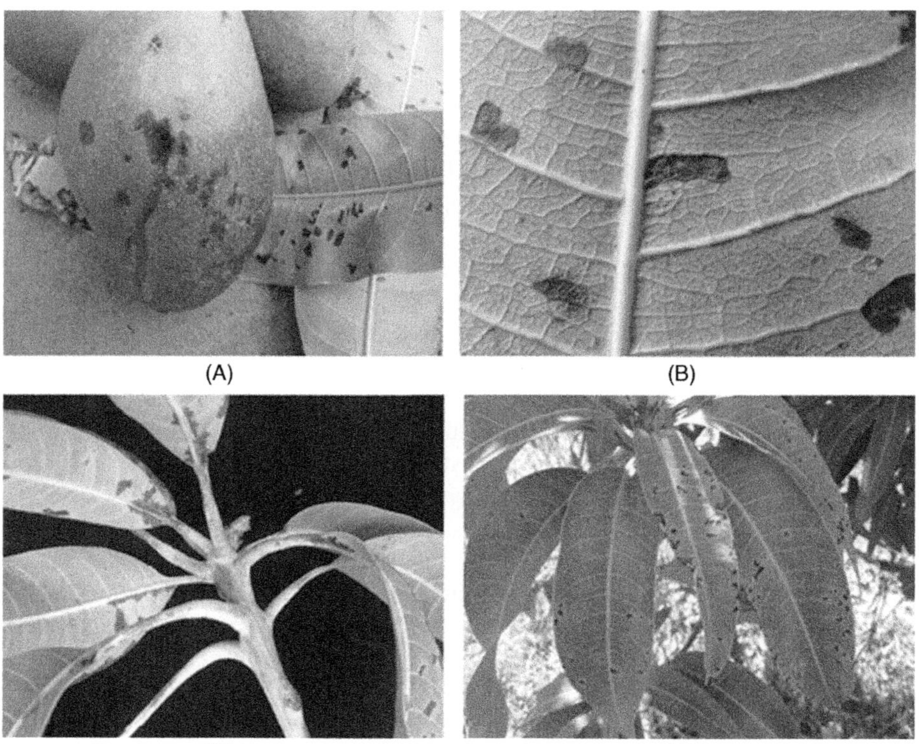

Figure 5.4 Mango bacterial canker: (A) spots on fruits and leaves; (B) lesions on leaves limited in the vicinity of the vein; (C) lesions on leaves and twigs; and (D) spots on leaves surrounded by chlorotic haloes.

The causal agent of MBC is a Gram-negative bacterium formally known as a pathovar of *Xanthomonas campestris.* Though studies applying multi-locus sequence analysis and AFLP showed that this pathogen is a pathovar of *X. citri,* it became identified as *X. citri* pv. *Mangiferaeindicae* (Ah-you *et al.* 2009). Another *Xanthomonas* species associated with MBC, reported as occurring in China, was *X. arboricola* (Qi *et al.* 2012). The pathogen can survive on asymptomatic mango leaves as primary endophitic (Pruvost *et al.* 2009) and in twig cankers that serve as an inoculum source (Ah-You *et al.* 2007). Bacterial infections occur through wounds and stomata, mainly on old leaves since young leaves have non-functional stomata (Gagnevin and Pruvost 2001). However, in Brazil, usually symptoms on young leaves are observed when the weather is favorable to infections. Long-distance dissemination occurs by infected propagation material, whereas short-distance contamination can be by rainwater splashing or wind-driven (Gagnevin and Pruvost 2001). Disease development is favored by high relative humidity (>90%) and moderate temperatures of 25–30°C (Pruvost and Luisetti 1991).

Mango cultivars can present low or partial resistance to MBC, but no cultivar has shown complete resistance. In Brazil, one of the most cultivated mango, the Palmer cultivar, is very susceptible to disease. The control of MBC must begin in the nursery by tool disinfection, destruction of any potential sources of inoculums, and prophylactic treatments with copper sprays. In the field, the recommended control of the disease is windbreaks planting around mango orchards to reduce wounds and dissemination of bacteria by wind-driven rainwater, and regular copper application (Gagnevin and Pruvost 2001).

5.6 Dieback and Stem-end Rot

Mango dieback and stem-end rot are two of the most serious diseases affecting mango production and postharvest losses worldwide. The causal agents of these diseases belong to a fungi complex of the Botryosphaeriaceae family (Costa *et al.* 2010; Ismail *et al.* 2012). Dieback symptoms result in the death of branches (Khanzada *et al.* 2004), while stem-end rot symptoms result in fruit losses during postharvest storage and marketing (Ni *et al.* 2012).

Among Botryosphaeriaceae fungi, *Lasiodiplodia theobromae, Neofusicoccum parvum, N. mangiferum,* and *Botryosphaeria dothidea* commonly cause stem cankers, dieback, and stem-end rot of mangoes (Sakalidis *et al.* 2011). In addition, *L. hormozganensis, L. iraniensis, B. scharifii* sp. nov., and *N. mediterraneum* are also associated with mango dieback and fruit rot in Iran (Abdollahzadeh *et al.* 2010, 2013). In Brazil, even though *L. theobromae* has been associated with mango dieback and stem-end rot for a long time, molecular studies revealed the presence of other Botryosphaeriaceae species, such as *N. parvum, Fusicoccum aesculi,* and *Pseudofusicoccum stromaticum* (Costa *et al.* 2010; Marques *et al.* 2012). Another study revealed more fungi species associated with mango diseases in Brazil, namely *L. crassispora, L. egyptiacae, L. hormozganensis, L. iraniensis, L. pseudotheobromae, L. theobromae, B. mamane, F. fabicercianum, N. brasiliense* sp. nov., and *Neoscytalidium dimidiatum* (Marques *et al.* 2013a,b).

Dieback symptoms are characterized by death of the twigs, which results in a scorched appearance. The young green twigs start withering first at the base that extends along the veins to the leaf's edges. Leaves turn brown, margins roll upwards and then scorch and fall, leaving a dead branch. Under severe conditions, branches start drying one after

Figure 5.5 Mango dieback and stem-end rot: (A) dieback symptoms and (B) fruit with stem-end rot lesion. *Source of B*: Ivan Herman Fischer, PhD (APTA, SP, Brazil).

another in a sequence, resulting in death of the whole tree. Infected plants show abundant gum secretion from branches, stem, and main trunk (Khanzada *et al.* 2004). The fungus stays dormant on the fruit until it begins to ripen. Blossom blight symptoms appear at the peduncle and pedicel tissues, evolve to rotting and can cover the entire fruit body in severe cases, with the flesh of the infected fruit developing an off flavor (Figure 5.5) (Alemu 2014).

The fungi live within branches (as endophytic) and colonize inflorescence tissues without causing symptoms, reaching the stem-end of fruit several weeks after flowering. The fungi colonize the fruit just after harvesting. Fruit can also be infected by soil contact at harvest, since some of the fungi may be harbored in leaf litter on the soil. High humidity and water are generally responsible for the release and dispersal of conidia. The fungus can also be spread to adjacent healthy fruit by physical contact (Alemu 2014).

The best control measure is to prevent wounding of trunk and twigs. Cankers, dead limbs, dead twigs, and old fruit should be removed from the tree. Wounds from pruning and removal of suckers should be protected with copper. Corrections of stress factors such as nutritional deficiency should be addressed appropriately. Sprays with copper and mancozeb fungicides and treatment of graft unions with fungicides are beneficial. Often, postharvest control of latent infections is needed, mainly if fruit is stored or shipped. Hot water treatments and fungicides dips are common. Basically, it consists of dipping the fruit in a hot water bath at temperatures ranging from 50–55 °C for 2–5 minutes, with or without fungicide. The treatment should be done soon after harvest, but no later than 2 days following harvest (Alemu 2014). Also, direct involvement of latex in fruit resistance and the possibility of its manipulation to protect ripe fruit from fungal rotting have been reported recently (Karunanayake *et al.* 2015).

5.7 Mango Wilt or Mango Blight

Mango wilt is a serious disease that results in plant death. The causal agent belongs to the *Ceratocystis* spp. genera, a typical vascular fungus that infects trees and causes internal necrosis of the stem tissues and the obstruction of the xylem vessels, resulting in wilting and death of plants (Al-Sadi *et al.* 2010). Among reported species, *C. fimbriata* is notable for its broad host range and its extensive geographic distribution (Baker *et al.* 2003). However, other species of *Ceratocystis* were associated with mango wilt, for example *C. omanensis* in Oman (Al-Subhi *et al.* 2006), *C. manginecans* sp. nov., occurring in Oman and Pakistan, closely related to but distinct from isolates of mango from Brazil (Van Wyk *et al.* 2007), *C. mangicola* sp. nov., and *C. mangivora* sp. nov. in Brazil (van Wyk *et al.* 2011).

A bark beetle (*Hypocryphalus mangiferae*) is known to be involved in transmission and introduction of *Ceratocystis* into healthy mango trees. The insect acts as a wounding agent and as a vector of spores (Masood and Saeed 2012). The fungus invades the xylem and subsequently induces tyloses production. Thus, the symptoms of the disease include discoloration of wood that becomes dark brown stained from the point of infection, gum exudation, wilting, and rapid death of trees (Figure 5.6) (Al-Sadi *et al.* 2010). The death of the tree can occur about six months after the first symptoms appear.

Some factors favor the disease development, such as wounding trees, water stress, extreme temperatures in summer, sun scorch, and nutritional deficiencies, mainly lack of manganese, iron, zinc, and copper (Masood *et al.* 2012). Cultural practices must be applied to control the disease. It is essential to seek such advice to avoid damage to the tree trunk. Portions of affected canopy should be pruned and the cut areas treated with Bordeaux mix, as infected trees serve as a source of inoculums (Saeed *et al.* 2012). Fungicide and insecticides for fungus and vector control, respectively, should be applied as soon as initial onset of wilting signs (Lozano *et al.* 2001; Masood *et al.* 2014; Saeed *et al.* 2011). Also, the use of resistant mango cultivars should be regarded, since there

(A) (B) (C)

Figure 5.6 Mango wilt disease: (A) gum exudation from an infected stem; (B) discoloration of wood (Galli *et al.* 2011); and (C) dead mango tree (Galli *et al.* 2011).

are susceptible and resistant basal cell responses of mango stem tissue to infection by *C. fimbriata* (Araujo *et al.* 2014).

References

Abd-Alla, M.A. and Haggag, W.M. (2013) Use of some plant essential oils as postharvest botanical fungicides in the management of anthracnose disease of mango fruits (*Mangi Feraindica* L.) caused by *Colletotrichum Gloeosporioides* (Penz). *International Journal of Agriculture and Forestry*, **3**, 1–6.

Abdollahzadeh, J., Javadi, A., Mohammadi, G.E., Zare, R. and Phillips, A.J.L. (2010) Phylogeny and morphology of four new species of *Lasiodiplodia* from Iran. *Persoonia: Molecular Phylogeny and Evolution of Fungi*, **25**, 1–10.

Abdollahzadeh, J., Zare, R. and Phillips, A.J.L. (2013) Phylogeny and taxonomy of *Botryosphaeria* and *Neofusicoccum* species in Iran, with description of *Botryosphaeria scharifii* sp. nov. *Mycologia*, **105**, 210–220.

Ah-You, N., Gagnevin, L., Chiroleu, F., Jouen, E., Rodrigues, N.J. and Pruvost, O. (2007) Pathological variations within *Xanthomonas campestris* pv. *mangiferaeindicae* support its separation into three distinct pathovars that can be distinguished by amplified fragment length polymorphism. *Phytopathology*, **97**, 1568–1577.

Ah-You, N., Gagnevin, L., Grimont, P.A.D., Brisse, S., Nesme, X. *et al.* (2009) Polyphasic characterization of *Xanthomonas* pathogenic to Anacardiaceae and their relatedness to different *Xanthomonas* species. *International Journal of Systematic and Evolutionary Microbiology*, **59**, 306–318.

Akem, C.N. (2006) Mango anthracnose disease: Present status and future research priorities. *Plant Pathology Journal*, **5**, 266–273.

Alemu, K. (2014) Dynamics and management of major postharvest fungal diseases of mango fruits. *Journal of Biology, Agriculture and Healthcare*, **4**, 1–10.

Al-Sadi, A.M., Al-Ouweisi, F.A., Al-Shariani, N.K., Al-Adawi, A.O., Kaplan, E.J. and Deadman, M.L. (2010) Histological changes in mango seedlings following infection with *Ceratocystis manginecans*, the cause of mango decline. *Journal of Phytopathology*, **158**, 738–743.

Al-Subhi, A.M., Al-Adawi, A.O., van Wyk, M., Deadman, M.L. and Wingfield, M.J. (2006) *Ceratocystis omanensis*, a new species from diseased mango trees in Oman. *Mycological Research*, **110**, 237–245.

Araujo, L., Bispo, W.M.S., Cacique, I.S., Moreira, W.R. and Rodrigues, F.A. (2014) Resistance in mango against infection by *Ceratocystis fimbriata*. *Phytopathology*, **104**, 820–833.

Arauz, L.F. (2000) Mango anthracnose: economic impact and current options for integrated and management. *Plant Disease*, **84**, 600–611.

Baker, C.J., Harrington, T.C., Krauss, U. and Alfenas, A.C. (2003) Genetic variability and host specialization in the Latin American clade of *Ceratocystis fimbriata*. *Phytopathology*, **93**, 1274–1284.

CABI/EPPO (2012) *Xanthomonas campestris* pv. *mangiferaeindicae*. [Distribution map]. Distribution Maps of Plant Diseases 2012 October Map 1135 (Edition 1). CABI Head Office, Wallingford, UK [Online]. Available: http://www.cabi.org/cpc/abstract/20123367489 [15April 2015].

Costa, V.S.O., Michereff, S.J., Martins, R.B., Gava, C.A.T., Mizubuti, E.S.G. and Câmara, M.P.S. (2010) Species of Botryosphaeriaceae associated on mango in Brazil. *European Journal of Plant Pathology*, **127**, 509–519.

Crespo, M., Arrebola, E.F.M., Maymon, C.M., Freeman, S., Torés, J.A. and Vicente, A. (2014) Characterization of *Fusarium mangiferae* isolates from mango malformation disease in Southern Spain. *European Journal of Plant Pathology*, **139**, 253–259.

Diedhiou, P.M., Diallo, Y., Faye, R., Mbengue, A.A. and Sene, A. (2014) Efficacy of different fungicides against mango anthracnose in Senegalese Soudanian Agroclimate. *American Journal of Plant Sciences*, **5**, 2224–2229.

Freeman, S., Katan, T. and Shabi, E. (1998) Characterization of *Colletotrichum* species responsible for anthracnose diseases of various fruits. *Plant Disease*, **82**, 596–605.

Freeman, S., Maymon. M., Biton, A., Levin, A.G. and Shtienberg, D. (2014) Management of mango malformation disease based on a novel strategy of timing of fungicide applications combined with sanitation. *Crop Protection*, **61**, 84–91.

Gagnevin, L. and Pruvost, O. (2001) Epidemiology and control of mango bacterial black spot. *Plant Disease*, **85**, 928–935.

Galli, J.A., Martins, A.L.M., Ito, M.F., Braghini, M.T., Narita N. and Rossetto, C.J. (2011) Mango wilt. vol. XXII: Survival of poli-embrionic varieties. *Revista Brasileira de Fruticultura*, **33**(4), 1119–1126.

Gamliel-Atinsky, E., Sztejnberg, A., Maymon, M., Vintal, H., Shtienberg, D. and Freeman, S. (2009) Infection dynamics of *Fusarium mangiferae*, causal agent of mango malformation disease. *Phytopathology*, **99**, 775–778.

Gupta, J.H. (1979) Influence of environmental factors on the development of powdery mildew of mango. *Indian Journal of Horticulture*, **36**, 96–98.

Hua, M., Yanga, D., Huberb, D.J., Jiangc, Y., Lia, M. *et al.* (2014) Reduction of postharvest anthracnose and enhancement of disease resistance in ripening mango fruit by nitric oxide treatment. *Postharvest Biology and Technology*, **97**, 115–122.

Ismail, A.M., Cirvilleri, G., Polizzi, G., Crous, P.W., Groenewald, J.Z. and Lombard, L. (2012) *Lasiodiplodia* species associated with dieback disease of mango (*Mangifera indica*) in Egypt. *Australasian Plant Pathology*, **41**, 649-660.

Karunanayake, K.O.L.C., Sinniah, G.D., Adikaram, N.K.B., Abayasekara, C.L. and Wijayasekara, D.S. (2015) Retention of latex at harvest, enhanced mango (*Mangifera indica* L.) fruit resistance and reduced anthracnose and stem-end rot. *Australasian Plant Pathology*, **44**, 113–119.

Khanzada, M.A., Lodhi, A.M. and Shahzad, S. (2004) Mango dieback and gummosis in Sindh, Pakistan caused by *Lasiodiplodia theobromae* [Online]. Available: https://www.plantmanagementnetwork.org/pub/php/diagnosticguide/2004/mango/ [24 March 2016].

Kongtragoul, P., Nalumpang, S., Miyamoto, Y., Izumi, Y. and Akimitsu, K. (2011) Mutation at codon 198 of tub2 gene for carbendazim resistance in *Colletotrichum gloeosporioides* causing mango anthracnose in Thailand. *Journal of Plant Protection Research*, **51**, 377–384.

Kumar, P., Misra, A.K. and Modi, D.R. (2011) Current status of mango malformation in India. *Asian Journal of Plant Sciences*, **10**, 1–23.

Lima, C.S., Monteiro, J.H., Crespo, N.C., Costa, S.S., Leslie, J.F. and Pfenning, L.H. (2009) VCG and AFLP analyses identify the same groups in the causal agents of mango malformation in Brazil. *European Journal of Plant Pathology*, **123**, 17–26.

Lima, C.S., Pfenning, L.H,, Costa, S.S., Abreu, L. and Leslie, J.F. (2012) *Fusarium tupiense* sp. nov., a member of the *Gibberella fujikuroi* complex that causes mango malformation in Brazil. *Mycologia*, **104**, 1408–1419.

Lima, N.B., Batista, M.V.A., Morais Jr, M.A., Barbosa, M.A.G., Michereff, S.J. *et al.* (2013) Five *Colletotrichum* species are responsible for mango anthracnose in northeastern Brazil. *Fungal Diversity*, **61**, 75–88.

Lima, N.B., Lima, W.G., Tovar-Pedraza, J.M., Michereff, S.J. and Câmara, M.P.S. (2015) Comparative epidemiology of *Colletotrichum* species from mango in northeastern Brazil. *European Journal of Plant Pathology*, **141**, 679–688.

Lozano, C., Pena, A., Sanchez-Raya, A.J. and Campos, M. (2001) Effectiveness of deltamethrin for protecting olives from attack by the beetle *Phloeotribus scarabaeoides* (Scolytidae). *International Journal of Pest Management*, **47**, 75–78.

Marasas, W.F.O., Ploetz, R.C., Wingfield, M.J., Wingfield, B.D. and Steenkamp, E.T. (2006) Mango malformation disease and the associated *Fusarium* species. *Phytopathology*, **96**, 667–72.

Marques, M.W., Lima, N.B., Michereff, S.J., Câmara, M.P.S. and Souza, C.R.B. (2012) First report of mango dieback caused by *Pseudofusicoccum stromaticum* in Brazil. *Plant Disease*, **96**, 144–145.

Marques, M.W., Lima, N.B., Morais Jr, M.A., Barbosa, M.A.G., Souza, B.O. *et al.* (2013a) Species of *Lasiodiplodia* associated with mango in Brazil. *Fungal Diversity*, **61**, 181–193.

Marques, M.W., Lima, N.B., Morais Jr, M.A., Michereff, S.J., Phillips, A.J.L. and Câmara, M.P.S. (2013b) *Botryosphaeria*, *Neofusicoccum*, *Neoscytalidium* and *Pseudofusicoccum* species associated with mango in Brazil. *Fungal Diversity*, **61**, 195–208.

Masood, A. and Saeed, S. (2012) Bark beetle, *Hypocryphalus mangiferae* Stebbing (Coleoptera: Curculionidae: Scolytinae) is a vector of mango sudden death disease in Pakistan. *Pakistan Journal of Botany*, **44**, 813–820.

Masood, A., Saeed, S., Mahmood, A., Malik, S.A. and Hussain, N. (2012) Role of nutrients in management of mango sudden death disease. *Pakistan Journal of Zoology*, **44**, 675–683.

Masood, A., Salman, M. and Saeed, S. (2014) Fungicide injection, an efficient management technique of mango sudden death disease in Punjab, Pakistan. *Pakistan Journal of Phytopathology*, **26**, 259–263.

Nasir, M., Mughal, S.M., Mukhtar, T. and Awan, M.Z. (2014) Powdery mildew of mango: A review of ecology, biology, epidemiology and management. *Crop Protection*, **64**, 19–26.

Nelson, S.C. (2008) Mango powdery mildew. *Plant Disease*, **46**, 1–6.

Newman, Z., Freeman, S., Biton, I., Sa'ada, D., Paz, T., Maymon, M. and Lavi, U. (2012) Molecular diagnosis of mango malformation disease and phylogeny of *Fusarium mangiferae*. *Phytoparasitica*, **40**, 287–297.

Ni, H., Yang, H., Chen, R., Lio, R. and Hung, T. (2012) New *Botryosphaeriaceae* fruit rot of mango in Taiwan: Identification and pathogenicity. *Botanical Studies*, **53**, 467–478.

Nishijima, W. (2009) Mango diseases and their control [Online]. Available: http://www.ctahr.hawaii.edu/oc/freepubs/pdf/hitahr_04-06-93_20-24.pdf [26 January 2016].

Nor, N.M.I.M., Salleh, B. and Leslie, J.F. (2013) *Fusarium* species associated with mango malformation in Peninsular Malaysia. *Journal of Phytopathology*, **161**, 617–624.

Otero-Colina, G., Rodríguez-Alvarado, G., Fernández-Pavía, S., Maymon, M., Ploetz, R.C. *et al.* (2010) Identification and characterization of a novel etiological agent of mango malformation disease in Mexico *Fusarium mexicanum* sp. nov. *Phytopathology*, **100**, 1176–1184.

Prusky, D., Kobiler, I., Miyara, I. and Alkan, N. (1997) Fruit diseases. In: *The Mango, Botany, Production and Uses* (ed. R.E. Litz), pp. 201–230. CAB International, Wallingford, UK.

Pruvost, O. and Luisetti, J. (1991) Effect of time of inoculation with *Xanthomonas campestris* pv. *mangiferaeindicae* on mango fruits susceptibility epiphytic survival of X. c. pv. *mangiferaeindicae* on mango fruits in relation to disease development. *Journal of Phytopathology*, **133**, 139–151.

Pruvost, O., Savelon, C., Boyer, C., Chiroleu, F., Gagnevin, L. and Jacques, M.A. (2009) Populations of *Xanthomonas citri* pv. *mangiferaeindicae* from asymptomatic mango leaves are primarily endophytic. *Microbial Ecology*, **58**, 170–178.

Pruvost, O., Boyer, C., Vital, K., Verniere, C., Gagnevin, L. and Traoré, Y.N. (2012) First report in Mali of *Xanthomonas citri* pv. *mangiferaeindicae* causing mango bacterial canker on *Mangifera indica*. *Plant Disease*, **96**, 581.

Pruvost, O., Boyer, C., Grygiel, P., Boyer, K., Verniere, C. *et al.* (2014) First report of *Xanthomonas citri* pv. *mangiferaeindicae* causing mango bacterial canker on *Mangifera indica* in Ivory Coast. *Plant Disease*, **98**, 1740.

Qi, Y.X., Pu, J.J., Zhang, X., Lu, Y., Zhang, H. *et al.* (2012) First report of bacterial leaf spot of mango caused by a *Xanthomonas arboricola* pathovar in China. *Journal of Plant Pathology*, **94**, 85.

Rivera-Vargas, L.I., Lugo-Noel, Y., McGovern, R.J., Seijo, T. and Davis, M.J. (2006) Occurrence and distribution of *Colletotrichum* spp. on mango (*Mangifera indica* L.) in Puerto Rico and Florida, USA. *Plant Pathology Journal*, **5**, 191–198.

Saeed, S., Masood, A., Sayyed, A.H. and Kwon, Y.J. (2011) Comparative efficacy of different pesticides against mango bark beetle, *Hypocryphalus mangiferae* Stebbing (Coleoptera: Scolytidae). *Entomological Research*, **41**, 142–150.

Saeed, S., Masood, A. and Khan, S.M. (2012) Diseased plants as a source of dissemination of mango sudden death disease in healthy mango plants. *Pakistan Journal of Phytopathology*, **24**, 21–25.

Sakalidis, M.L., Ray, J.D., Lanoiselet, V., Hardy, G.E. St J. and Burgess, T.I. (2011) Pathogenic Botryosphaeriaceae associated with *Mangifera indica* in the Kimberley Region of Western Australia. *European Journal of Plant Pathology*, **130**, 379–391.

Souza, A., Carboni, R.C.D., Wickert, E., Lemos, E.G.M. and Goes, A. (2013) Lack of host specificity of *Colletotrichum* spp. Isolates associated with anthracnose symptoms on mango in Brazil. *Plant Pathology*, **62**, 1038–1047.

Thinh, D.C. and Kunasakdakul, K. (2013) Inhibition of *Colletotrichum gloeosporioides* and control of postharvest anthracnose disease on mango fruit using propionic acid combined with bee-carnauba wax emulsion. *Journal of Agricultural Science*, **5**, 110–116.

Tucho, A., Lemessa, F. and Berecha, G. (2014) Distribution and occurrence of mango anthracnose (*Colletotrichum gloesporioides* Penz and Sacc) in humid agro-ecology of Southwest Ethiopia. *Plant Pathology Journal*, **13**, 268–277.

Van Wyk, M., Al Adawi, A.O., Khan, I.A., Deadman, M.L., Al Jahwari, A.A. *et al.* (2007) *Ceratocystis manginecans* sp. nov., causal agent of a destructive mango wilt disease in Oman and Pakistan. *Fungal Diversity*, **27**, 213–230.

Van Wyk, M., Wingfield, B.D., Al-Adawi, A.O., Rossetto, C.J., Ito, M.F. and Wingfield, M.J. (2011) Two new *Ceratocystis* species associated with mango disease in Brazil. *Mycotaxon*, **117**, 381–404.

Zhan, R-L., Yang, S-J., Ho, H-H., Liu, F., Zhao, Y-L. *et al.* (2010) Mango malformation disease in South China caused by *Fusarium proliferatum*. *Journal of Phytopathology,* **158**, 721–725.

Zhang, Z., Yang, D., Yang, B., Gao, Z., Li, M. *et al.* (2013) β-Aminobutyricacid induces resistance of mango fruit to postharvest anthracnose caused by *Colletotrichum gloeosporiodes* and enhances activity of fruit defense mechanisms. *Scientia Horticulturae,* **160**, 78–84.

6

Harvesting and Postharvest Technology of Mango

Jeffrey K. Brecht[1] *and Elhadi M. Yahia*[2]

[1] *Horticultural Sciences Department, University of Florida, Gainesville, Florida, USA*
[2] *Horticultural Sciences Department, Universidad Autonoma de Queretaro, Queretaro, Mexico*

6.1 Introduction

Mangoes are produced in over 80 countries with different marketing scenarios present in each country, therefore harvesting and postharvest handling practices vary considerably. However, the basic considerations and underlying physiology of mango fruit are the same worldwide. As a climacteric fruit (Brecht and Yahia, 2009; Yahia 2011), mangoes can be harvested when physiologically mature, but unripe, and then ripened postharvest. In practice, however, this leads to ripe fruit of inferior quality. Therefore, mangoes are almost always harvested after initiation of ripening on the tree. The extent to which ripening on the tree is allowed to progress prior to harvest is dependent on the intended market for the fruit – local markets allowing almost fully ripe fruit to be harvested and distant markets requiring the fruit to be progressively closer to ripening initiation, or even prior to ripening initiation, at the time of harvest.

Similarly, the postharvest technologies that are used for mangoes vary according to the intended utilization of the fruit. Various handling practices, such as sizing, application of fruit coatings, packaging and palletization, cooling, and use of modified and controlled atmospheres (MA and CA) are tailored to the market requirements. Export to countries with quarantines in effect against microbe or insect pests requires application of disinfestation treatments (Johnson and Hofman, 2009; Yahia *et al.* 2011). The choices that are made regarding the specific technologies to be used must correspond to mango physiology and quality requirements. This chapter will cover harvesting and postharvest technology of mangoes in relation to the above considerations.

6.2 Harvest Maturity Indices

Mango fruit take approximately 3 months to reach maturity, with maximum fruit size attained about 2–3 weeks before horticultural maturity. Harvesting mangoes at the optimum horticultural maturity for their intended purpose is arguably the most

Handbook of Mango Fruit: Production, Postharvest Science, Processing Technology and Nutrition,
First Edition. Edited by Muhammad Siddiq, Jeffrey K. Brecht and Jiwan S. Sidhu.

important practice contributing to mango quality after harvest. In actuality, horticultural maturity for mango corresponds to initiation of ripening. Mango is a climacteric fruit (i.e., it can ripen off the tree), exhibiting a climacteric pattern of respiration and an increase in ethylene production (Brecht and Yahia, 2009; Subramanyam *et al.* 1975; Yahia 2011). Changes associated with mango fruit ripening include:

- peel color changes from green to yellow in most cultivars;
- flesh color changes from greenish yellow to yellow to orange in all cultivars;
- chlorophyll content decreases, while carotenoid content increases;
- flesh firmness decreases and juiciness increases;
- starch is converted into sugars;
- soluble solids content increases;
- titratable acidity decreases;
- characteristic aroma volatiles production increases;
- respiration (carbon dioxide production) rate increases from 40–50 to 160–200 mg/kg hr at 20 °C; and
- ethylene production rate increases from 0.2–0.4 to 2–4 µl/kg·hr at 20 °C.

Since mangoes are typically allowed to at least begin ripening on the tree prior to harvest, most harvest indices are related to the changes occurring during fruit ripening; other indices are related to physical and chemical changes during fruit development (Brecht *et al.* 2014; Singh and Singh 2012; Yahia 2011). The change in fruit shape resulting in full cheeks and prominent shoulders, increasing fruit density, and change in skin color of some cultivars from dark green to light-green or yellow are common horticultural maturity indices for mango. Red blush on the skin is not a measure of fruit maturity, being related only to sun exposure, which stimulates synthesis of the red pigments. There is no single index that can be used successfully for all cultivars to select mangoes with uniform maturity.

Maturity at harvest determines both the postharvest life of the fruit and its quality. Generally, tree-ripe fruit attain much better quality than fruit harvested when less ripe and ripened off the tree. However, tree-ripe fruit will not withstand prolonged storage and transport periods, being more susceptible to bruising, decay, and over-ripening. Therefore, it is recommended that fruit destined for a local market (immediate consumption) and for processing should be harvested at later stages of ripening than fruit intended for storage or to be transported to distant markets. Mango fruit should be harvested after they reach physiological maturity and before they are overripe. Fruit harvested before physiological maturity will either never ripen or never reach optimum quality. Mangoes harvested at the mature and half-mature stages ripen to good-quality fruit, while immature fruit do not ripen normally and are also more susceptible to chilling injury and heat injury than mature fruit (Brecht and Yahia 2009; Johnson and Hofman 2009; Mohammed and Brecht 2002).

Maturity indices should ideally fulfill several requirements, including:

- simple to use;
- not influenced by environmental conditions;
- should relate to ultimate quality of the fruit;
- objective; and
- be preferably non-destructive.

Table 6.1 Examples of the use of computation to predict harvest time in different ASEAN countries.

Cultivar	Age (days)	Computation	Country
Arumanis	90	Full bloom	Indonesia
Arumanis	91	Fruit set	Malaysia
Blencong	86	First bloom	Indonesia
Carabao	84	Full bloom	Philippines
Carabao	116	Flower	Philippines
Cempora	101	First bloom	Indonesia
Golek	78	Full bloom	Indonesia
Golek	84	Fruit set	Malaysia
Kam Daeng	71	Full bloom	Thailand
Malgoa	108	First bloom	Indonesia
Malgoa	112	Fruit set	Malaysia
Nam Dorkmai	100	Full bloom	Thailand
Nam Dorkmai	102	Full bloom	Philippines
Nam Dorkmai	93	Fruit set	Thailand
Nang Klarngwun	115	Full bloom	Thailand
Tok Boon	105	Fruit set	Malaysia
Tongdum	102	Full bloom	Thailand
Yampulu	101	First bloom	Indonesia

Source: Yahia (1999, 2011).

6.2.1 Fruit Development Period

Harvest maturity in mango can be reached in about 10–16 weeks after fruit set, depending on the cultivar and the growing environment. Table 6.1 contains examples of typical fruit development periods calculated from full bloom or fruit set to full maturity for different countries in the ASEAN (Association of Southeast Asian Nations) region. However, the specific number of days must be calculated separately by each grower, since it may vary even for the same cultivar grown in different places. There can also be seasonal differences in the same growing region. For example, growers in the Philippines are accustomed to harvest mangoes during April–June, that are 100–110 days from fruit induction, but fruit picked in January–March that develop in cooler conditions are 120–125 days from induction. It has been found that Carabao mangoes grown in different regions of the Philippines reach maturity after 1,000 heat units (degree-days in excess of 17.9 °C), even though this may be reached after different time periods in different seasons and different regions (Yahia 2011).

6.2.2 Fruit Shape

On the basis of shape and form, four maturity stages during mango fruit development can be defined (Figure 6.1):

1) shoulders in line with the stem end and green olive color;
2) shoulders outgrowing the stem end and olive-green color;

Figure 6.1 Immature (left) and mature (right) mango fruit, as shown by shoulder development and fullness of the cheeks. *Source:* Brecht *et al.* (2014), with permission.

3) shoulders outgrowing the stem end and light color; and
4) flesh becoming soft and blush developing.

Fruit are recommended to be harvested at stages 3 and 4. However, the change in the shape of the shoulder may not apply to all cultivars.

6.2.3 External and Internal Color

Loss of external green color is an obvious indication of fruit ripening in some, but not all, mango cultivars. The skin color may change with maturation from dark green to a lighter shade or to an olive green color. This is related to loss of chlorophyll in the peel. Further ripening related color changes typically are to a yellow or yellowish-orange color and those are due to conversion of chloroplasts to chromoplasts along with synthesis of carotenoid and xanthophyll pigments. Red "blush" on the fruit surface is related to stimulation of anthocyanin synthesis by exposure of the fruit to sunlight and is not indicative of maturity or ripeness. The external color of some green mango varieties does not change at all. For most varieties, the loss of green color may occur too late in development to be a useful indicator of harvest maturity, especially for mangoes intended for export markets with associated long shipping times.

Fruit position on the tree in terms of shade or sun exposure can affect peel color development. Nitrogen (N) fertilization significantly affects the color development, with green color enhanced and yellow and red colors retarded when N levels are excessive (Garcia-Martinez *et al.* 2015; Nguyen *et al.* 2004). Therefore, peel color is not an optimum maturity index for many mango cultivars. On the other hand, in most mango cultivars, flesh color changes are somewhat uniform as the fruit ripening proceeds. Unfortunately, this is a destructive index, but it is more consistent and more utilized than skin color change. Flesh color is commonly used as a harvest maturity index in many mango growing regions where mangoes are being produced for export. Typically, there are five stages that are defined by the appearance and progression of yellow to

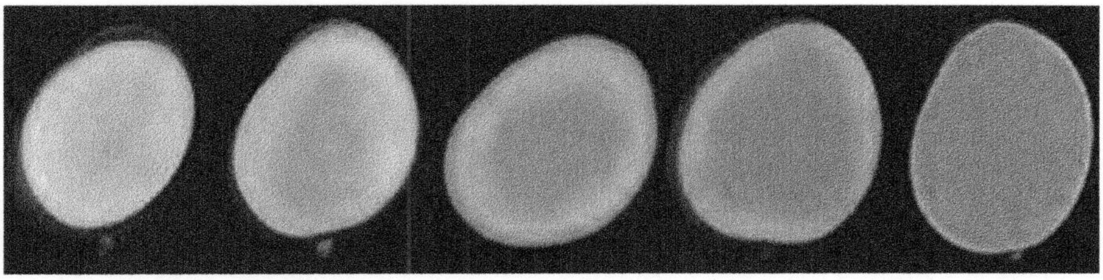

Figure 6.2 Internal flesh color development stages (1 to 5, left to right) in Tommy Atkins mango fruit. *Source:* Brecht *et al.* (2014), with permission.

orange color, from 0–100% of the flesh in 25% increments starting near the pit and progressing outwards (Figure 6.2). The procedure is described in the *Mango Maturity and Ripeness Guide* provided by the National Mango Board in the USA (NMB 2010)

6.2.4 Dry Matter Content

Changes in dry matter during mango fruit maturation are due primarily to starch accumulation. Dry matter content can be measured directly by dehydration of fruit flesh to determine the dry weight as a percentage of the fresh weight or indirectly by determining the fruit specific gravity [density or mass per unit volume (g/cm^3) of the fruit divided by the density of pure water at the same temperature]. As mango fruit mature, they accumulate dry matter and thus become denser. Dry matter/specific gravity changes are rapid in young fruit and slow as maturity is reached. The specific dry matter and specific gravity values need to be determined separately for each variety and region. In Egypt, physiological maturity of Tommy Atkins, Kent, and Keitt mangoes was reached at 113, 122, and 122 days after fruit set, respectively, at which time the specific gravity of the fruit was 1.013–1.014 for Tommy Atkins and 1.021–1.022 for the other varieties (Abourayya *et al.* 2011). The advantage of measuring specific gravity over dry matter is that specific gravity measurement is non-destructive. However, although specific gravity changes during fruit maturation, the change is small, which reduces its reliability as a practical maturity index.

6.2.5 Combined Indices

Due to differences among mango types, diversity of cultivars, and diversity of production conditions, there is no consensus on maturity indices. It is important that maturity indices be established for cultivar, growing region, and purpose of harvest. Mango fruit transported to distant markets by sea should generally be harvested firmer and with less external color change than fruit for domestic marketing. The export fruit are picked, treated, packed, and shipped while still mostly green and with a firm texture, so that they will arrive at the importer more yellow or red than green, but still firm. It is necessary that several indices be used together to make the best harvest decision. For example, in Australia, flesh color is combined with skin color and dry matter content (Johnson and Hofman 2009). In Mexico and Central and South America, flesh color is used to determine when to harvest an orchard, while fruit shape (shoulder development) and skin color are used to determine which fruit to pick (Brecht *et al.* 2014).

6.3 Harvesting Practices

6.3.1 Fruit Bagging

Paper bags are used in some places to cover and protect individual mango fruit from insects and diseases. Fruit are commonly bagged when about half-developed or after the period of heavy fruit drop and up to about 30–45 days before harvest. Bagging decreases the incidence of insects, anthracnose, and stem-end rot. However, bagging may also affect fruit maturity by changing the micro-environment around the fruit.

6.3.2 Harvest Management

The time of day at which mangoes are harvested should be selected carefully. There are benefits from picking in the coolest hour of the day due to lower field heat, respiration, and heat damage. Harvesting during warmer parts of the day will increase the rate of fruit deterioration until cooling can be done, increase the energy requirement for cooling, and negatively affect worker comfort. However, latex flow can also be higher in the earliest part of the day, due to higher water potential in the trees.

Harvesting should be conducted by experienced pickers. The harvest operation should be conducted carefully to avoid mechanical damage to the fruit such as peel abrasion and flesh fracturing. All of the fruit on a mango tree do not typically mature at the same time, necessitating multiple harvests, especially when harvesting is done at earlier stages of maturity, to ensure that only mature fruit are harvested. For young trees, and in the lower part of larger trees, picking can be done by hand. It is preferable that picking be done by hand whenever possible. However, for fruit on higher branches, some harvesting aids are needed to assist in picking.

6.3.3 Harvest Aids

Several types of harvest aids are used in different regions, but the most common is a half-elliptical basket attached to a long pole, with the leading edge of the basket frame fitted with a cutting blade (Figure 6.3). The basket should be big enough to hold no more than 3 to 5 fruits, so that the fruit will not be injured, either mechanically or by the latex that exudes from the cut pedicel and can injure the peel ("sap burn"). Baskets should be made of a material that facilitates cleaning. It is important that these devices be designed so that about 2 cm of the peduncle of the fruit remains attached after harvest. Fruit are cut by the blade and fall into the bag. When the bag is full, the pole is lowered and the fruit transferred to baskets, buckets, or crates. Ladders or three-wheeled motorized picking platforms (known in some parts as cherry pickers) are employed in some regions to pick fruit from tall trees.

6.3.4 Mechanical Harvesting

Mechanical harvesting of mangoes is not very commonly employed. In a system used in Australia, fruit are pulled from the tree using hooks that separate the fruit from the panicle flush with the fruit. The spurting sap is allowed to squirt away from the fruit while it is falling. Fruit are then caught on suspended plastic tarpaulins that are constantly being sprayed with water and a detergent solution. The solution coats the fruit

Figure 6.3 A mango harvest aid being used to remove mango fruit from the tree by cutting the pedicel and catching the severed fruit in the basket.

as they roll down the tarpaulin, protecting the fruit from the sap. It is very important that care is taken to reduce mechanical injury. In some countries in Africa and Asia, detached fruit are allowed to drop to the ground, where they are collected. This practice causes major injury and fruit losses. To avoid fruit injury, branches should not be vigorously shaken, and fruit should not be allowed to impact tree branches or the ground. Mechanical injury during harvest is, unfortunately, common and causes major quantitative and qualitative losses.

6.3.5 Latex (sap) Staining and Control

When a mango fruit is detached from the tree, or when the skin of an unripe fruit is cut, transparent liquid oozes or squirts out. The sap contained in the fruit is under considerable pressure. When the pedicel is broken, the sap is exuded towards the open end. Latex (sap) exudates, with low pH and high oil content, can stain the fruit, burn the skin, and thus reduce fruit quality. Immature fruit usually produce more latex than mature fruit. Since the flow of latex is higher during the morning than during the afternoon due to fruit turgidity, it is recommended that fruit be picked after about 9.00 am. The flow of latex also increases immediately after rainfall due to increased turgidity of the fruit.

The susceptibility of mangoes to latex staining depends on several factors, including cultivar, maturity stage at harvest, time of harvest, growing region, season, and age of tree, among others. It has been reported that skin damage by sap burn is particularly severe with Kensington, and less serious in Florida cultivars (Yahia *et al.* 2006; Yahia 2011). Areas of the mango fruit surface damaged by latex may be more susceptible to development of bacterial and fungal decay lesions.

Several strategies have been suggested to combat the problem of latex (Robinson *et al.* 1993). Some of the suggested methods include:

- Harvesting the fruit with 2–10 cm of pedicel or longer attached, later reducing the pedicel length to about 0.5 cm or less in the packinghouse, often the following day. Latex commonly does not exude from the longer pedicels, because there is no continuity between the fruit and the pedicel resin ducts.
- Inversion of fruit picked with 0.5 cm or shorter pedicel immediately upon removal from the tree in a shady spot for 30 minutes to allow the sap to drain and the pedicel to dry. This should be done on racks because, in addition to the food safety considerations, inversion of fruit on the ground can cause a significant increase in rot, resulting from soil-borne inoculums.
- Use of detergents and emulsions after harvest that reduce the effects of the sap, such as de-sapping in a 1% solution of calcium hydroxide or washing fruit in 1% aluminum potassium sulfate.
- Applying a surface coating to the fruit prior to eliminating the latex (de-sapping).
- Trimming of pedicels and de-sapping at the packinghouse followed by inversion on a stationary rack or a roller-conveyer running below water or water plus detergent sprays for 20 minutes.

6.3.6 Fruit Transport to the Packinghouse

Harvested fruit should be transported to the packinghouse or to the market as soon as possible. Exposure to the sun in the field, on the road, at the packinghouse, or in the market should be avoided. Fruit should be transported with care, so that mechanical injury is minimal.

6.4 Fruit Grades and Standards

Mangoes should be uniform in size and shape, with no scars or other surface defects such as abrasions or bruising. High-quality mangoes at destination should have appealing color and shape, firm texture, tender flesh, sweet taste, and be highly aromatic. Fiber content should be minimal. Acceptable quality characteristics for ripe (flesh color stage 5) Ataulfo, Francis, Haden, Keitt, Kent, and Tommy Atkins mangoes were considered to include flesh firmness (measured with an 8-mm diameter probe) ranging from 0.5–2.75 kg-force and soluble solids from 12–18% (National Mango Board, *Mango Maturity and Ripeness Guide*) (Kader, 2008). Flesh firmness of less than 0.5 kg-force is indicative of over-ripe fruit.

The Organization for Economic Cooperation and Development (OECD) provides guidelines defining international marketing requirements. Minimum requirements for mangoes intended for the international market are that the fruit should be intact, firm, fresh in appearance, sound, clean, free from black stains (i.e. latex staining) and bruising, free from damage caused by low temperature, free from pests and pest damage, carefully picked at the stage that allows transport and handling, and continuation of the ripening process so as to arrive in a satisfactory condition at its destination. Class standards are defined as "extra", "good quality" (class I), and "marketable" (class II). Fruit sizes defined are A: 200–350 g, B: 351–550 g, and C: 551–800 g. The maximum permissible differences allowed within each size group are 75, 100, and 125 g, respectively.

Mangoes for the export market must comply with consumer requirements and with the quality standards of the importing market in terms of size, color, appearance, freedom from defects, ripeness stage, and uniformity, absence of insect damage and disease, and phytosanitary regulations. The fruit must arrive at the market at a ripeness stage that can allow reasonable shelf life for retail marketing. Depending on the importing market, the mangoes may need to be subjected to a legal disinfestation treatment. The most common insect disinfestation treatment accepted by most importing countries is the use of heat treatments. Some countries accept the use of irradiation.

Import markets have established quality standards. The three important quality standards developed are:

1) *Codex Alimentarius*: Worldwide Codex Standard for Mangoes, Codex Stan 184-1993;
2) *Mexico*: NOM-FF-58-1985. Mexican Official Standard for Mango (Norma official Mexicana); and
3) *Europe*: UN/ECE standard FFV-45. Concerning the Marketing and Commercial Quality Control of Mangoes.

Generally, import markets require that mangoes, on arrival, be:

- physiologically mature;
- 30–50% development of color;
- in the case of red cultivars, a significant amount of red area on the fruit shoulder;
- firm;
- a minimum of 10% soluble solids content;
- uniform in shape;
- free from diseases, insects, latex stains, soil burns, and mechanical injury; and
- in compliance with specific size and weight requirements.

Prices in import markets are usually set on the basis of:

- *the presence of red color*: red varieties are usually more expensive than green or yellow varieties;
- *maturity stage*: immature and over-ripe fruit are less desirable;
- *uniformity*: fruit should be very uniform in size, weight, and color; and
- *absence of diseases and defects*: fruit should be free from disease (internal or external) and physical and physiological defects.

Regarding fruit uniformity, which is a very important characteristic, a package with a capacity of 4 kg, for example, should contain: 6, 8, 10, 12, or 14 fruits, each of 666 g, 500 g, 400 g, 350 g, or 290 g, respectively.

6.4.1 Export Market Mangoes

An export plan should be prepared for each market, and should include:

1) Fruit must be screened in the field to insure export quality. This operation can be done by the exporter or by agents working for the exporter association. In this way, exported fruit are assured to be of high quality and suitable for export markets.

Table 6.2 Number of packages that need to be sampled during the inspection of mangoes.

Total number of packages	Packages to be sampled
Up to 100	5
101 to 300	7
301 to 500	9
501 to 1,000	10
More than 1,000	15 (minimum)

2) A detailed historical registry of fruit is needed and it should include all aspects of the operation, such as origin of fruit, date of harvest, date and hour of packing, and special treatments applied. This registry will help identify the cause of any problems that appear during marketing.
3) All packed fruit should be inspected (fruit and package) before leaving the packinghouse to assure their quality. It is important that all required information should be printed on the package. Fruit should be removed from the package and inspected individually and the grade specified according to a set of standards. The sample size is established by the International Standardization Office (ISO874-1980) according to Table 6.2.
4) Specifications of export mangoes are set according to the requirements and the standards of the importing country. Some of the general specifications for the markets in England, Europe, and Canada are as follows:
   a) *Varieties*: The most popular varieties are Ataulfo, Tommy Atkins, Haden, Keitt, and Kent.
   b) *Color*: Peel color should be partly red (except Ataulfo), with green and yellow. Pulp color should be yellow to orange.
   c) *Appearance*: Size and ripeness should be uniform. Fruit should be free from defects such as misshapen, scars, healed mechanical damage, or healed insect damage.
   d) *Size*: Size requirements for Haden and Tommy Atkins are as in Table 6.3:
   e) *Condition*: Fruit should not show latex staining, mechanical damage such as abrasions, cuts and bruises, or insect and disease damage.
   f) *Phytosanitary regulations*: Disinfestation treatments for fruit flies are not needed for the EC, despite the large production of temperate fruit in regions free of fruit fly. Fly infestation has never been perceived as a threat, because freezing winter temperatures throughout much of the region effectively prevent establishment

Table 6.3 Size and weight requirements for mangoes imported by England, Europe, and Canada.

	Haden	Tommy Atkins
Minimum weight	300 g	350 g
Minimum longitude	9.0 cm	10 cm
Minimum width	8.0 cm	9.0 cm
Minimum thickness	8.0 cm	9.0 cm

of the insects. Canada also does not require fruit fly disinfestation of tropical products for the same reason. The USA requires that mangoes be disinfested using treatments such as vapor heat, hot air, hot water, or irradiation.

Some importing countries allow inspection at the exporting site (packinghouse). This is the case, for example, in Mexico for mangoes exported to the USA. Official inspectors (through agreement between the US Department of Agriculture and the Mexican Secretary of Agriculture) assure that phytosanitary regulations are correctly applied and fruit are strictly secured from re-infestation. Quality Inspection can be done either at the export site, at the port of exit, or at the port of entry.

It is important to note that prior approval to import is required by some countries. These import permits may cover multiple importations, but usually require renewal every 3–12 months. Phytosanitary certificates must be issued by a government agent based on an agreement commonly done between export and import countries. Consignments that are found to contain pests subject to quarantine will be rejected, and will either be returned or destroyed.

6.4.2 Establishment of a Quality Control System

Due to competition in the international mango market, it is essential to develop a quality control (QC) system following the standards established by the International Organization for Standardization (ISO; www.iso.org). Due to the fact that the term "quality" is very ambiguous and difficult to define, it is important to first establish a "certification system" that can be developed and agreed upon by an association of producers and exporters. This system should include the following:

- Maturity indices should be developed and a harvesting system should be established.
- Each packer/producer/exporter should make sure that maturity indices and harvest techniques used in his/her establishment are ideal. A "seal of quality" should be established, indicating the certification of the fruit.
- A certification system for fresh fruit should be developed by producers/exporters. A "quality manual" should be developed.
- A certification team (technicians) should be trained and used to establish the system. These "quality promoters" will be the responsible agents to certify the fruit received in the packinghouse/market, decide its grade, price, and whether it is adequate for export or not.
- Not all fruit will pass through the system, nor will all producers and probably not all exporters be part of this system. However, the seal of quality should be the factor that distinguishes between "certified" and "not certified" fruit. Producers/exporters associated with the system should be the only entities that receive the service of certification, and thus the seal of quality. Mangoes with the seal of quality should be promoted and distinguished.
- The seal of quality should be awarded when:
  a) the QC is done by an inspector approved by the association, and the inspection follows the "quality manual" developed by the association;
  b) an initial visit is made to the establishment to install the procedure of the QC system; and
  c) random visits should be done to the establishment to monitor the process.

6.5 Field and Packinghouse Sanitation

It is important to follow good sanitation practices during harvesting and handling of mangoes, in order to minimize the opportunities for plant decay or human pathogen microorganisms to inoculate or contaminate the fruit. Sanitation practices, as well as handling conditions, especially temperature and humidity, can influence the survival of microorganisms on mangoes and on various surfaces involved in mango handling.

In the mango orchard, the main procedure for reducing microbial contamination is pruning or collecting of diseased plant parts and debris that may harbor pathogens, such as mummified fruit, decay-affected leaves, twigs and panicles, and burning the material. These practices have been shown to reduce incidence of anthracnose and stem-end rot (Saaiman 1997). With regard to food safety and human pathogens, field sanitation centers on preventing mango fruit from coming in contact with soil, which can be assumed to be contaminated with animal and possibly human waste, and avoiding contact with dirty hands, tools, and containers. Contact of the fruit with the soil risks the spread of soil-borne plant diseases as well as human pathogens and should be avoided at all times. Harvested fruit should not be allowed to fall to the ground, or be picked up from the ground. Workers in the field must also practice good personal hygiene, especially proper handwashing, and also guard against transferring soil to fruit. If ladders are used, hands should not contact the steps or rungs, which will have come in contact with shoes with dirt on them.

All equipment and implements that come in contact with mangoes during harvesting and handling operations should be subject to a regular schedule of cleaning and sanitation to minimize cross-contamination. All the packinghouse equipment and surfaces with which mangoes may come in contact, or that workers may touch and subsequently transfer microorganisms to fruit, should also be cleaned and sanitized regularly. It is necessary to clean equipment and surfaces with appropriate detergents or other cleaners and sanitize the equipment, surfaces, and fruit with chlorine or other approved sanitizers. It must be noted that hot water and vapor heat quarantine treatments and hot water decay control treatments do not, by themselves, sanitize mangoes for food safety purposes. Food Safety Training Kits have been developed for the National Mango Board in the USA and are available in English, Spanish, Portuguese, and Creole at www.mangofoodsafety.org.

6.6 Postharvest Handling

Mango fruit are susceptible to physical damage throughout the postharvest handling chain. The types of physical damage that can occur in mangoes include abrasion, compression, vibration, and impact bruising. Abrasion occurs when fruit surfaces are abraded by stems of other fruit, the sides of rough or dirty picking containers, dirty packing line conveyers, and worn or stiff packing line brushes. Mangoes can also suffer vibration bruising in transit when individual fruit are allowed to rub against each other. Careful handling should be performed to reduce all forms of physical injury.

6.6.1 Handling at the Packinghouse

At the packinghouse, if there will be a delay before packing, the fruit should be collected under cover so as to be protected from the sun, high temperatures, wind, and rain. Fruit should ideally be packed within no more than 3–4 hours of arrival. It is preferable that the peduncle be trimmed at this point to leave about 0.5 cm. Samples should be taken to evaluate the quality of the fruit. This inspection point is very important to determine the condition of the fruit, determine acceptance, the suitability for a desired market, the treatments that the fruit should receive in the packinghouse, and decide the price on the basis of quality evaluation. The results of this initial QC inspection also can form the basis for feedback to the harvest crews, whether or not there are any problems noticed with the incoming fruit.

It is recommended that mango fruit be received on the packing line in water using a dump tank to minimize physical injury and then transported on a roller conveyer to the other steps of the packinghouse. This water can contain up to 200 ppm chlorine adjusted to pH 7 and the fruit should be exposed for at least 3–5 seconds. Chlorine is degraded easily and thus its concentration should be monitored regularly. Detergents are commonly added to the dump tank water to also serve as a washing step. Otherwise, washing of the fruit in water with detergent, often sprayed on the fruit as they pass over a brush roll conveyor, is a separate step that is taken prior to any other packing line steps. After washing, the fruit should be rinsed with clean water, then dried by air using fans and/or sponges before continuing to other packinghouse operations.

There are many types of designs of packing lines and packinghouses; some are very simple, while others are computerized. However, the most important consideration is that the packing line should not be complicated, should not damage the fruit, and the process should be clean and organized.

6.6.2 Sizing and Grading

When quarantine treatment is required, a pre-sizing operation is usually necessary in order to separate the fruit into size categories requiring different treatment durations according to the controlling treatment protocol. Quarantine treatments (see below) are usually applied to fruit that have been returned to the field crates after washing and pre-sizing, prior to further packing line operations. Mangoes are introduced to the packing line either manually or mechanically, pouring the fruit from the field crates onto an inclined surface or a conveyor belt. On the packing line, fruit should be selected for uniformity of size and appearance and freedom from defects, and only good-quality fruit should be packed. The purpose of this sequence is to sort fruit into uniform categories (according to size, shape, color, and absence of defects), and to divert low-quality fruit to another use such as a low-quality market or for processing. Fruit can be classified into different grades according to the requirements of the market and all the defective fruit should be eliminated.

The purpose of sizing is to categorize fruit into different and uniform groups on the basis of size or weight. Mango fruit can be classified manually, but uniformity will be much better when sizing is done mechanically – either by size or weight according to the requirements of the importing market. Sizing can be done by using diverging rollers, in which the distance is minimal at the beginning and increases continuously. Distances between rollers are adjustable so that fruit sizes can be modified depending on the cultivar and the requirements of the market. Smaller fruit are dropped at the beginning and largest fruit are dropped at the end.

Mangoes can also be classified by weight. Weight sizers consist of cups controlled electronically. Each cup carries one fruit. When a fruit is dropped into a sizing cup, its weight is recorded, and the fruit will be dropped automatically onto the packing line conveyor to which its size corresponds.

Quality grading of mango fruit is usually done by manual inspection and observation of individual fruit by trained persons. Fruit are classified according to uniformity, maturity stage, and color, and for absence of injuries and defects, latex stain, disease, and insect infestation.

6.6.3 Waxing

"Waxing" is the common term used for coating the fruit surface with a thin layer of some material that improves visual appearance of the fruit and is beneficial as a barrier to mass transfer of water vapor and gases; restriction of gas diffusion means that such coatings can provide an internal MA in the fruit tissues. Waxing of mangoes is most commonly used by some exporters to reduce water loss of fruit treated with hot water for insect or disease control, because when mangoes are treated with hot water it can deteriorate the "bloom" (natural waxy coating) on the fruit surface. Aqueous wax emulsions, consisting of vegetable (sisal, sugarcane, and carnauba) waxes and mineral petroleum (paraffin) with and without shellac and emulsifiers, can increase the storage life of mangoes. It is important to investigate the regulations of the importing country regarding types of coating materials allowed, but it is always preferable that these be of natural origin. The coating should be applied in a thin uniform film, using preferably roller brushes, or a coating applicator, or by very light hand application.

Improper application of fruit coatings can inhibit gas exchange between the fruit tissue and the environment, to the extent that ripening can be inhibited and fermentation and off-flavor development can occur in the fruit. Thus, the use of light waxes is preferred to heavy waxes that can hamper ripening and associated color development. Dipping of fruit in a wax emulsion is not recommended. The coating can be applied together with the fungicide. The fruit should be completely dry before applying the wax, otherwise foaming of water-emulsion waxes may occur.

6.6.4 Packing

Packing into shipping cartons is done manually. Mangoes for export are commonly packed in a single layer in corrugated cartons. Size and weight of packages depend on the requirements of the importer, but mangoes are commonly packed in cartons with a capacity of 4–6 kg. Packaging can be a marketing tool, and thus should portray an excellent marketing image. Packaging should meet several requirements:

- It should protect the fruit from mechanical damage and from contaminants, and not cause injury.

- It should have sufficient strength to survive the marketing system. The cartons will commonly be in contact with a humid atmosphere, and that should not debilitate the structure of the carton.
- Cartons should have sufficient openings/vents for ventilation. Vents should be oriented with the direction of air circulation in the transport container – horizontal for truck trailers and vertical for marine containers.
- Packaging should provide sufficient information, including promotional data. Required information includes product name, source, and class. Optimum temperatures for transport/storage (12.5 °C) and ripening (20–22 °C) should be indicated on the packaging.
- The package should be easily opened and closed, and preferably, packages should be recyclable.
- Regulations of countries should be checked before using the packages. Special attention should be given to the preference for recycling in Europe, and the special material and size requirements in Germany. It is also important to use packages that are compatible with the pallet to be used (e.g. 1.0 × 1.2 m for Europe or 40 × 48" for the USA).

There is a strong movement toward recycling and standardization, especially in Europe. Packages made of mixed products are to be avoided. Also to be avoided are stapled packages. Waxed packages are not acceptable in many markets, because they are not recyclable. It is important to communicate with the importer regarding the requirements of each country.

6.6.5 Palletizing

Palletizing allows much easier, faster, and uniform handling, lowers costs, and better utilizes space in the storage rooms or transport containers. Pallet dimensions can be different in different export markets, and thus should be investigated before shipping. The dimensions of shipping cartons must be compatible with the pallets being used. Ideally, the carton dimensions should allow full coverage of the pallet surface for efficiency.

6.6.6 Cooling

Mangoes, especially those that are heat treated, should be pre-cooled, which is very important to remove field heat. Pre-cooling also decreases the refrigeration demand during cold storage or refrigerated transport. Pre-cooling is needed to reduce possible high temperature damage, and also to delay fruit ripening and to maintain fruit quality for prolonged periods.

Mangoes should be pre-cooled to about 2 °C above the desired storage or transport temperature, with the remainder of cooling to take place in the storage room prior to transit or in the transit container. This is because the cooling curve (Figure 6.4) is asymptotic, meaning that the rate of cooling continuously decreases over time. This makes removal of the last few degrees slow and inefficient. Delay in cooling mango fruit results in accelerated ripening, shortened postharvest life, and lower quality. The ideal pre-cooling method for mangoes is forced-air cooling. With this

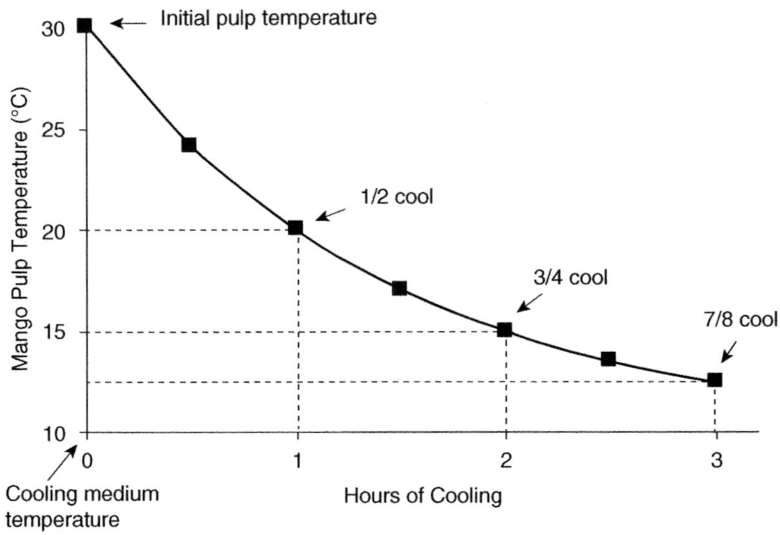

Figure 6.4 Fruit temperature change over time during forced-air cooling of mangoes with an initial temperature of 30 °C in a refrigerated room at 10 °C to a 7/8 cooling temperature of 12.5 °C.

method the temperature of the fruit can be reduced from about 35 °C to about 14 °C in a room set at 10–12.5 °C in 2–5 hours. The chilling threshold temperature for mangoes is typically considered to be about 12.5 °C for most mango cultivars, meaning that the fruit will not be injured by that temperature, no matter how long the exposure.

6.7 Storage and Transport Technologies

6.7.1 Storage

Mangoes are not commonly stored for prolonged periods. However, after precooling, the fruit should be moved immediately to a cold room or refrigerated transport container. It is common (in cases that fruit quantity is insufficient to fill a transport container, or the container is not ready) that fruit be held in the cold storage at 10–12.5 °C for short periods. It is important that the cold chain should not be interrupted (fast cooling, prompt transfer to a cold room or pre-cooled transport container, no exposure to high temperature). Fruit should be picked, packed, precooled, and placed into cold storage or cold transport containers within no more than 24 hours.

The ideal postharvest temperature for long-term storage or transport of mangoes is the chilling injury threshold temperature of 12.5 °C. However, in most export operations, mango cold storage rooms are maintained at 10–14 °C, depending on the cultivar and holding period, with the relative humidity (RH) maintained in the range of 85–90% RH. Cold storage rooms should be equipped with adequate systems of air circulation and air exchange, and should be clean. Fruit should be stacked in a way that can permit adequate circulation of cold air. Harvesting at optimum maturity, prompt cooling, the use of optimum temperature, and optimum storage conditions, can maintain most mango cultivars for 2–4 weeks.

The ideal holding temperature for mangoes depends on several factors such as cultivar, maturity stage, and holding period, which all affect expression of chilling injury. A temperature of 12.5 °C is generally considered to be optimum for unripe mango storage and transport. Ripened fruit can be kept at as low as 8 °C for up to 21 days, because chilling sensitivity decreases as mango fruit ripen, but the fruit will deteriorate rapidly after removal from storage.

Shelf life decreases markedly with increasing delay in time of fruit placement in cool storage. Delay in time from picking until placement of the fruit under refrigeration increases the rate of ripening, especially for fruit picked at later stages of ripening. If the delay is prolonged, refrigerated storage/shipping may become totally ineffective in prolonging the life of the fruit.

6.7.2 Modified and Controlled Atmospheres (MA and CA)

Modified or controlled atmospheres (lower levels of O_2 and/or higher levels of CO_2 compared to those found in normal air) have been shown to be beneficial in delaying ripening and controlling diseases and insects of mangoes (Yahia and Singh 2009). The difference between MA and CA lies in their precision: CA involves active generation and feedback control of the atmosphere, while MA is designed to roughly balance fruit respiration with restricted gas exchange of the storage space or packaging to reach a desired O_2 and CO_2 equilibrium. Optimum O_2 and CO_2 concentrations for mangoes depend on the cultivar, storage period, and the purpose of using MA and CA (Yahia 1997; Yahia and Singh 2009). While CA is much more effective than MA, it is also more expensive. Optimum CA storage conditions for different mango cultivars are summarized in Table 6.4.

Information on the response of mangoes to CA and the benefits that can be gained are not yet very well established. This is mostly due to the diversity of cultivars and the variability between growing regions, but also to the limited amount of research and the little control in the type of research done on this fruit. Optimum atmosphere composition for long-term transport and storage of mature-green mangoes is usually considered to be 3–5% O_2 plus 5–10% CO_2. Ripening mangoes can benefit from higher CO_2 levels (up to 15%) and lower storage or transport temperature, but O_2 cannot be lowered very much due to risk of initiation of fermentative metabolism in the fruit (Bender and Brecht, 1994; Bender *et al.* 2000a,b). These moderate concentrations of O_2 and CO_2 atmospheres are not very effective for mango disease or insect control. The ideal atmosphere for any particular purpose should be decided on the basis of the fruit maturity or ripeness, cultivar, temperature, shipping period, and treatments given to the mangoes prior to MA/CA application (i.e. heat treatment). Improper use of MA and CA can result in physiological disorders and quality deterioration.

There is no current widespread use of CA storage for mangoes. However, long-term marine shipping in MA and CA containers has seen limited commercial use in different countries, including Mexico (Yahia 1993; 1998; 2009). Use of CA and MA can be beneficial in delaying fruit ripening during long-distance marine transport for 2 weeks or more, but use is limited by the expense and by MA/CA container availability.

Table 6.4 Optimum controlled atmosphere storage conditions for different cultivars of mango.

Cultivars	Country	Recommended CA O_2 kPa	CO_2 kPa	Storage temperature (°C)	Storage life (days)
Alphonso	India	5	5	8–13	30–45
Amelie	Senegal	5	5	10–12	28
Banganapalli	India	5	3	8–13	35–45
Carlota	Brazil	6	10	8	35
Chok Anan	Malaysia	2 or 5	0	15	28
Delta R2E2	Australia	3	6	13	34
Haden	Brazil	6	10	8	30
Irwin	Japan	5	5	8–12	28
Irwin (Tree ripe)	Japan	5	5–10	5–15	30
Jasmin	Brazil	6	10	8	35
Julie	Senegal	5	5	10–12	28
Kensington Pride	Australia	2–4	4–6	13	30–35
Rad	Thailand	6	4	13	25
Sao Quirino	Brazil	6	10	8	35
Tommy Atkins	USA, Chile	3–5	0–5	12–15	21–31

Source: Narayana *et al.* (2012).

6.7.3 Ripening

Mangoes are often harvested early in the ripening process and ripened to eating ripe stage after harvest. The quality of ripened mangoes will depend on maturity at harvest and on ripening conditions, especially temperature. Temperature management is very important during ripening of mango fruit and ripening rooms are a specialized type of storage room. The ideal temperature for mango ripening is 20–22 °C; ripening at 15–18 °C may result in the most attractive skin color, but flavor becomes tart; flavor development is best at 27 °C, but those fruit will require about 2–3 days at 21–24 °C to attain sweet flavor; ripening at 27–30 °C results in mottled skin and an unpleasantly strong flavor; optimum RH during ripening is 90–95% (Brecht *et al.* 2014).

Ethylene is not commonly used for mango ripening, because mangoes are usually picked after initiation of ripening and thus initiation of ethylene biosynthesis has already occurred on the tree. However, in certain cases, ethylene treatment may result in more uniform ripening by increasing the ripening rate of the less advanced fruit in a non-uniform lot, while having less effect on the more advanced fruit. Exposure to 100 ppm ethylene at 20–22 °C and 90–95% RH for 12–24 hours can result in accelerated and uniform ripening within 5–9 days, depending on cultivar and maturity stage. Carbon dioxide concentration should be kept below 1% in the ripening room to minimize potential interference with ethylene action. Once fully ripened, mangoes can be kept at 8–10 °C and 90–95% RH for up to 1 week.

6.7.4 Transport

Mangoes can be transported by road, air, or sea. The method to be selected depends on the type of market, distance, cost of transportation, and the stage of maturity of the fruit. Major losses occur during transport of mango fruit in developing countries. This is due to improper and rough handling, improper containers, poor road and vehicle conditions, and high transit temperatures. In international marine shipping, chilling injury due to low maturity fruit being transported over long distances at below 12.5 °C is a major problem. It is recommended that a sample of shipped fruit be maintained at the origin to evaluate its quality during the shipping and marketing period. This sample should be maintained at the same conditions (temperature and RH) as the transported fruit.

Trailers and containers for land and sea transport should be:

- clean, not damaged, and with adequate and well-functional air ducts;
- pre-cooled before loading the cooled fruit;
- refrigeration capacity should be adequate for the quantity of fruit to be transported;
- containers should be equipped with a temperature registry system. These should be preferably more than one, and should be placed in representative locations in the container, preferably inside the fruit pallet; and
- transport containers are designed for maintenance of temperature of the load and not to pre-cool it (i.e. not to remove field heat).

Overland, non-refrigerated transport to local markets may be feasible if distances are short, and weather is not very warm. In warm weather, transport can be done at night when the temperature is much lower. Refrigerated land transport containers should be checked before loading to insure proper conditions – clean, same temperature and RH settings as for storage rooms. Fruit should be loaded after cooling, and containers should also be pre-cooled before loading.

Many of the exported mangoes worldwide are shipped by sea in refrigerated containers. Several treatments can be applied on board marine containers, including MA and CA, quarantine treatments, and ripening treatments. Marine transport is less expensive than air transport, but slower, and thus fruit need to be handled adequately to ensure a long postharvest life. Marine containers are characterized by vertical or bottom-air circulation, in which cold air is circulated from below, in contrast to the horizontal air circulation normally used in trailers for land transport. Cold air is forced through the packages to the upper part of the load. However, the package and the loading pattern must be compatible with this system.

A very limited amount of tree-ripe mangoes are shipped by air, typically without refrigeration. Air transport is fast, but expensive. It should only be used when the fruit are very perishable, or the price in the export market is very high. Air cargo containers are normally closed containers and usually are made of aluminum or fiberglass. They are not ventilated or significantly insulated and thus there can be an increase or decrease in temperature, especially if these are left on the airport tarmac or elsewhere at ambient temperature. Some of these containers are cooled with dry ice or with refrigerated gels.

6.8 Quarantine Treatments for Insects

In the case that mangoes are to be shipped to a market that restricts the entry of certain insects and requires a quarantine treatment, the fruit should be treated with a legal system established through an agreement between export and import authorities. Handling of the fruit must follow the protocol of the quarantine system. The protocol usually defines pre-harvest treatments and precautions, restrictions of type of fruit that can be treated, and fields where fruit for treatment can be harvested, traps for insects in the field, integrated pre-harvest treatments, control systems, and possibly other requirements. Upon arrival at the packinghouse, fruit that have been harvested according to the quarantine protocol should be sampled to assure the absence of insects. Fruit found to be infested should not be treated or packed. Following application of an insect quarantine treatment, the treated fruit must be kept in an isolated, insect-proof area to prevent re-infestation.

The quarantine systems most commonly used for mangoes in many countries are treatment with hot water or vapor heat. Some importing countries allow the use of low doses of irradiation as an insect sterilization treatment. Quarantine systems using fumigants such as ethylene dibromide (EDB) and methyl bromide (MB) are not acceptable by most importing countries.

6.8.1 Hot Air Treatment

Vapor heat was approved in Japan in 1986 for the importation of mangoes from the Philippines. This treatment requires that mangoes be treated with vapor heat until the fruit surface temperature is 46 °C and the seed temperature reaches 46.5 °C, and held that way for 10 minutes (Mitcham and Yahia 2009). A vapor heat treatment has also been approved for the control of the Queensland fruit fly (*Bactrocera tryoni*) in Kensington mangoes exported from Australia to Japan (Mitcham and Yahia 2009). The treatment consists of raising the pulp temperature to 46.5 °C and holding it for 10 minutes. The USA also approved a quarantine vapor heat treatment for the control of Mexican fruit fly (*Anastrepha ludens*) and other Anastrepha species in Manila mangoes from Mexico (APHIS-PPQ 2015a), but the hold time is 6 hours and the treatment is not used commercially; and for mangoes from Taiwan infested with the oriental fruit fly (*Bactrocera dorsalis*) (APHIS-PPQ 2015a).

In forced hot-air treatment, the hot air is forced over the surface of the fruit, which will slowly heat the pulp. In forced hot-air treatment, the fruit surface temperature remains cooler than during vapor heat treatment due to evaporative cooling on the fruit surface. This results in lower incidence of surface injury with forced hot-air than vapor heat treatment. When air at 50 °C was forced over mango fruit surfaces, all stages of mango fruit fly (*Ceratitis cosyra*), West Indian fruit fly (*Anastrepha obliqua*), and *Anastrepha serpentina* were killed when the seed surface temperature reached 48 °C (Mangan and Ingle 1992; Yahia *et al.* 2011). Mango fruit weighing over 700 g cannot be treated with this system.

6.8.2 Hot Water Treatment

The use of hot water treatment as a quarantine system was intensified after the elimination of EDB. The use of this quarantine system in Mexico started in 1988 for the control of fruit flies (*A. ludens* and *A. obliqua*) for fruit exported to the USA and Japan. The treatment was also found to be effective in disinfestation of insects other than the genus Anastrepha. It consists of immersion in water at 46.1 °C for 65, 75, 90, or 110 minutes, depending on fruit type (shape) and weight (APHIS-PPQ 2015b). This system is now widely used in different mango growing regions in Mexico, Central and South America, and the West Indies. After heat treatment, fruit are allowed to be cooled with water at no less than 22.2 °C. However, the fruit must either remain at ambient air temperature for 30 minutes prior to hydrocooling or, in order for the fruit to be hydrocooled immediately after hot water treatment, the hot water treatment duration must be extended by 10 minutes. The fruit are subsequently packed and cooled with forced-air cooling.

6.8.3 Irradiation

In 1986, the US Food and Drug Administration (FDA) approved the use of food irradiation treatment of up to 1 KGy (1,000 Gy, 100 Krad). Irradiation is effective at killing, sterilizing, or preventing further development of a wide variety of insect pests. The dose required for sterilization of most insects is below 0.50 KGy, but doses required to kill the insect are usually higher than those tolerated by the fruit (Yahia *et al.* 2011). All immature stages of a fruit fly will become effectively non-viable upon being irradiated at a minimum dose of 0.25 KGy. Irradiation has been used since 2004 to disinfect mangoes shipped from Australia to New Zealand (Torres-Rivera and Hallman 2007), and currently Mexico is shipping irradiated mango to the USA. Fruit damage by irradiation is a function of cultivar, irradiation dose, and fruit maturity/ripeness at the time of treatment. Symptoms of irradiation stress on fruit include accelerated softening, uneven ripening, and surface damage, and it is additive to other types of stress such as physical, chilling, and water stresses.

6.9 Nutritional and Quality Losses

Mango quality is strongly related to the occurrence of the normal progression of ripening related changes in the fruit. In order to ripen to optimum quality, mango fruit must be harvested when they are physiologically mature, ideally after initiation of ripening. This ensures that the fruit have accumulated the maximum amount of dry matter, primarily starch, which determines the eventual sweetness of the fruit, as well as providing the precursors for all other ripening-related compounds in the fruit, such as organic acids, pigments, and aroma compounds. Fruit softening is among the principal changes associated with fruit ripening. In this sense, quality losses in mangoes during the postharvest period are related to over-ripening, which results in fruit that are too soft for marketing and more prone to decay. Over-ripening is avoided primarily by proper temperature management with possible supplementation by MA or CA.

Physical injury stimulates respiration and ethylene production in mango fruit. Therefore, in addition to quality losses related to development of unsightly injury symptoms, mangoes that have been abraded, cut, or bruised, ripen faster and cause surrounding fruit to do the same. Physical injuries also serve as avenues of infection for decay microorganisms, such as *Rhizopus oryzae*, *Botryodiplodia theobromae*, and those organisms responsible for stem end rot [*Lasiodiplodia theobromae* (Pat.) Griff and Maubl (syn. *B. theobromae* Pat.), *Phomopsis mangiferae* Ahmad, and *Dothiorella dominicana* (Sacc.) Petr. and Syd.].

Mango fruit are a good source of vitamin C (ascorbic acid), containing about half the vitamin C content of oranges or strawberries (USDA 2015). Significant amounts of vitamin B_1 (thiamine), vitamin B_2 (riboflavin), and folic acid are also found in mangoes. The mango fruit is a rich source of carotenoids, which include pro-vitamin A (ß-carotene) and act generally as antioxidants in promoting human health. Ripe mangoes are ten times richer in carotene than partially ripe fruit, while unripe green mangoes contain only trace amounts. Of these nutrients, vitamin C is the most prone to losses during postharvest handling, primarily as a result of poor temperature management.

Mangoes harvested when they are immature do not have the ability to ripen to good quality. Since immature mangoes are more susceptible to chilling injury than mature mangoes, they can be injured during storage and transport at temperatures that would normally be safe for mature or ripening fruit. Chilling injury in mango is manifested initially as inhibition of aroma volatile synthesis, which can result in fruit with normal appearance but little flavor. More severe chilling injury results in lenticel discoloration, uneven ripening (especially in terms of color changes), then brown, scald-like peel discoloration, often accompanied by pitting. In more severe cases, the peel discoloration becomes more pronounced and then the flesh is also affected and the fruit become prone to decay. Immature mangoes are also more susceptible to heat injury than mature fruit. Heat injury symptoms include lenticel spotting, skin browning ("scald"), and flesh discoloration; hot water injury can also result in development of internal cavitation and stem-end collapse.

6.10 Future Research Needs

There is a need for further understanding of mango postharvest physiology, in order to overcome important problems such as chilling injury, internal breakdown, and premature and uneven ripening. There is also a need for research to develop better maturity and harvesting indices. Mango fruit possess a highly attractive appearance, taste, and aroma when ripened on the tree; mangoes harvested pre-ripe and ripened off the tree do not possess the same potential sensory appeal. Unfortunately, many of the mangoes being introduced in new markets in temperate regions of the world are being picked early and shipped at low (chilling) temperatures in order to arrive with good appearance and firmness, which do not translate into good taste and aroma. This situation is most acute for mangoes from the most distant exporting countries. Future expansion of mango international marketing on a year-round basis depends on attracting consumers who have yet to taste a mango. Achieving a good sensory experience for those consumers will require the development and use of technology that makes it possible to ship riper fruit. Basically this means slowing mango fruit ripening.

Mangoes do not produce much ethylene during ripening compared with other climacteric species. It may be possible to scrub ethylene sufficiently from the postharvest environment to impede mango ripening. Blocking ethylene action, either through application of competitive ethylene binding inhibitors like 1-methylcyclopropene, or transgenic efforts to alter expression of binding proteins, could have similar results as ethylene scrubbing. CA technology is proven to effectively slow mango ripening and extend postharvest life, but is expensive and there is a limited supply of CA marine containers to use for mango shipping. Use of MAP, perhaps in conjunction with ethylene scrubbing, may be a more practical approach to be able to ship tree-ripe mangoes internationally. In short, efforts to develop technologies that allow delivery of fully ripe, sweet, juicy, fragrant mangoes to consumers will lead to greater sales and consumption of this fruit.

References

Abourayya, M.S., Kassim, N.E., El-Sheikh, M.H. and Rakha, A.M. (2011) Fruit physical and chemical characteristics at maturity stage of Tommy Atkins, Keitt and Kent mango cultivars grown under Nubariya conditions. *Journal of American Science*, **7**, 228–233.

APHIS-PPQ [Animal and Plant Health Inspection Service, Plant Protection and Quarantine]. (2015a) Treatment manual. T106-Vapor heat. Various commodities from Mexico: Mango (Manilla variety only; T106-a-3); Mango from Taiwan, for *Bactrocera dorsalis* (T106-d), and Mango from the Philippines (the island of Guimaras only) (T106-d-1), for *Bactrocera occipitalis, Bactrocera cucurbitae,* and *Bactrocera philippinensis.* In: *Plant Protection and Quarantine Treatment Manual.*, pp. 5-2-74 to 5-2-76. US Department of Agriculture, Hyattsville, MD.

APHIS-PPQ [Animal and Plant Health Inspection Service, Plant Protection and Quarantine]. (2015b) Treatment manual. T102-Water treatment. Mango (T102-a). In: *Plant Protection and Quarantine Treatment Manual*, pp. 5-2-58 to 5-2-60. US Department of Agriculture, Hyattsville, MD.

Bender, R.J. and Brecht, J.K. (1994) Responses of Kent and Tommy Atkins mangoes to reduced O_2 and elevated CO_2. *Proceedings of the Florida State Horticultural Society*, **107**, 274–277.

Bender, R.J., Brecht, J.K., Baldwin, E.A. and Malundo, T.M.M. (2000a) Aroma volatiles of mature-green and tree-ripe Tommy Atkins mangoes after controlled atmosphere vs. air storage. *HortScience*, **35**, 684–686.

Bender, R.J., Brecht, J.K., Sargent, S.A. and Huber, D.J. (2000b) Mango tolerance to reduced oxygen levels in controlled atmosphere storage. *Journal of the American Society for Horticultural Science*, **125**, 707–713.

Brecht, J.K. and Yahia, E.M. (2009) Postharvest physiology. In: *The mango: Botany, production and uses*, 2nd edition, (ed. R. Litz), pp. 484–528. CAB International, Wallingford, UK.

Brecht, J.K., Sargent, S.A., Kader, A.A., Mitcham, E.J., Maul, F. *et al.* (2014) *Mango Postharvest Best Management Practices Manual* – HS1185 (revised), (ed. J.K. Brecht). National Mango Board and UF/IFAS, Gainesville, FL.

Garcia Martinez, R., Lopez Jimenez, A., Saucedo Veloz, C., Salazar-Garcia, S. and Suarez Espinosa, J. (2015) Ripening and fruit quality of mango Kent with three levels of fertilization. *Revista Mexicana de Ciencias Agricolas*, **6**, 665–678.

Johnson, G.I. and Hofman, P.J. (2009) Postharvest technology and quarantine treatments. In: *The mango: Botany, production and uses*, 2nd edition, (ed. R. Litz), pp. 529–605. CAB International, Wallingford, UK.

Kader, A.A. (2008) Mango Quality Attributes and Grade Standards: A review of available information and identification of future research needs (National Mango Board) [Online]. Available: http://www.mango.org/Mangos/media/Media/Documents/Research%20And%20Resources/Research/Industry/Post-Harvest/Mango_Grade_Standards_Final_Report_Eng.pdf [23 March 2016].

Mangan, R.L. and Ingle, S.J. (1992) Forced-hot air quarantine treatment for mangoes infested with West Indians fruit fly (Diptera: Tephritidae). *Journal of Economic Entomology*, **85**, 1860–1864.

Mitcham, E.A. and Yahia, E. (2009) Alternative Treatments to Hot Water Immersion for Mango Fruit. Report to the National Mango Board [Online]. Available: http://www.mango.org/Mangos/media/Media/Documents/Research%20And%20Resources/Research/Industry/Post-Harvest/Alternatives_Hot_Water_Treatment_Exec_Summary_Eng.pdf [23 March 2016].

Mohammed, M. and Brecht, J.K. (2002) Reduction of chilling injury in Tommy Atkins mangoes during ripening. *Scientia Horticulturae*, **95**, 297–308.

Narayana, C.K., Rao, D.S. and Roy, S.K. (2012) Mango production, postharvest physiology and storage. In: *Tropical and Subtropical Fruits: Postharvest physiology, processing and packaging* (ed. M. Siddiq), pp. 259–276. Blackwell Publishing, Ames, IA.

Nguyen, H., Hofman, P., Holmes, R., Bally, I., Stubbings, B. and McConchie, R. (2004) Effect of nitrogen on the skin colour and other quality attributes of ripe Kensington Pride mango (*Mangifera indica* L.) fruit. *The Journal of Horticultural Science and Biotechnology*, **79**, 204–210.

NMB [National Mango Board] (2010) Mango Maturity and Ripeness Guide [Online]. Available: http://www.mango.org/Mangos/media/Media/Documents/Retail-Quality%20Assessment/Mango-Maturity-and-Ripeness-Guide.pdf [23 March 2016].

Robinson, S.P., Loveys, B.R. and Chacko, E.K. (1993) Polyphenol oxidase enzymes in the sap and skin of mango fruit. *Functional Plant Biology*, **20**, 99–107.

Saaiman, W.C. (1997) Orchard sanitation as a means of controlling postharvest diseases. *South African Mango Growers Association Yearbook*, **17**, 73–74.

Singh, Z. and Singh, S.P. (2012) Mango. In: *Crop Post-harvest: Science and technology* (eds D. Rees, G. Farrell and J. Orchard), pp. 108–142. Blackwell Publishing, Oxford, UK.

Subramanyam, H., Krishnamurthy, S. and Parpia, H.A.B. (1975) Physiology and biochemistry of mango fruit. *Advances in Food Research*, **21**, 223–305.

Torres-Rivera, Z. and Hallman, G.J. (2007) Low dose irradiation phytosanitary treatment against Mediterranean fruit fly (Diptera: Tephritidae). *Florida Entomologist*, **90**, 343–346.

USDA [US Department of Agriculture] (2015) USDA National Nutrient Database for Standard Reference, Release 28 [Online]. Available: http://www.ars.usda.gov/ba/bhnrc/ndl [23 March 2016].

Yahia, E.M. (1993) Modified/controlled atmosphere storage in Mexico. In: *Proceedings of the Sixth International CA Conference*, 15–17 June 1993. Northeast Regional Agricultural Engineering Service, NRAES-71, vol. 2, pp. 570–578. Cornell University, Ithaca, NY.

Yahia, E.M. (1997) Modified/controlled atmospheres for mango. In: *CA '97 Proceedings, Volume 3: Fruits other than Apples and Pears* (ed. A.A. Kader), pp. 110–116. University of California, Davis, CA.

Yahia, E.M. (1998) Modified and controlled atmospheres for tropical fruits. *Horticultural Reviews*, **22**, 123–183.

Yahia, E.M. (1999) Maturity indices, grades for quality, and postharvest handling and packing of mango, In: *Postharvest Treatments for Mango* (ed. A. Lizana), Proceedings of a short course. 28–29 July, 1988, Cairo, Egypt. RONCO/ATUT, Giza, Egypt.

Yahia, E.M. (ed.) (2009) *Modified and Controlled Atmospheres for Transportation, Storage And Packaging of Horticultural Commodities*. CRC Press (Taylor and Francis), Boca Raton, FL.

Yahia, E.M. (2011) Mango (*Mangifera indica* L.). In: *Postharvest Biology and Technology of Tropical and Subtropical Fruits*, vol. 3 (ed. E.M. Yahia), pp. 492–565. Woodhead Publishing, Oxford, UK.

Yahia, E.M. and Singh, S.P. (2009) Tropical fruits. In: *Modified and Controlled Atmospheres for the Storage, Transportation, and Packaging of Horticultural Commodities* (ed. E.M. Yahia), pp. 397–461. CRC Press, Boca Raton, FL

Yahia, E.M., Ornelas-Paz, J.J. and Aziza, F.R. (2006) *The Mango*. Editorial Trillas, Mexico City. (in Spanish)

Yahia, E.M., Jones, R.W. and Thomas, B.D. (2011) Quarantine pests of tropical and subtropical fruits and their control. In: *Postharvest Biology and Technology of Tropical and Subtropical Fruits*, vol. 1 (ed. E.M. Yahia), pp. 224–287. Woodhead Publishing, Oxford, UK.

7

Packaging of Fresh Mangoes and Processed Mango Products

Aman Ullah Malik[1], *Farihah Siddiq*[2] *and Muhammad Siddiq*[3]

[1] *Institute of Horticultural Sciences, University of Agriculture, Faisalabad, Pakistan*
[2] *Food Packaging Associate, East Lansing, Michigan, USA*
[3] *Department of Food Science and Human Nutrition, Michigan State University, East Lansing, Michigan, USA*

7.1 Introduction

Recent technological advancements and improvements in logistics have increased the fresh mango trade manifold, both in remote domestic markets of its growing countries, as well as in global markets. Supermarkets in advanced countries are now offering mangoes almost year round, by sourcing them from different countries in the Northern Hemisphere (Mexico, Thailand, Philippine, Pakistan, India) and in the Southern Hemisphere (Australia, Brazil, Peru, Ecuador), due to difference in seasonality. However, delivering quality with consistency is the greatest challenge in mango marketing and the role of appropriate packaging is one of the most crucial aspects of mango logistics. Studies of international markets show that bruised mangoes are often either totally rejected or sold at lower market prices. Furthermore, if the packaging of mangoes is improved according to the requirements and demands of the wholesalers, they will be willing to pay the expected price (ASLP 2012). Good packaging and cushioning prevents bruising and fruit losses in the mango supply chain (Chonhenchob and Singh 2003). The term "right" packaging for mango is relevant to particular supply chains (air, sea, road) and intended markets; however, the very basic requirements remain proper fruit containment, protection, communication, and convenience in distribution and marketing (Chonhenchob *et al.* 2012; FAO 2011), along with their positive impact on quality and shelf life.

Likewise, along with fresh mangoes, processed products are also becoming more diverse, from traditional mango pickles and tetra-packed juices to minimally processed fruit chunks, mango leather, and dried chips, with increasing emphasis on food safety and quality. Hence, packaging is in an ever state of transformation, both for fresh and processed products. This chapter focuses mainly on the applied aspects of mango packaging (fresh fruit and processed products), in the light of current and best practices in the global mango industry, as well as its future research directions.

Handbook of Mango Fruit: Production, Postharvest Science, Processing Technology and Nutrition,
First Edition. Edited by Muhammad Siddiq, Jeffrey K. Brecht and Jiwan S. Sidhu.

7.2 Current Packaging Systems

Mature mangoes are firm when harvested, but as the ripening process advances, the fruit become soft and have limited shelf life under ambient conditions. Any kind of injury, abrasion, or compression damage will induce ethylene, and accelerate ripening (reduce shelf life) and fruit decay later in the supply chain. Hence, apart from good packaging, careful handling, and maintaining optimal conditions during packaging, storage, transport, and retail marketing are essential to ensure the desired shelf life.

7.2.1 Packaging of Fresh Mangoes (Bulk/Retail)

Harvested mangoes go through de-sapping, washing, sizing, and grading prior to packing. In most developing countries, fruit are bulk packed (multilayer) in wooden crates normally lined with paper, thermoformed containers, bamboo baskets, gunny bags, and plastic woven polypropylene sacks of different sizes (Defraeye *et al.* 2015; FAO 2011, 2014). Traditional bulk packs have various logistic issues (pre-cooling, food safety, environmental, loading, unloading), and do not integrate into modern cold chain systems (Defraeye *et al.* 2015). Mango supply chain studies in Pakistan showed that over-packing mangoes in wooden crates and overloading in trucks were significant contributors to high postharvest losses (Malik *et al.* 2013). However, wooden crates have been gradually replaced, mainly with cardboard boxes in high-end domestic markets as well as in major export markets (USA, EU), excluding the Middle East, where mango export in wooden crates is still allowed. The Australian mango industry uses 7-kg mango trays in domestic as well as for many export markets. In general, the packaging design, size, and footprints are largely dictated by the wholesalers (Defraeye *et al.* 2015).

Retail packaging of fresh produce is generally developed by the consumer's requirements (Chonhenchob *et al.* 2012). Selected pictures of retail mango packaging in different countries are shown in Figure 7.1. Cardboard boxes of varying sizes (1, 2, 5 kg) are the most commonly used, while there is an increasing trend of using plastic trays, and trays being overwrapped or heat sealed films and clamshells (Figure 7.2) (Chonhenchob *et al.* 2012). However, unsuitable plastic films and lack of temperature control at retail sometime causes fogging, although plastic films with high permeability to gases and with antifogging properties are available (Boonruang *et al.* 2012). Cell or tray packing provides produce protection from stacking pressure, as well as from impact with each other; however, due to loose packing density and headspace in each compartment, produce may be damaged due to vibration during transportation (FAO 2011). In-package dividers, liners (cardboard, plastic, foam), and cushioning material (shredded paper, soft paper, bubble wrap) are often used to protect mangoes from bruising during shipping and distribution (FAO 2011; Malik *et al.* 2013). Individual wrapping of mangoes with colorful soft papers, fruit stickers, and use of additional colored ribbons, are also popular in Chinese markets.

7.2.1.1 General Considerations for Fruit Packaging

Uniformity of fruit in package, freedom from bruising, count, and weight specifications are the major considerations in mango packing. Fruit should be packed tight enough to avoid vibration during transport, but not be over-packed to cause pressure damage. Various studies on bruise control during transportation show that use of foam net

Figure 7.1 Mango packing styles for the retail markets in selected countries.

provides the best protection (ASLP 2012; Chonhenchob and Singh 2004). However, another study concluded that molded paper as a cushioning material performs better than foam net (Wongsuriyasak and Srichandr 2012).

Fruit should be short stemmed (1 inch) to avoid physical injury to other fruit. A mango is stronger at the shoulders and should be best packed lying at the shoulders and not at its base, which becomes soft over time and often results in compression damage and discoloration, usually called "soft nose". While packing, the tray should be held at an angle, starting from the distal end toward the packer. Carton height should be sufficient to avoid compression of fruit when cartons are stacked. During closing of boxes, ventilation holes of the tray and the lid must be aligned. Since mangoes lose weight during ripening, carton weight at packing may need to be at least 5% greater than shown on the package label (Malik *et al.* 2013).

7.2.1.2 Packaging – Labeling and Phytosanitary Regulations

Minimum label information for most of the domestic wholesale markets in developing countries include the cultivar, class (local criteria), gross weight, and grower/suppliers name. However, for export markets, labeling information usually includes country of origin, cultivar, and grower, packing company, market agent, count (fruit per carton and

Figure 7.2 Clamshell containers (top) and plastic liner cups (bottom) used for mango packaging.

weight range), and class (Johnson and Hoffman 2009). Currently, mango supplies to supermarkets require a third-party audited quality assurance system, of which complete traceability is an integral part. Both electronic bar codes as well as use of numerical codes are in practice for maintaining traceability. The barcode uniquely allows tracking of the product's physical location at any point in the supply chain, and also has several other stock recording advantages. There are several types of bar codes available in the market, but the Universal Product Code (UPC) barcode is the most common. More recently, work on wireless based radiofrequency identification (RFID) labels, tags, and chips in food product traceability are also underway (Lee and Rahman 2014).

Mango packaging is also subjected to strict phytosanitary regulations by many importing countries, mostly aiming at avoiding contamination with quarantine pest or regulated articles (e.g., trash, soil, weed seeds), unprocessed packing material (vector pest), etc., preventing post-treatment infestation, and to ensure traceability. For instance, Biosecurity Australia requires mangoes to be packed in new cartons sealed with a sticker (by mandated government agency) or a secure seal placed across the carton opening. Unprocessed packing material of plant origin is not allowed. Any openings in cartons are to be either screened (mesh no greater than 1.6 mm diameter) or taped closed. All cartons are to be marked "For Australia", and labeled with a packing date, packinghouse name or number, and registered treatment (for fruit fly disinfection), center name or number (Anonymous 2011).

7.2.2 Bulk Shipment by Air, Sea, and Land

For export, mangoes are mostly air freighted in a two-piece cardboard box of varying sizes (more commonly 2, 3, 4, 5 kg), packed as single layers, and delivered in two to three days to most destinations around the world. Waxed boxes are no longer allowed in Europe, due to difficulties in recycling. Boxes need to be strong and lightweight, with ventilation holes. It is generally recommended that holes should be away from the vertical corners and covers less than 5% of the total box wall area to reduce strength loss (Kader 2002). In airplanes, fruit can be loaded in metallic closed containers (called AVEs) with a capacity of about 1.3 ton, with boxes individually stacked inside. Strapping of 2–3 boxes is preferred in such un-palletized loads. Another option is the air pallet with a capacity of 3.5 ton and is generally preferred (considering airflow and heat transfer), but not necessary (Malik *et al.* 2013). Mangoes intended for air freighting should be conditioned to 20–22 °C before loading, to maintain better shelf life. Trans-shipment of mangoes often has higher risk of damage to boxes during loading and offloading, causing fruit bruising and physical losses.

For sea shipping, mangoes are usually packed in 4-kg single-piece open trays, stacked on pallets varying from region to region (Europe mostly: 1000 (W) × 1,200 mm (L); some Asian countries: 1,100 × 1,100 mm; and the USA (40 × 48 in), and held in refrigerated containers under low temperature (10–13 °C) and high relative humidity (85–90%). The boxes need to be much sturdier than air freight boxes, to withstand high RH, for extended time in transit. The minimum requirement for the packaging material is a C-flute corrugated board with 200 lb burst strength (Singh *et al.* 2013). A simulated shipping study on closed top versus open top packaging concluded more marketable fruit outrun in open top compared to close top packaging of the two mango cultivars studied (Hafeez *et al.* 2012), possibly due to better cooling and air flow.

Mangoes for the US market are usually shipped in 4-kg boxes (8.8 lb), with 14-down footprint on pallets. Importers prefer the Grocery Manufacturers Association (GMA) standard pallet (40 × 48 inch). Controlling package weight is an issue. A US National Mango Board study proposed two 15-down tray designs: 3.75 kg (8.3 lb): 13.2 × 9.5 × 4 inches designs (12-down tray designs: 5 kg (11 lb): 12.87 × 11.75 × 4 inches) with 12- and 15-down footprints, respectively, fitting the GMA pallet in pattern for clear airflow through the trays. The outside dimensions of the 12- and 15-down tray design are 12.87 × 11.75 × 4 inches and 13.2 × 9.5 × 4 inches, respectively (Singh *et al.* 2013). The box strength mainly depends on the strength of its walls and upright edges; perfectly aligned column stacking provides best strength by creating a beam-like support structure within the pallet load. Pallet overhang and wide deck board gaps reduce stacking strength (Pathare and Opara 2014). Generally, importers avoid re-packing of fruit to save on high labor costs and time constraints, and sea shipped mangoes usually require repacking to remove damaged/diseased fruit and to have uniformity of fruit color in pack. Figure 7.3 shows mango boxes used for air/sea freight.

7.2.3 Packaging of Fresh-cut Mango

Minimally-processed or fresh-cut fruits have been one of the fastest growing segments of the food industry in North American and European markets. In comparison to 1990, the per capita consumption of mangoes has increased 4-fold to about 2 lbs in recent years (Siddiq *et al.* 2012). The mango, due to its unique flavor and aroma, is suitable for the ready-to-eat market. Fresh-cut mangoes are processed as either sliced or diced chunks and packaged in rigid plastic trays, also referred to as "clamshell" containers (Figure 7.4). Polyethylene terephthalate (PET) is the material of choice for packaging

Air-freight Box

Sea-freight Box

Shipment Ready Pelletized Boxes in Cold Room

Export Pack with Mesh (qurantine requirement)

Figure 7.3 Mango boxes used for air/sea freight, shipment ready palletized mango, and a display ready fancy export pack.

fresh-cut fruits, including mango. PET trays are closed with snap-on lids, which do not provide an air-tight seal, thus the product is at atmospheric conditions. Application of anti-browning and texture firming pre-treatments (e.g., ascorbic and citric acid, calcium chloride) is done prior to packaging (Siddiq *et al.* 2013). Fresh-cut mangoes are marketed under refrigerated conditions and must be kept at 4 °C to preserve quality and ensure food safety prior to consumption, that is, the "use by" date. The typical consumer package size is 10–12 oz.

Modified atmosphere packaging (MAP) has been investigated for extending the shelf life and preserving quality of fresh-cut mangoes by many researchers, using polypropylene or polystyrene plastic trays. Trays are thermo-conglutinated (heat sealed) with a lidding film having appropriate thickness and oxygen transmission rate (OTR), for

(a) Mango Slices

(b) Mango Chunks

Figure 7.4 Fresh-cut mango slices and chunks/dices packaged in plastic trays with snap-on lids.

example, Cryovac LDX-5406 film (Beaulieu and Lea 2003; Chonhenchob *et al.* 2007; González-Aguilar *et al.* 2000; Martinez-Ferrer *et al.* 2002). MAP used in combination with low-temperature storage to extend the shelf life of other fresh-cut fruits can also be used for fresh-cut mangoes, since these condition reduce the respiration rate and minimize surface browning and damage (González-Aguilar *et al.* 2008).

Packaging plays an important role in preserving quality through marketing channels. Siddiq *et al.* (2011) reported that from a consumer's perspective, color/appearance is the most important quality attribute that is readily visible. If the color or appearance of a fresh-cut product is not of acceptable quality, the consumer is less likely to purchase it regardless of excellent texture, flavor, taste, or other quality attributes it may possess. Therefore, in addition to acceptable physico-chemical properties, a high level of sensory quality is a prerequisite for successful marketing of fresh-cut mangoes.

7.2.4 Packaging of Processed Mango Products

7.2.4.1 Mango Juice, Nectar, and Beverage

Among processed mango products, juice, nectar, and beverages are the most commonly available and consumed widely. In recent years, these products have gained popularity in North American and European countries. These products are packaged in a family pack (32 oz or 1 liter size) composite packages or "brick" packages, due to their brick-like shape and single-serve (12–16 oz or 340–450 ml) juice. Beverage products are packaged in PET or glass bottles, and aluminum cans (Figure 7.5). It is to be noted that the type of packages shown in the figure are not all inclusive, as a number of variations in material, shape, size, and lid/seal closure are available in the market. Cost consideration plays a role in the selection of packaging material, especially for small processors in developing countries with less stringent food safety regulations and quality control practices.

Composite package is a fabricated structure made with multi-layers of homogeneous materials (Figure 7.5a). In order to impart specific properties to such packages, the sub-layers of the composite material are carefully chosen. The sub-layers are composed of polyolefin (polyethylene (PE)), aluminum foil, and paperboard, and are bound with selected adhesives. Besides juice and beverages, brick-type packages can also be used for aseptically-processed mango purees. In the case of retorting (thermal processing) conditions, the heat-sealing layer of choice is polypropylene to withstand high temperatures (Pascall 2010). The right selection of packaging material(s) for mango juice products is critical to preserve physico-chemical and sensory quality and ensure microbial safety (Alaka *et al.* 2003; Morris *et al.* 2007; Vásquez-Caicedo *et al.* 2007a). For best preservation and nutrients retention, materials with appropriate barrier properties against oxygen are selected. Glass bottles offer the best barrier properties, while material used in PET bottles possesses excellent structural stability, significantly less weight, and no breakage as compared to glass. PET bottles also a offer good barrier to moisture and gases, e.g., carbon dioxide.

7.2.4.2 Canned Mango Slices, Chunks, and Pulp

Metal cans of various forms have been used in commercial food preservation for well over a century and continue to be a cost-effective, environmentally friendly, and the safest packaging choice. The most commonly processed and packaged mango products

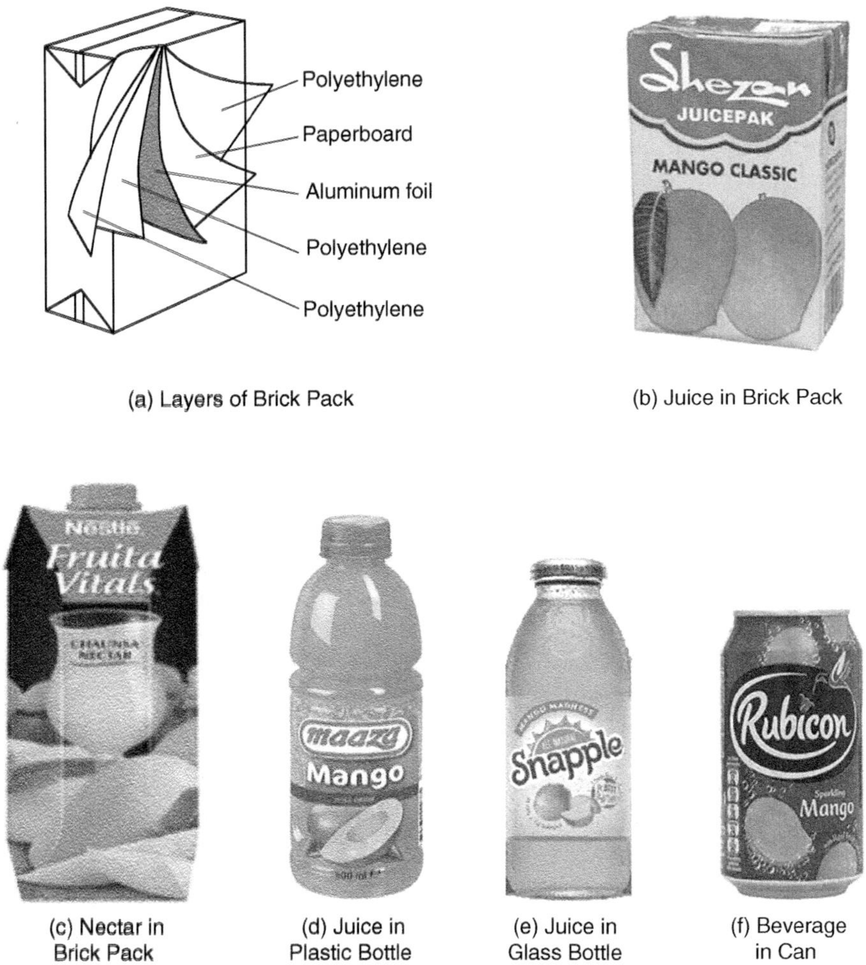

Figure 7.5 Composite layers of brick-type package (a), with juice/nectar (b,c), juice in plastic and glass bottles (d,e), and beverage in aluminum can (f).

in cans are sliced and diced mangoes, mango pulp, and mango puree (Hansawasdi *et al.* 2006; Ledeker *et al.* 2014; Vásquez-Caicedo *et al.* 2007b). Figure 7.6 shows a selection of processed mango products in metal cans. The appropriate selection of temperature and holding time for canning is critical to achieve commercial sterility and ensure public safety. Some processors market mixed mango pickles in metal cans, but these do not require thermal processing.

Marsh and Bugusu (2007) reported that metal cans offer corrosion resistance, possess good barrier properties to gases, water vapor, light, and odors, and can be easily recycled. Cans of different shapes and sizes can be heat treated and sealed hermetically (suitable for sterile products). The weight of metal cans has been reduced significantly in the last two decades and new innovations in metal package design include easy-open cans with pull-tab lids, and easy-to-grasp cans. Packaging of mango products in metal cans offers the most efficient way of long distance transportation without any quality

Figure 7.6 Metal cans used for mango products – diced mangoes, mango slices in pull-tab lid can, and mango pulp.

loss due to temperature and humidity fluctuations, which are detrimental to dried or frozen products during transit. This is of particular advantage, since most mangoes are produced and processed in tropical and subtropical regions.

7.2.4.3 Mango Products in Glass Containers

Mango jam, jelly, spread, pickle, and chutney (relish) are the most commonly packaged products in glass containers, typically in jars of various shapes and sizes. A selection of mango products packaged in glass is shown in Figure 7.7. These shelf stable products are hot-filled into glass jars and closed using lug type twist lids. Once the product is cooled to room temperature, a vacuum is created in the headspace, thus lowering the amount of oxygen, which helps minimize the quality loss. According to the Glass Packaging Institute (GPI 2015), the life cycle of glass has set the environmental standards for food

Figure 7.7 Mango products in glass jars.

(a) Frozen, 1-lb (b) Frozen, 5-lb (c) Dried, 3.5 oz

Figure 7.8 Frozen mango chunks (a,b) and dried mango slices (c) in flexible pouches.

packaging. Glass has a proven 400 years record of safety and, besides being made from all natural materials, it is infinitely recyclable.

Robertson (2012) reported that, besides being recyclable, reusable, and recloseable, glass containers have very low interaction with the food product, are impermeable to gases and vapors, and have the ability to withstand high processing temperatures. The disadvantage of glass containers is that they are heavy and brittle, have higher transportation costs, and are susceptibility to breakage from impact and mishandling.

7.2.4.4 Frozen and Dried Mango Product

Frozen mangoes are packaged in flexible pouches of retail and bulk sizes, while dried mangoes are mostly packaged in individual-serving size bags (Figure 7.8). Bulk-packaged frozen mango chunks are generally used in the food service sector, such as restaurants, fast food establishments, and institutional use. Dried mangoes, usually in sliced form, are packaged in low-density polyethylene (LDPE) and high-density polyethylene (HDPE) bags with suitable moisture and oxygen barrier properties. Robertson (2012) reported that the choice of barrier layers, sealant layers, and food contact layers depends on the processing conditions, product application, and desired shelf life. These pouches are made of materials that can withstand frozen storage at low temperatures (−18° C) for 3–6 months. A broad range of materials is used for manufacturing of flexible pouches, including PET, PE, oriented polypropylene (OPP), HDPE, LDPE, cast polypropylene (CPP), and metalized films of variable thicknesses.

Flexible pouches for packaging of frozen fruit require high tear/rupture resistance to withstand shock and vibration during transportation and retail handling and storage. Plasticizers are generally added to increase the flexibility of polymeric films used for frozen products. Chonhenchob *et al.* (2012) recommended technologies for tracking packaged products throughout the supply chain. since temperature is one of the critical factors affecting the quality and safety of frozen fruits. In this regard, intelligent packaging, such as RFID, is increasingly being used in the supply chain.

(a) 4lb (b) 7-oz (c) 4-oz

Figure 7.9 Ready to eat mango chunks in extra light syrup packaged in plastic jar (a) and cups (b,c).

7.2.4.5 Mango Chunks in Syrup in Plastic Packages

Ready-to-eat diced mangoes in syrup are packaged in 1–4 lb plastic jars and 4–7 oz individual-serve cups, and typically made of PET or CPP materials (Figure 7.9). Fill media is extra light syrup made with 10% sugar, which helps maintain the product color. These products are shelf stable, but must be kept refrigerated after the package is opened. Individual-serve cups offer added convenience of use when on-the-go for a healthy snack.

7.3 Innovative Packaging Technologies

7.3.1 Modified Atmosphere Packaging (MAP)

MAP is a simple and economical packaging technology, that modifies or alters the gas composition around the products in food packages from normal air (20.95% O_2, 78.09% N_2, 0.93% argon, and 0.038% CO_2) to provide an atmosphere for increasing shelf life and maintaining the quality (Zhang *et al.* 2015). The principal techniques for maintaining the quality of mangoes under MAP are based on high gas permeability and an ethylene-removing mechanism of film (Boonruang *et al.* 2012). The decreased O_2 and increased CO_2 levels reduce respiration rate, prevent water loss, and alleviate chilling injury in mango (Chidtragool and Ketsa 2013; Pesis *et al.* 2000). At a level of 2% O_2, anaerobic respiration may result in the development of off flavors and off odors (Sandhya 2010); however, low oxygen tolerance varies with produce type and cultivars, and mango cv Sindhri can be successfully shipped using an atmosphere of 2% O_2 and 4% CO_2 (Malik *et al.* 2013). Among several types of commercially available MAP films (Stepac, Activepk, Bio-fresh, Intellipack, Everfresh, etc.) (Malik *et al.* 2012), the two more commonly tested for mangoes are Xtend bags and Active Pack (Githiga *et al.* 2014; Malik *et al.* 2012; Pesis *et al.* 2000).

Effect of MAP on mangoes and other fresh produce depends upon a number of factors, including produce type, film (thickness, perforations, strength), crop maturity, and processing and storage conditions (temperature, RH, duration) (Brecht and Yahiya 2009; Kader 2002; Yahiya 2009). The in-package gas levels for mangoes are reported from 3–7% O_2, 5–8% CO_2, and 85–92%RH (Sandhya 2010). However, a recent study on mango, comparing Xtend versus Biofresh packaging films, show that gas concentrations significantly varied between the two film types (Figure 7.10).

A study on Nam Dok Mai mangoes, using four different films (i.e. non-perforated, highly gas-permeable film; non-perforated, ethylene-absorbing, highly gas-permeable film; micro-perforated, highly gas-permeable film; and common non-perforated polyethylene film), showed that shelf life of mangoes was extended to 40 days with the non-perforated, highly gas-permeable film, as compared to 20 days with the control. Furthermore, film incorporated with fine particles of mordenite framework inverted-type zeolite with ethylene-absorbing abilities, increases the prospects of further extension of shelf life in mangoes (Boonruang *et al.* 2012). For effective use of MAP technology, mangoes must be pre-cooled before sealing of bags. MAP limitations

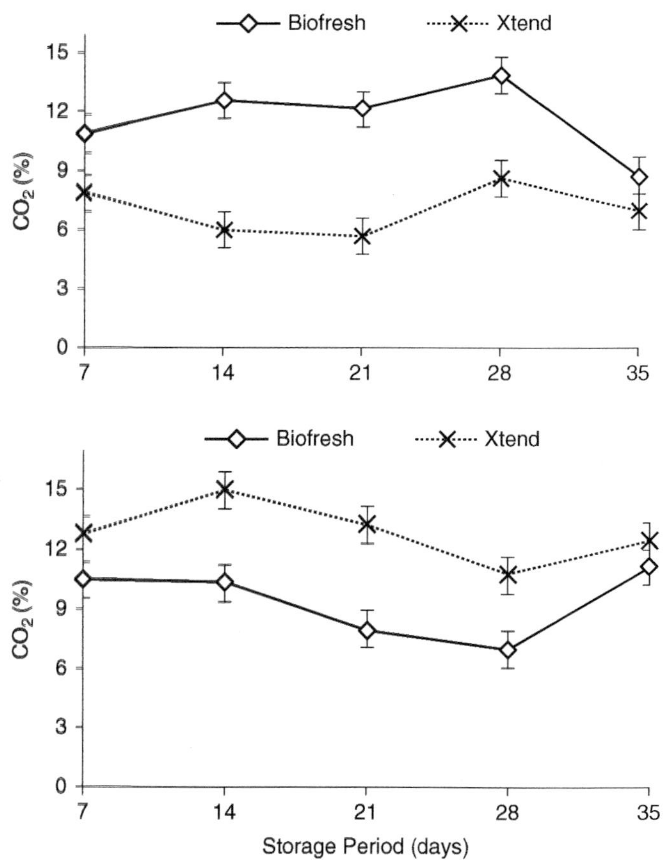

Figure 7.10 In-package CO_2 and O_2 concentrations during storage of *Samar Bahisht Chaunsa* mango. *Source:* Zohaib and Malik (*unpublished data*).

include their use in conjunction with good cold chain, increased cost (cost of bag and packing /unpacking), and occasional production of off-flavor (McCollum *et al.* 1992). More recently, Temperature Responsive Films (Landec Labs) have been developed which increase gas permeability in response to temperature as well as increase in respiration, thus stabilizing the modified atmosphere under various temperatures (Sandhya 2010). The new TR-film technology can provide some opportunities for mango shelf life extension under ambient conditions, since previous MAP films could only be effective under good cool chain system.

Individual shrink wrap films have been studied with earlier reports of negative impact on fruit quality of mangoes, mainly on account of off odor and decay incidence (McCollum *et al.* 1992). A study using food grade cling film resulted in retarding ripening, lower physiological weight loss, increased (>70–80%) marketable fruit, and shelf life (10–15 days), after 21 days of mango storage (Janave and Sharma 2005).

7.3.2 Active Packaging

Active packaging is defined as the incorporation of certain compounds into packaging systems to extend shelf life and maintain or enhance food quality. The chemicals have a diverse range, including organic acids, enzymes, bacteriocins, fungicides, natural extracts, ions, and ethanol, as well as the materials in which they are incorporated, for example papers, plastics, metals, or mixtures of these materials as mentioned in different studies (Dainelli *et al.* 2008; Lee *et al.* 2015). In mangoes, the commercially important scavengers are for gases such as ethylene (removal), CO_2 (addition/removal), and O_2 (removal), along with prospects of antimicrobial agents.

7.3.2.1 Ethylene Controllers

Ethylene is the natural gaseous hormone, which can also be applied exogenously to trigger fruit ripening (Kader 2002). Ethylene removal from storage atmosphere or its in-package absorption can help improve the shelf life of produce including mangoes (Illeperuma and Jayasuriya 2002). These chemicals include potassium permanganate, aluminum oxide, activated carbon, and silicon dioxide under various trade names such as Ethysorb (StayFresh Ltd.), Ageless C (Mitsubishi Gas Chemical Company), Freshkeep (Kurarey), Acepack (Nippon Greener), and Peakfresh (Klerk Plastic, and Chantler Packaging, Inc.) (Sandhya, 2010). Potassium permanganate is more common and removes ethylene through potassium permanganate-generated oxidation via the breakdown of ethylene's double bond (Wills and Warton, 2004), which can cause abnormal fruit ripening and off flavor. Better alternatives like zeolites (natural and synthetic) are incorporated into the films, as ethylene-absorbing additives (Boonruang *et al.* 2012). Sachets of ethylene-absorbing chemicals can be placed into MAP bags along with pre-cooled mangoes, before sealing off the bags. It is important to understand that this technique is helpful only when applied timely (as soon as mangoes are harvested, at the pre-climacteric stage), and in conjunction with other ethylene management strategies, including injury prevention and better temperature management.

7.3.2.2 Scavengers (CO_2 and O_2)

Maintaining low levels of O_2 and higher CO_2 during transportation/shipping of mangoes are desirable for extending shelf life, by reducing rate of respiration and ethylene production. However, during the process of respiration, mangoes continuously produce CO_2 and accumulation of its high levels over time may have negative impacts on fruit quality, for example lack of ripening or color development, etc. Therefore, CO_2 levels are to be maintained within desirable limits by removing it from the atmosphere of the produce. Commercial technologies (FreshLock, Mitsubishi Gas Chemical Co.) based on calcium hydroxide are used in shipping containers to create a modified atmosphere for shipping mangoes to distant destinations. Calcium hydroxide reacts with CO_2 to form calcium carbonate and water (Lee *et al.* 2015):

$$Ca\left(OH\right)_2 + CO_2 \rightarrow CaCO_3 + HO$$

Other scavengers include Verifrais (Codimer Tournessi, Gujan-Mestras) (Sandhya 2010), iron powder/calcium hydroxide, and ferrous carbonate/metal halide, as summarized by Lee *et al.* (2015). Oxygen absorbing sachets are not normally used in fresh mango packs, but are more widely used in processed products such as dried and dehydrated fruits (Chonhenchob *et al.* 2012).

7.3.2.3 Antimicrobial Substances

Active packaging with antimicrobial properties is extensively used for processed food products to reduce spoilage and contamination (Le *et al.* 2015; Lee and Rahman 2014); however, studies on fresh mangoes are limited. Controlling postharvest decay is one of the main challenges in extending the shelf life of mangoes. Active sachets incorporated with either oregano or lemongrass essential oils showed reduced growth of the fungi (*Colletotrichum gloeosporides, Lasiodiplodia theobromae, Xanthomonas campestris* pv. mangiferae indica, and *Alternaria alternate*) in Tommy Atkins mangoes stored under ambient condition (25 ° C ± 2 °C, RH 80% ± 5%) for 9 days; lemongrass being more efficient by decreasing about 2 log cycles both the mesophilic aerobes, molds, and yeast counts compared to the control (Antonio *et al.* 2011). However, there is still long way to go for achieving commercially effective control of mango postharvest diseases without using fungicides.

7.3.2.4 Edible Films

Mango fruit is harvested commercially over a range of maturities, for example immature green (dark green), mature green (lighter green), and tree ripe with color breaking to red or orange-yellow. Edible coatings that retard ripening, prevent water loss, and reduce decay can be used on mangoes to extend shelf life and improve appearance (Baldwin *et al.* 1999). Edible films and coatings can be considered one class of active packaging, owing to their functions of edibility and biodegradability (Chonhenchob *et al.* 2012). Generally, edible films and coatings can be produced from three major types of biopolymers – polysaccharides (carbohydrates and gums), proteins, and lipids (Lin and Zhao 2007), and food-grade additives, such as plasticizers. In addition to permeability and mechanical integrity, wettability should also be considered for ensuring effective film application, that is, it is dependent upon the spreading coefficient (or wettability) versus the surface properties of the fruit (Cerqueira *et al.* 2009).

Many studies have reported the use chitosan films and other coatings for whole mangoes (Baldwin *et al.* 1999; Dang *et al.* 2008; Srinivasa *et al.* 2002) and coatings of carboxymethycellulose containing maltodextrin to retain the visual quality and flavor of fresh-cut mangoes (Plotto *et al.* 2004). Edible coatings offer a practical route to the incorporation of functional additives such as antimicrobials, antioxidants, and texture improving compounds (Siripatrawan and Harte 2010). Campos *et al.* (2011) indicated that these films are not a replacement for traditional packaging. Instead, they provide an additional stress factor to be applied to food preservation and can also help to reduce the cost and amount of traditional packaging used.

7.3.3 Intelligent Packaging

Tracking information of the packaged products is increasingly becoming an important part of modern trade. To achieve this, intelligent packaging technologies are being researched and commercialized and can provide the information about the quality and/or safety of the packaged food products by monitoring during handling, distribution, and storage (Chonhenchob *et al.* 2013). Puligundla *et al.* (2012) recommended that ideal indicators should be safe, rapid, reliable, cost-effective, and non-destructive. The major intelligent packaging systems are time-temperature indicators (TTIs)/time-temperature biosensors (TTBs), leak detectors, ripeness and spoilage indicators, and RFID. Such indicators can be used for mangoes and mango products to enhance product reliability. Ideal indicators should be rapid, accurate, and reliable as well as cost-effective, non-destructive, and safe (Puligundla *et al.* 2012). The RFID technology has been used for trackability and traceability from the grower to consumer throughout the supply chain, to ensure the quality and safety of products (Chonhenchob *et al.* 2013).

There is increasing consumer interest in buying ready-to-eat fruit, and mangoes can become popular in many international markets if consumers are able to know exactly when fruit is at optimum ripening stage with best taste and flavor. Such a scenario is becoming a reality with the introduction of novel technologies like RipeSenses (Jenkins Group, Auckland, New Zealand; www.ripesense.com). This is the first of its kind sensor label that can calorimetrically indicate the degree of preferred ripeness of fruits including mangoes. Its working principle is based on detecting the natural aroma compounds as ripening progress. It has novel features such as a unique container design, to capture the aroma and protecting fruits from bruising, along with allowing retailers to sell ripe, ready-to-eat fruit without excessive shrinkage (Lee and Rahman 2014).

7.4 Future Directions

In recent years, novel packaging materials such as bio-based polymers or biopolymers, biodegradable plastics, compostable plastics, oxo-degradable plastics, and UV-degradable plastics have gained attention from the food industry, consumers, and government (Almenar *et al.* 2012). The future of fresh mango packaging is expected to see changes in carton design and strength, for improved transportation, especially under sea freight conditions. There is still improvement to be made in designing more suitable in-package liners with better, flexible material, in order to gently, but effectively hold fruit and reduce bruises. More efficient, temperature responsive films provide opportunities for reducing postharvest losses and extension of shelf life of mangoes, particularly in developing countries, where cold chain facilities are limited, often inaccessible, and costly.

References

Alaka, O.O., Aina, J.O. and Falade, K.O. (2003) Effect of storage conditions on the chemical attributes of *Ogbomoso* mango juice. *European Food Research and Technology*, **218**, 79–82.

Almenar, E., Siddiq, M. and Merkel, C. (2012) Packaging for processed food and the environment. In: *Handbook of Food Process Design* (eds, M.S. Rahman and J. Ahmed), pp. 1369–1405. Blackwell Publishing, Oxford.

Antonio, E., Medeiros, A., de Fatima, N., Soars, F., de Olivier, T. *et al.* (2011) Antimicrobial sachets post-harvest mango fruits. *Revista Brasileira De Fruticultura*, **33**, 363–370.

Anonymous (2011) Non-regulated Analysis: Extension of existing policy for fresh mango fruit from Pakistan. Biosecurity Australia, Australian Government [Online]. Available: http://www.daff.gov.au/SiteCollectionDocuments/ba/plant/2011/BAA_2011-03_Draft_extension_of_policy.pdf [14 August 2015].

ASLP [Agriculture Sector Linkages Program] (2012) Mango trial shipments to China. Mango value chain improvement project. Australia-Pakistan Agriculture Sector Linkages Program. University of Agriculture, Faisalabad, Pakistan.

Baldwin, E.A., Burns, J.K., Kazokas, W., Brecht, J.K., Hagenmaier, R.D. *et al.* (1999) Effect of two edible coatings with different permeability characteristics on mango (*Mangifera indica* L.) ripening during storage. *Postharvest Biology and Technology*, **17**, 215–226.

Beaulieu, J.C. and Lea, J.M. (2003) Volatile and quality changes in fresh-cut mangos prepared from firm-ripe and soft-ripe fruit, stored in clamshell containers and passive MAP. *Postharvest Biology and Technology*, **30**, 15–28.

Brecht, J.K. and Yahia, E.M. (2009) Postharvest physiology. In: *The Mango: Botany, production and uses* (ed, R.E. Litz), 2nd edition, pp. 484–514, CAB International, Wallingford, UK.

Boonruang, K., Chonhenchob, V., Singh, S.P. Chinsirikul, W. and Fuongfuchat, A. (2012) Comparison of various pack aging films for mango export. *Packaging Technology and Science*, **25**, 107–118.

Campos, C.A., Gerschenson, L.N. and Flores, S.K. (2011) Development of edible films and coatings with antimicrobial activity. *Food and Bioprocess Technology*, **4**, 849–875.

Cerqueira, M.A., Lima, Á.M., Teixeira, J.A., Moreira, R.A. and Vicente, A.A. (2009) Suitability of novel galactomannans as edible coatings for tropical fruits. *Journal of Food Engineering*, **94**, 372–378.

Chidtragool, S. and Ketsa, S. (2013) Modified atmosphere reduces chilling injury of mango fruit during low temperature storage. *Acta Horticulturae*, **992**, 513–519.

Chonhenchob, V. and Singh S.P. (2003) Comparison of corrugated boxes and reusable plastic containers for mango distribution. *Packaging Technology and Science*, **16**, 231–37.

Chonhenchob, V and Singh, S.P. (2004) Testing and comparison of various packages for mango distribution. *ASTM International Journal of Testing and Evaluation*, **32**, 69–72.

Chonhenchob, V., Chantarasomboon, Y. and Singh, S.P. (2007) Quality changes of treated fresh-cut tropical fruits in rigid modified atmosphere packaging containers. *Packaging Technology and Science*, **20**, 27–37.

Chonhenchob, V., Chinsirikul, W. and Singh, S.P. (2012) Current and innovative packaging technologies for tropical and subtropical fruits. In: *Tropical and Subtropical Fruit Processing and Packaging* (ed, M. Siddiq), pp. 115–134. John Wiley & Sons, Ames, IA.

Chonhenchob, V., Saha, K., Singh, S.P. and Siddiq, M. (2013) Packaging technologies for dates and date products. In: *Dates – Postharvest Science, Processing Technology and Health Benefits* (ed, M. Siddiq), pp. 137–156. John Wiley & Sons, Oxford, UK.

Dainelli. D., Gontard, N., Spyropoulos, D., Zondervan-van den B.E. and Tobback, P. (2008) Active and intelligent food packaging: legal aspects and safety concerns. *Trends in Food Science and Technology*, **19**(Suppl. 1), S103–S112.

Dang, K.T., Singh, Z. and Swinny, E.E. (2008) Edible coatings influence fruit ripening, quality, and aroma biosynthesis in mango fruit. *Journal of Agricultural and Food Chemistry*, **56**, 1361–1370.

Defraeye, T., Paul, C.P., Berry, T., Opara, U.L., East, E. *et al.* (2015) Towards integrated performance evaluation of future packaging for fresh produce in the cold chain. *Trends in Food Science and Technology*, **44**, 201–225.

FAO (2011) Packaging in fresh produce supply chains in South Asia, RAP Publications 2011/20. FAO-UN Regional Office for Asia and the Pacific. Available: http://www.fao.org/docrep/015/ba0135e/ba0135e00.htm [14 August 2015].

FAO (2014) Appropriate Food Packaging Solutions for Developing Countries [Online]. Available: http://www.fao.org/docrep/015/mb061e/mb061e00.pdf [18 August 2015].

González-Aguilar, G.A., Wang, C.Y. and Buta, J.G. (2000) Maintaining quality of fresh-cut mangoes using anti-browning agents and modified atmosphere packaging. *Journal of Agricultural and Food Chemistry*, **48**, 4204–4208.

González-Aguilar, G.A., Celis, J., Sotelo-Mundo, R.R., De La Rosa, L.A., Rodrigo-Garcia, J. and Alvarez-Parrilla, E. (2008) Physiological and biochemical changes of different fresh-cut mango cultivars stored at 5° C. *International Journal of Food Science and Technology*, **43**, 91–101.

Githiga, R., Ambuko, J., Margaret, H. and Willis, O. (2014) Effect of Activebag® modified atmosphere packaging on the postharvest characteristics of mango fruits, *Mangifera indica L. cultivar Tommy Atkins. Journal of Applied Biosciences*, **83**, 7535–7544.

GPI [Glass Packaging Institute] (2015) Recycling Glass [Online]. Available: http://www.gpi.org/recycling [14 August 2015].

Hafeez, O., Malik, A.U., Khan, A.S., Rehman, A. and Javaid, Q.A. (2012) Impact of different packaging types and low temperature shipping durations on fruit quality and marketability of Pakistani mangoes. *International Journal of Agriculture and Biology*, **14**, 47–54.

Hansawasdi, C., Kem Kaow, M. and Chaiprasart, P. (2006) The production of canned mango slices in syrup from *Mangifera indica* Linn. cv. *Nahm-dawg-mai. VIII International Mango Symposium*, **820**, 769–774.

Illeperuma, C.K. and Jayasuriya, P. (2002) Prolonged storage of "Karuthacolomban" mango by modified atmosphere packaging at low temperature. *Journal of Horticultural Science and Biotechnology*, **77**, 153–157.

Janave, M.T. and Sharma, A. (2005) Extended storage of gamma-irradiated mango at tropical ambient temperature by film wrap packaging. *Journal of Food Science and Technology*, **42**, 230–233.

Johnson, G.I. and Hofmann, P.J. (2009) Postharvest technology and quarantine treatments. In: *The Mango: Botany, Production and Uses* (ed, R.E. Litz), 2nd edition, pp, 530–569. CAB International, Wallingford, UK.

Kader, A. (2002) Postharvest Biology and Technology, An overview. In: *Postharvest Technology of Crops* (ed, A.A. Kader.), Publication No. 3529. University of California, Agriculture and Neutral Resources, Oakland, CA.

Ledeker, C.N., Suwonsichon, S., Chambers, D.H. and Adhikari, K. (2014) Comparison of sensory attributes in fresh mangoes and heat-treated mango purées prepared from Thai cultivars. *LWT-Food Science and Technology*, **56**, 138–144.

Lee, S.J. and Rahman, A.T.M. (2014) Intelligent packaging for food products. In: *Innovations in Food Packaging* (ed, J.H. Han), 2nd edition, pp. 171–209. Academic Press, Cambridge, MA.

Lee, S.Y., Lee, S.J. Choib, D.S. and Hur, S.J. (2015) Current topics in active and intelligent food packaging for preservation of fresh foods. *Journal of the Science of Food and Agriculture*, **95**, 2799–2810.

Lin, D. and Zhao, Y. (2007) Innovations in the development and application of edible coatings for fresh and minimally processed fruits and vegetables. *Comprehensive Reviews in Food Science and Food Safety*, **6**, 60–75.

Malik, A.U., Khan, A.S., Malik, O.H. and Hameed, R. (2012) *Controlled and Modified Atmosphere Storage of Fruits and Vegetables.* Postharvest Research and Training Centre, Institute of Horticultural Sciences, MAS Printers, University of Agriculture, Faisalaba.

Malik, A.U. Campbell J., Hofman P., Amin, M. and Collins, R. (2013) *ASLP Best Practice Mango Supply Chain Management Manual.* Mango Value Chain Improvement Project, University of Agriculture, Faisalabad, Pakistan.

Marsh, K. and Bugusu, B. (2007) Food packaging: roles, materials, and environmental issues. *Journal of Food Science*, **72**, R39–R55.

Martinez-Ferrer, M., Harper, C., Pérez-Muñoz, F. and Chaparro, M. (2002) Modified atmosphere packaging of minimally processed mango and pineapple fruits. *Journal of Food Science*, **67**, 3365–3371.

McCollum, T.G., D'Aquino, S., Miller, W.R. and McDonald, R.E. (1992) Individual shrink film warping of mangoes. *Proceedings Florida State Horticultural Society*, **105**, 103–05.

Morris, C., Brody, A.L. and Wicker, L. (2007) Non-thermal food processing/preservation technologies: a review with packaging implications. *Packaging Technology and Science*, **20**, 275–286.

Pascall, M. (2010) Packaging for fresh vegetables and vegetable products. In: *Handbook of Vegetables and Vegetable Processing* (ed, N. K. Sinha), 1st edition, pp. 405–422. Blackwell Publishing, Ames, IA.

Pathare, P.B and Opara U.L. (2014) Structural design of corrugated boxes for horticultural produce: A review. *Biosystems Engineering*, **125**, 128–140.

Pesis, E., Aharoni, D., Aharon, Z., Ben-Arie, R., Aharoni, N. and Fuchs, Y. (2000) Modified atmosphere and modified humidity packaging alleviates chilling injury symptoms in mango. *Postharvest Biology and Technology*, **19**, 93–101.

Plotto, A., Goodner, K.L. and Baldwin, E.A. (2004) Effect of polysaccharide coatings on quality of fresh cut mangoes *(Mangifera indica). Proceedings of the Florida State Horticultural Society*, **117**, 382–388.

Puligundla, P., Jung, J. and Ko, S. (2012) Carbon dioxide sensors for intelligent food packaging applications. *Food Control*, **25**, 328–333.

Robertson, G.L. (ed) (2012) *Food Packaging: Principles and Practice, 3rd edition.* CRC Press, Boca Raton, FL, 733 p.

Sandhya (2010) Modified atmosphere packaging of fresh produce: Current and future needs. *LWT-Food Science and Technology*, **43**, 381–392.

Siddiq, M., Harte, J.B., Singh, S.P., Khan, A.A., Dolan, K.D. and Saha, K. (2011) Effect of processing, packaging and vibration treatment on the sensory quality of fresh-cut apple slices. *Packaging Technology and Science*, **24**, 309–315.

Siddiq, M., Akhtar, S. and Siddiq, R. (2012) Mango processing, products and nutrition. In: *Tropical and Subtropical Fruit Processing and Packaging* (ed, M. Siddiq), pp. 277–297. John Wiley & Sons, Ames, IA.

Siddiq, M., Sogi, D.S. and Dolan, K.D. (2013) Antioxidant properties, total phenolics, and quality of fresh-cut "Tommy Atkins" mangoes as affected by different pre-treatments. *LWT-Food Science and Technology*, **53**, 156–162.

Singh, S.P., Singh, J. and Saha, K. (2013) Development of New Standardized Package System and Interfacing GMA Pallet that Allows a Paper Corrugated Tray with a 4 kg Weight Capacity for Imported Mangos [Online]. Available: http://www.mango.org/sites/default/files/download/Research%20Studies/Industry/Packaging%20Report%20_Final_ENG.pdf [26 July 2015].

Siripatrawan, U. and Harte, B.R. (2010) Physical properties and antioxidant activity of an active film from chitosan incorporated with green tea extract. *Food Hydrocolloids*, **24**, 770–775.

Srinivasa, P., Baskaran, R., Ramesh, M., Prashanth, K.H. and Tharanathan, R. (2002) Storage studies of mango packed using biodegradable chitosan film. *European Food Research and Technology*, **215**, 504–508.

Vásquez-Caicedo, A.L., Schilling, S., Carle, R. and Neidhart, S. (2007a) Effects of thermal processing and fruit matrix on β-carotene stability and enzyme inactivation during transformation of mangoes into puree and nectar. *Food Chemistry*, **102**, 1172–1186.

Vásquez-Caicedo, A.L., Schilling, S., Carle, R. and Neidhart, S. (2007b) Impact of packaging and storage conditions on colour and β-carotene retention of pasteurised mango purée. *European Food Research and Technology*, **224**, 581–590.

Wills, R.B.H. and Warton, M.A. (2004) Efficacy of potassium permanganate impregnated into alumina beads to reduce atmospheric ethylene. *Journal of the American Society for Horticultural Science*, **129**, 433–438.

Wongsuriyasak, S. and Srichandr, P. (2012) Novel packaging material for mango transportation. In: *3rd International Conference on Manufacturing Science and Engineering (ICMSE2012)* (eds W.Z. Chen, X.P. Xu, P.Q. Dai *et al.*), Xiamen, Peoples Republic of China.

Yahia, E.M. (ed) (2009) *Controlled and Modified Atmospheres for the Storage, Transportation and Packaging of Horticultural Commodities*. CRC Press, London, 608 p.

Zhang, M., Meng, X., Bhandari, B., Fang, Z. and Chen, H. (2015) Recent application of modified atmosphere packaging (MAP) in fresh and fresh-cut foods. *Food Reviews International*, **31**, 172–193.

8

Processing and Quality of Fresh-cut Mangoes

Blanca Salinas-Roca[1], *Jorge Welti-Chanes*[2], *Olga Martin-Belloso*[1] *and Robert Soliva-Fortuny*[1]

[1] *Departament de Tecnologia d'Aliments, Universitat de Lleida, Lleida, Spain*
[2] *Tecnológico de Monterrey, Escuela de Ingeniería y Ciencias, Centro de Biotecnología FEMSA, Monterrey, Mexico*

8.1 Introduction

In recent years, an increasing emphasis is put on the consumption of fruit and vegetables as part of a healthy diet. The changes in the life styles of the consumers have driven the development of ready-to-eat food products, including fresh-cut fruits and vegetables. The demand of fresh-cut fruits with extended shelf life and fresh-like characteristics is a challenge for academia and industry. According to the International Fresh Produce Association (IFPA), fresh-cut fruits are those peeled, cut into pieces, and ready-to-eat. These products have also been referred to as lightly processed or minimally processed (Sinha *et al.* 2010).

Among the most tropical fruits, mango seems to be highly attractive for consumers due to its special flavor, characteristic aroma, texture, and high concentration of bioactive compounds. However, organoleptic and nutritional properties are different among the mango varieties grown worldwide. Mango varieties such as Tommy Atkins, Haden, and Palmer are commonly found in global markets for fresh consumption, as they retain their yellowish color, texture, and fruity flavor throughout at least 2–3 weeks of storage at 5–10 °C. Apart from these cultivars, Manila, Alphonso, and Irwin, among others, are important in local markets, since they show excellent hedonic properties as fleshy texture, juicy, sweetness, and highly aromatic flavor, although they have shorter storage life than other varieties (Ribeiro *et al.* 2008; Sinha 2012).

Mango is a climacteric fruit and therefore its respiration rate rises as it reaches full maturity and then begins to over-ripen (Siddiq 2012). After reaching its climacteric peak, respiration behavior tends to decline until cell death occurs. Once detached from the plant, mangoes, as living tissues, suffer a cascade of metabolic reactions that can result in accelerated ripening, physical changes and, finally, senescence.

Thus, mango undergoes physiological changes during handling and processing, increasing respiration rate, oxidation reactions, color change, and loss of nutrients because of diverse biochemical and enzymatic reactions. Furthermore, metabolic changes are accelerated when the natural barrier of skin is removed by peeling and

Handbook of Mango Fruit: Production, Postharvest Science, Processing Technology and Nutrition, First Edition. Edited by Muhammad Siddiq, Jeffrey K. Brecht and Jiwan S. Sidhu.

subsequent cutting, which results in the microbial load being modified. Moreover, when handling is not adequate, the resulting microbial growth could further reduce the shelf life of fresh-cut mango. Consequently, during minimal processing of mango, two main objectives are sought: to maintain the fresh-like characteristics of the flesh (color, flavor, texture, and nutritional value) and to assure the safety of the product as well as to extend shelf life long enough to allow marketing of fresh-cut mango.

The integration of all aspects from field to fork, including postharvest handling operations, can determine the final quality of fresh-cut products. According to this, minimally processed mango has to keep the attributes of fresh fruit with minimal variations throughout the storage. This chapter focuses on postharvest mango requirements and processing conditions needed to get high-quality fresh-cut mango with fresh-like attributes through the storage period.

8.2 Mango Fruit and Postharvest Handling

8.2.1 Fruit Characteristics

The quality of fresh-cut fruits depends directly on the raw material, therefore mango characteristics prior to fresh-cut processing must be controlled. Mango undergoes chemical reactions, which can be triggered by endogenous metabolism and also by exogenous factors during postharvest. Therefore, it is important to control those variables that affect whole fruit quality, such as harvest maturity index, temperature, and air humidity.

Generally, for an appropriate quality and subsequent processing of mango, the fruit should be harvested at pre-climacteric or very early climacteric stages (Narayana *et al.* 2012; Sinha 2012). Although the ripening of whole mango depends on the variety, this fruit is best harvested in a mature but unripe stage, about 2–4 months from blooming and usually transported in a firm green pre-climacteric stage. Moreover, whole mangoes could be carefully stored under refrigeration (5–10 °C) to retain acceptable characteristics for up to 2–3 weeks. The optimum relative humidity (RH) during storage is between 85 and 90%, although for short storage periods (5–7 days), values of 90–95 % RH could be also accepted (Siddiq 2012).

During postharvest, an optimal quality regarding physical aspects is needed. Thus, fruit with 72–86% of moisture, 14–23 °Brix, and pH of 3–4.5 would be sufficient for further fresh-cut mango processing, although differences among varieties should be taken into consideration (Siddiq 2012).

8.2.2 Postharvest Handling

The handling during postharvest, which leads to physiological changes affecting the quality of the final product, is affected by the climacteric respiratory behavior. In this sense, postharvest handling, including pre-treatments and sanitation, is key to maintaining suitable physicochemical characteristics of raw mango and to avoid microbial contamination during its processing.

A variety of postharvest stress-type treatments and interventions have been developed to preserve fruits (Figure 8.1), with many of these treatments also applicable to fresh-cut mangoes. González-Aguilar *et al.* (2010) reported that to prevent

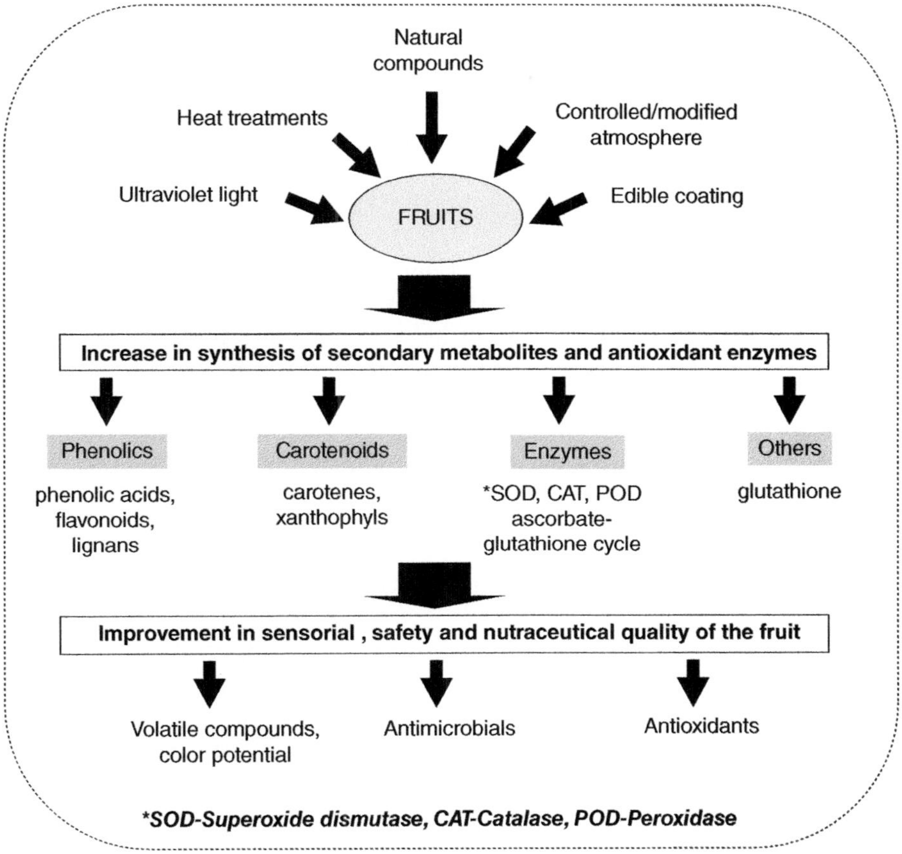

Figure 8.1 Controlled postharvest stress type applied to tropical fruit which induces the synthesis of secondary metabolites and antioxidant enzyme system increasing their sensorial, safety, and nutraceutical quality. *Source:* González-Aguilar *et al.* (2010).

injuries, the plant cell has developed mechanisms involving some secondary metabolic compounds (flavonoids, lignans, carotenoids, ascorbate, and glutathione) and some enzymes (supeoxide dismutase, catalase, peroxidase, ascorbate peroxidase, and glutathione reductase).

8.2.2.1 Pre-treatments

Mangoes have a limited shelf life under ambient conditions and the quality of the fruit rapidly decreases once they are fully ripe. Pre-treatments, which are applied to whole fruit, have a considerable impact on the microbiological and sensorial quality of mango as well as the fruit metabolism. In order to assure an acceptable raw material for fresh-cut processing, different treatments on whole fruit are applied; the most common pre-treatments involve the use of 1-methylcyclopropene (1-MCP), hot water, ethanol vapor, or waxing. The 1-MCP is an ethylene inhibitor and thereby prevents ethylene-dependent responses. The application of 1-MCP used for partially ripe fruit generally suppresses firmness and color loss in fresh-cut fruit (Martin-Belloso and Soliva-Fortuny 2010). Also, the use of hot water dipping (HWD) applied to whole mango improves physical, physiological, and biochemical quality of the final product. Whole mangoes subjected to hot water dipping (HWD) at 50 °C for 30 min, as well as ethanol vapor, have been recommended to retard ripening (Djioua *et al.* 2009). On the other hand, waxing mango with carnauba-based wax or polysaccharide coatings reduces moisture loss. The ultimate goal of all pre-treatments is to reduce the respiration rate of mango, generating a delay in ripening.

8.2.2.2 Washing and Sanitation

Adequate hygienic conditions are required during postharvest handling to assure a high-quality raw material for fresh-cut processing. Sanitizing the whole mango would not assure complete product safety, but it helps in removing pesticide residues and reduces the load of deteriorative and pathogenic microorganisms.

Mango is generally immersed in cold tap water containing sanitizing agents, such as chlorine to reduce the microbial loads on its surface; a concentration of 100 mg/L of chlorine at 11 °C together with brushing reduces surface contamination. A lower concentration of chlorine could be used (0.01 mg/L) by dissolving chlorine in hot water (50 °C) for 10–15 minutes. Thus, the destruction and removal of microorganisms on the peel, which cause damage, can be achieved. Nevertheless, it must be noted that after HWD, storage at low temperatures is highly recommended to reduce fruit surface damage (Ngarmsak *et al.* 2006b).

Various decontamination ways has been explored in other fruits, although no application in fresh-cut mango has been reported. In this regard, hydrogen peroxide, chlorine dioxide, peroxyacetic acid, and organic acids have been tested for washing whole fruits. Their efficacy for extending shelf life and reducing the native and pathogenic microbial population has been demonstrated in grapes, prunes, apples, oranges, melons, tomatoes, and fresh-cut melons (Sinha 2012). Alternatives to traditional sanitizing agents are the use of ozone, UV light, and pulsed light (González-Aguilar *et al.* 2001; Oms-Oliu *et al.* 2008). Furthermore, electron beam irradiation up to 1.5 kGy was assayed in Tommy Atkins mangoes, ensuring pathogen decontamination and maintaining sensory attributes of the fruit during 21 days of storage at 12 °C (Moreno *et al.* 2007).

8.3 Processing of Fresh-cut Mangoes

Minimal processing is defined as the treatment to achieve minimal changes in final product quality characteristics as well as during storage and distribution (Ohlsson and Bengtsson 2002). Figure 8.2 shows typical processing steps for the production of minimally processed mango, which have an influence on the fresh-cut fruit attributes. Fresh-cut mango processing includes three basic mechanical operations: peeling, cutting, and seed removal. All of these operations provoke an increase in the respiration rate and the production of ethylene after one hour of processing, accelerating the maturity and senescence of fresh-cut mango (Souza *et al.* 2004). Moreover, the wounding effect causes the degradation of tissues by the cell wall enzymes, mainly pectin methylesterase and polygalacturonase, which consequently induces physicochemical changes. Also, the phenolic production and wound healing metabolisms are initiated during the mechanical operations of fresh-cut mango processing (Siddiq 2012).

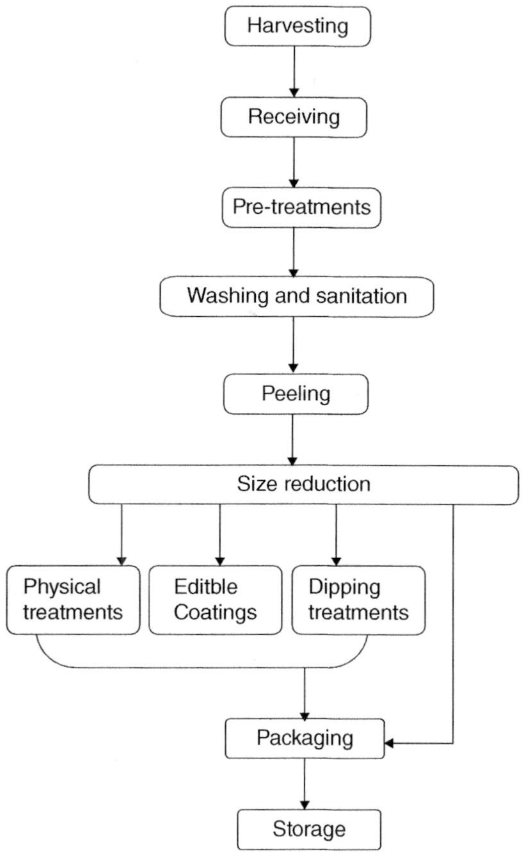

Figure 8.2 Flow diagram for fresh-cut mango production. *Source:* Sinha *et al.* (2012).

The deleterious processes such as the wounding effect and fresh-cut mango metabolism acceleration have an influence on the final product. Thus, quality changes in fresh-cut mango, including softening, loss of sugars, bioactive compounds content, moisture and weight, appearance of undesired flavor and color, occur. Furthermore, fresh-cut mango could suffer chilling as well as water and heat stress during processing. To avoid the undesired changes and obtain a high quality final product, the cold chain during whole processing must be maintained. The temperature conditions during mechanical operations (10–20 °C) and storage (4–5 °C) need to be controlled. Refrigerated storage enables a reduction in physiological deterioration of fresh-cut fruit, which can then be marketable after 9 days (Gil *et al.* 2006). On the other hand, preservation methods and package conditions could also reduce the deleterious reactions triggered by processing operations and extend the shelf life of fresh-cut mango (Rojas-Graü *et al.* 2007).

8.3.1 Preservation Methods

In recent years, dipping treatments with antimicrobials, anti-browning agents, and texture preservatives as well as edible coating application, among other preservative methods, have been studied to assure the quality of minimally processed mango. The combination of different preservation methods could also be a feasible way to achieve better fresh-cut product quality. In this sense, the combination of dipping, edible coating, and physical treatments have been successful in achieving a higher reduction of microbial load in fresh-cut mango, keeping fresh-like quality parameters, thus leading to an increase in the shelf life of the product (Salinas-Roca *et al.* 2014).

8.3.1.1 Dipping Treatments

The incorporation of chemical agents in dipping solutions is the most common way to control deleterious impact of cutting such as browning. The antimicrobial action of organic acids and essential oils generally regarded as safe (GRAS) allow extending shelf life of fresh-cut fruits (Oms-Oliu *et al.* 2010). Organic acids can also be effective in preventing fresh-cut mango browning (González-Aguilar *et al.* 2000). The use of citric acid is widely extended to preserve color and the pH reduction in fresh-cut mango, which extends its shelf life by controlling microbial growth (Chiumarelli *et al.* 2011). Nevertheless, they observed a 9% weight loss in citric acid dipped fresh-cut mangoes. Also, vanillic acid has been tested on fresh-cut mango against spoilage microorganisms, it was shown that the use of 80 mM of vanillic acid delayed the growth of microorganisms by up to 14 days (Ngarmsak *et al.* 2006a). On the other hand, other chemicals such as calcium chloride can be incorporated into dipping treatments to maintain the turgidity of the cell membrane, allowing a texture similar to the fresh fruit throughout storage (Oms-Oliu *et al.* 2010). In addition, the combination of chemical agents could even enhance the final quality in fresh-cut mango. Siddiq *et al.* (2013) determined that ascorbic acid, citric acid, and calcium chloride at 1%, 1%, and 0.5%, respectively, maintained good color, firmness, and soluble solids content in fresh-cut mango during 12 days at 4 °C.

8.3.1.2 Edible Coatings

Edible coatings (EC) are materials such as polysaccharides, proteins. and lipids that are either applied to or formed directly on fresh-cut fruit surface. ECs used in minimally-processed mango can act as a barrier to external elements, protecting the fruit and reducing the changes to its characteristic aroma and weight. Common coatings used in fresh-cut mango are sodium alginate (Robles-Sánchez *et al.* 2013), carboxymethylcellulose, chitosan, and carrageenan (Plotto *et al.* 2010), and cassava starch (Chiumarelli and Hubinger 2012).

The use of sodium alginate in mango pieces using calcium chloride as a cross-linker maintains fresh-like characteristics, such as texture and color (Chiumarelli *et al.* 2011). Other ECs, such as chitosan, carrageenan, carboxymethylcelluloses, and pectin, can also preserve texture, color, pH, and acidity of mango pieces at concentrations between 0.5 and 2% (w/v) (Djioua *et al.* 2009; Plotto *et al.* 2010). Although no antimicrobial effects have been observed for most ECs, chitosan used as a coating in fresh-cut mango could reduce microbial growth. Hence, chitosan maintains both physicochemical and microbiological quality, extending the shelf life of mango pieces by up to 10 days (Campaniello *et al.* 2008; Chien *et al.* 2007). Souza *et al.* (2015) reported that chitosan-alginate nano-multilayer EC extended the shelf life of fresh-cut mangoes for up to 8 days. In order to assure high quality fresh-cut fruit, the incorporation of antimicrobial essential oils or organic acids in EC is recommended (Rojas-Graü *et al.* 2009).

Vanillin, oregano, and lemongrass are the most common antimicrobials incorporated as essential oils.

The use of EC for fresh-cut mango preservation can significantly increase the retention of bioactive compounds, because less deleterious reactions occur. In addition, coated fresh-cut mango has a barrier effect that avoids the deleterious reactions that reduce the content of polyphenols and carotenoids present in fresh mango (Robles-Sánchez *et al.* 2013; Rojas-Graü *et al.* 2009). Furthermore, other ingredients with health-related functions, such as vitamins, can be incorporated into EC too. Consequently, coated fresh-cut mango could have similar and even higher nutritional value than whole fruit.

8.3.1.3 Physical and Combined Treatments

The application of intense light pulses (ILP), which consists of the application of light (λ = 100–1,800 nm) on the fruit tissue surface, can assure safety and maintain fresh-like characteristics of the fruit, while avoiding the undesired effects caused by handling and processing in dipping treatments (González-Aguilar *et al.* 2007a,b). The ILP treatment has two main benefits, that is, the decontamination and the activation of the bioactive compounds pathway. The decontamination is attributed to a damage of the cell membrane of the microorganism, hence the growth of the microbial population is limited and the shelf life extended (Oms-Oliu *et al.* 2008). In this regard, the full spectrum light (λ = 100–1,800 nm) achieved a 4.5 log reduction of *L. innocua* population after 20 pulses in fresh-cut mango (González-Aguilar *et al.* 2007a,b). On the other hand, by using light at a lower wavelength (λ = 100–280 nm), the preservation of phenolic content in fresh-cut mango was improved from 0.34 mg/g fw at day 0 to 0.46 mg/g fw after 15 days (González-Aguilar *et al.* 2007a,b).

Ultrasound is another technology that has been explored as a post-processing treatment in fresh-cut mangoes. Santos *et al.* (2015) sonicated fresh-cut mangoes at 25 kHz for 30 min, followed by storage for 7 days at 4 °C. However, most of the quality attributes were impacted negatively by ultrasound treatment. Additional research is needed on the use of ultrasound technology for fresh-cut applications.

Salinas-Roca *et al.* (2016) studied the combined effect of pulsed light (PL), edible coating of alginate (ALC), and malic acid (MA) dipping on the safety and quality of fresh-cut mangoes. Microbial stability and quality parameters (color, pH, soluble solids, and firmness) of fresh-cut mango were examined throughout 14 days of storage at 4 °C. Results showed that MA-PL and PL-ALC-MA treatments additionally reduced *L. innocua* counts by additional 4.5 and 3.9 log cycles, respectively; and the microbial population remained less than 6 log CFU/g over 14 days. All the treatments showed a marked difference in the firmness values of fresh-cut mangoes, as compared to untreated samples.

8.3.1.4 Storage and Packaging

Fresh-cut mango can be affected by internal and external factors, as both respiration rate and storage conditions cause loss of quality and microbial growth throughout storage. On the one hand, the wounding effect triggers an increase in the respiration rate, but it can be regulated when fresh-cut mangoes are preserved with different treatments. Table 8.1 shows the effect of dipping and the use of edible coatings as preservation methods that enhances shelf life by reducing the respiration rate of fresh-cut mango by up to

Table 8.1 Respiration rate of fresh-cut mangoes subjected to different treatments.

Treatment	Respiration rate x 10^{-3} [mL CO_2/kg/s]
Control	2.41 ± 0.12
Citric acid	1.41 ± 0.15
Citric acid + cassava starch coating with glycerol	2.08 ± 0.03
Citric acid + alginate coating	1.68 ± 0.03
Citric acid + alginate coating with glycerol	1.85 ± 0.14

Source: Adapted from Chiumarelli *et al.* (2010, 2011).

58% compared with untreated mangoes (Chiumarelli *et al.* 2010, 2011). Temperature and atmosphere composition in fresh-cut mango packaging also affect the quality of fresh-cut mango during storage. Refrigeration temperatures can induce chilling injury in whole mango; however, temperatures from 5–10 °C in minimally processed mango can avoid deleterious reactions during 11 days of storage (Hodges and Toivonen 2008; Souza *et al.* 2004).

The atmosphere gases composition at appropriate levels can also contribute to a better preservation of minimally processed mango. Overall, the content of CO_2 tended to diminish during storage, although differences were observed, depending on the initial headspace atmosphere and packaging material (Martínez-Ferrer *et al.* 2002). Modified atmosphere packaging (MAP) is capable of inducing substantial reductions in respiration rates, oxidative stress, and tissue senescence, thus consequently an enhanced shelf life is achieved (Hodges and Toivonen 2008). A MAP system for mango pieces using an initial gas mixture of moderate to low O_2 and high CO_2 was shown to result in the longest shelf life compared to other treatments, including vacuum packing. Thus, microbial growth, texture, and color of fresh-cut mangoes were also maintained significantly better using MAP (González-Aguilar *et al.* 2000; Sothornvit and Rodsamran 2010). According to Sinha (2012), less quality deterioration is observed in minimally processed mango when O_2 and CO_2 values are 3–5% and 5–10%, respectively. In contrast, Rattanapanone *et al.* (2001) observed that the marketable period of mango cubes of Tommy Atkins and Kent varieties was extended when held in 4% O_2, 10% CO_2, and 86% N_2 atmosphere, but refrigeration temperature was the most influencing factor.

The fresh-cut mangoes have a high risk of microbial spoilage, since the fruit tissue surface is an appropriate growth substrate for spoilage microorganisms (Ngarmsak *et al.* 2006a,b). The microbial growth in fresh-cut mangoes could be minimized under refrigerated storage, thus the load of yeasts and molds remain stable at 2 $\log_{10}$ CFU/g throughout an 8-day storage at 5 °C (Rattanapanone *et al.* 2001). In this regard, the use of pre-cutting treatments, refrigeration temperatures, preservation methods, and/or MAP can retard the microbial growth, thus extending the fresh-cut mangoes' shelf life (Djioua *et al.* 2010). For instance, the use of HWD, as a pre-treatment on the whole mango, followed by a chitosan coating on fresh-cut mango, had a higher reduction of microbial load compared to their individual application (Table 8.2).

Table 8.2 Effect of hot water dipping and chitosan coating on aerobic psychrotrophic bacteria and mold and yeast count in fresh-cut mangoes after day 1 and 9 storage at 6 °C.

	Aerobic psychrotrophic bacteria (CFU g^{-1})		Molds and yeast (CFU g^{-1})	
Treatment	**Day 1**	**Day 9**	**Day 1**	**Day 9**
Control	$<1.0 \times 10^1$	2.5×10^6	2.5×10^2	6.0×10^2
Chitosan coating (CC)	1.5×10^2	9.0×10^5	1.0×10^2	2.0×10^2
Hot water dip (HWD)	$<1.0 \times 10^1$	7.2×10^5	$<1.0 \times 10^1$	1.8×10^4
HWD + CC	$<1.0 \times 10^1$	4.2×10^3	$<1.0 \times 10^1$	2.0×10^2

Source: Djioua *et al.* (2010).

8.4 Quality Attributes of Fresh-Cut Mangoes

8.4.1 Physical Attributes

8.4.1.1 Color

Mango fruit has a characteristic yellow color, although differences in the degree of yellowness can be observed among varieties. Studies in Tommy Atkins, Ataulfo, and Kent fresh-cut varieties described very similar L* values around 60 and b* values ranging from 30 to 40 (González-Aguilar *et al.* 2008). The browning phenomena, which promote the loss of characteristic color, appears when an increase of L* and a* parameters occur simultaneously with a decrease of b* values. For instance, in fresh-cut Ataulfo mango, a* and b* values changed from 4.02–6.59 and 69–49 after 15 days, respectively; hence, resulting in a loss of yellow color (Robles-Sánchez *et al.* 2009). Enzymatic browning, which is caused by enzymes liberated from tissues due to mechanical operations, is mainly caused by polyphenol oxidase (PPO) that, in the presence of oxygen and a substrate, converts phenolic compounds into dark-colored pigments (Martin-Belloso and Soliva-Fortuny 2010). Also, light and high oxygen availability has an effect on the oxidative reactions occurring on the surface of fresh-cut mango, leading to negative changes in color. Nevertheless, the characteristic yellowish color was maintained throughout 9 days in minimally processed mango under refrigerated storage (5 °C) (Gil *et al.* 2006; Robles-Sánchez *et al.* 2009).

The color of the fresh-cut produce, like mangoes, has a greater impact on the marketability of the product. From the consumer's perspective, color or appearance of fresh-cut produce is one of the most important quality attributes. If the color of a fresh-cut product is not attractive and of acceptable quality, the consumer is less likely to purchase it, regardless of its excellent flavor, taste, or other quality attributes (Siddiq *et al.* 2012).

8.4.1.2 Texture/Firmness

Mango has been described as a fibrous fruit with firmness values generally between 5 and 8 N (Salinas-Hernández *et al.* 2013). Texture is a quality attribute that changes

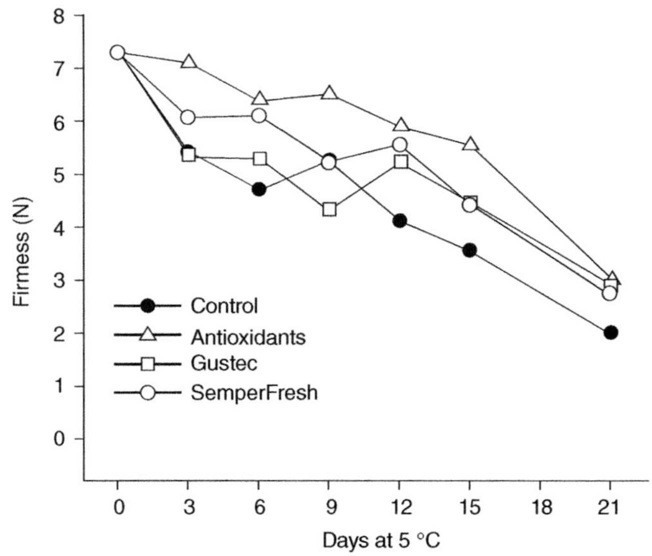

Figure 8.3 Firmness of fresh-cut Ataulfo mangoes treated with antioxidants or edible films during storage at 5 °C. (Antioxidants: 2 % ascorbic acid + 2 % citric acid; edible coatings: 2% Gustec or SemperFresh). *Source:* Gonzalez-Aguilar *et al.* (2008).

easily in fresh-cut mango because of wounding effects, which trigger the increase in enzymatic activity (González-Aguilar *et al.* 2000). Generally, fresh-cut mango firmness is reduced significantly during two weeks of storage, which is a typical storage period. It has been shown that pectinolytic and proteolytic enzymes may be the main factors affecting softening when they are delivered from bruised cells during slicing operations. The cutting operation increases the activity of pectin methylesterase, among other pectinolytic enzymes, which leads to a reduction in pectin content and loss of cell turgor (Skibsted *et al.* 2010). On the other hand, mechanical operations cause moisture losses in cut mango pieces, which also lead to loss of turgidity and an unacceptable appearance.

González-Aguilar *et al.* (2008) studied the effect of antioxidants, Gustec, and SemperFresh on fresh-cut mangoes (cv. Ataulfo, Kent, and Keitt) and reported that the softening effect can be delayed by the use of preservation treatments; the firmness results of fresh-cut Ataulfo mango are shown in Figure 8.3. Furthermore, the use of calcium dipping and low storage temperatures (5–10 °C) minimized the effects of mechanical injury, because they reduce enzymatic activity and cause water loss. Siddiq *et al.* (2013) studied the effect of different treatments on the quality attributes, including firmness, of fresh-cut Tommy Atkins during a 12-day storage at 4 °C. The combined treatment (1% ascorbic acid + 1% citric acid + 0.5% $CaCl_2$) was found most effective (Figure 8.4) in retaining the firmness of fresh-cut mango cubes. The control samples lost about 20% firmness at the end of 12-day storage.

8.4.1.3 Flavor

Naturally ripened mangoes have a characteristic taste and flavor. Munafo *et al.* (2014) reported that ethyl butanoate, 3-(methylthio)propanal and 4-hydroxy-2,5-dimethyl-3(2H)-furanone are the major volatile components in five different mango varieties.

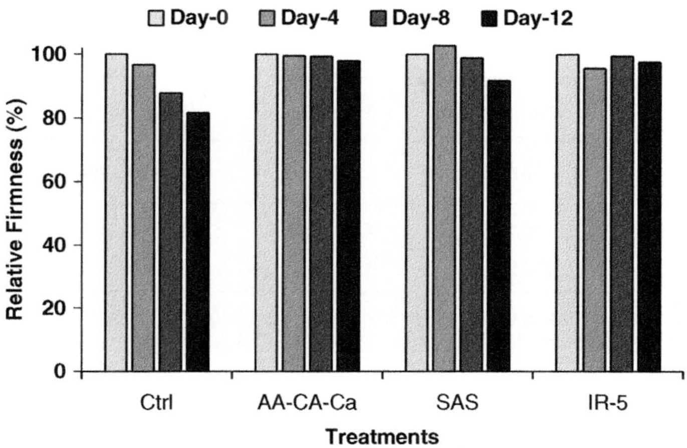

Figure 8.4 Effect of different treatments on the firmness of the fresh-cut Tommy Atkins mango cubes, stored at 4 °C. (Ctrl = Control; AA-CA-Ca = 1% ascorbic acid + 1% citric acid + 0.5% $CaCl_2$; SAS = 1% Sodium acid sulfate; IR-5 = Infrared heat, 5 min). *Source:* Adapted from Siddiq *et al.* (2013).

The flavor in fresh-cut mangoes could be affected by oxidative reactions leading to an off odor. Overall, fresh-cut mangoes retain a better aroma and appearance (closer to that of fresh mango) when the loss of water is diminished. The water can contain aromatic compounds, therefore the dehydration effect during inadequate storage results in the loss of the characteristic mango flavor. Plotto *et al.* (2010) observed an improvement of the sensory profile (sweetness, tartness, and fruity flavor) of fresh-cut mango with different preservation treatments, such as edible coatings containing organic acids. These researchers noted that the sweet taste was reduced after 5 days, although the fruity flavor was increased. Preservation treatments and adequate packaging help in maintaining the volatile compounds that give the characteristic mango flavor.

8.4.2 Bioactive Compounds

Mango contains phenolic compounds, carotenoids, and ascorbic acid that have antioxidant capacity and, therefore, they are a powerful defense against degenerative oxidative diseases in the fruit (Watson and Preedy 2009). Nutritional composition of mangoes varies with cultivar, harvest conditions, maturity index, postharvest handling, and processing; during minimal processing, a decrease of antioxidant bioactive content can occur due to the wounding response and other deleterious reactions (Sinha *et al.* 2012). Although there is a slight loss of the content in bioactive compounds when comparing whole and fresh-cut mangoes, processed mango remains a rich source of polyphenolic compounds, carotenoids, and vitamins. Table 8.3 shows the content of total phenols, β-carotene, vitamin C, and vitamin E in whole and fresh-cut mango stored at 5 °C for 10 days (Robles-Sánchez *et al.* 2011; Sarkiyayi 2013).

8.4.2.1 Phenolic Compounds

Phenolic compounds make an important contribution to the antioxidant capacity of mango (Sogi *et al.* 2012). Mangiferin, gallic acid, quercetin, ellagic acid, and

Table 8.3 Total phenols), β- carotene, vitamin C and vitamin E in whole and fresh-cut mangoes stored at 5 °C for 10 days.

	Total phenolics	β-carotene	Vitamin C	Vitamin E
Samples	mgGAE/100g FW		mg/100g FW	
Whole mango	110.7 ± 1.1a	4.47 ± 0.31b	93.5 ± 4.5b	0.92 ± 0.02a
Fresh-cut mango	116.0 ± 1.2b	3.03 ± 0.08a	86.7 ± 6.3a	0.93 ± 0.03a

Values in columns with different letters (a,b,c) in columns differ significantly ($p \leq 0.05$).
Source: Adapted from Robles-Sánchez *et al.* (2011).

β-glucogallin are among the polyphenolic compounds identified in mango flesh (Masibo and He 2008). Phenolic compounds are affected by peeling, and cutting can trigger their degradation. Wounding response stimulates two main enzymes: phenylalanine ammonia-lyase (PAL) and PPO, which are directly involved in phenolic compounds pathways (Martin-Belloso and Soliva-Fortuny 2010). The PAL activation can elicit reactive oxygen species (ROS) with further production of phenolic compounds. On the other hand, the decompartmentalization of fruit membrane cells allows the oxidation of phenolic compounds by PPO, thus browning of cut mango surface can occur. Moreover, phenolics are degraded by exposure to light, oxygen, and high temperatures during storage.

Generally, an increase in the total phenolics concentration in fresh-cut mango compared to whole fruit is observed, which is due to the release of the compounds from the cells after mechanical operations (Table 8.3). Nevertheless, during the processing and storage, phenolic compounds loss can occur. Preservation techniques can maintain phenolic compounds concentration in fresh-cut mango after 10 days under refrigeration (Robles-Sánchez *et al.* 2011). However, fresh-cut mango showed a slight decrease from 11 to 9 mg/100 g, fresh-wt basis, in the total phenolics content from day 0 to day 3 respectively, but the content was maintained thereafter for 9 days under refrigerated storage (Gil *et al.* 2006).

8.4.2.2 Carotenoids

The predominant carotenoid in mango is β-carotene, which is affected by variety, ripening, cutting processing, and storage conditions (Mercadante and Rodriguez-Amaya 1998; Pierson *et al.* 2014). Mango ripening produces an increase in β-carotene content, which is more significant at room temperature than under refrigeration, and may be due to an increase in mevalonic acid and geraniol syntheses, which lead to higher levels of total carotenes (González-Aguilar *et al.* 2008). Therefore, minimally processed mango preserves the original carotenoid concentration between 4 and 5 mg/100 g, fresh-wt, for at least 5 days at 5 °C (Robles-Sánchez 2009), although other studies have shown longer stability of up to 2 weeks (Gil *et al.* 2006; Robles-Sánchez *et al.* 2011). The loss of carotenoids during storage is typically influenced by temperature and oxygen concentration within the package, variables that allow the acceleration of this metabolism.

8.4.2.3 Vitamins

Vitamin C is present in high concentration in fresh-cut mango. The concentration in ready-to-eat Ataulfo mango (75 mg/100 g, fresh-wt) does not differ significantly from the fresh flesh (77 mg/100 g, fresh-wt) during the first week of storage (Gil *et al.* 2006). Nevertheless, a decrease, which could be promoted by processing, has been observed during storage of 15–20 days at 5–10 °C by Robles-Sánchez *et al.* (2009). In addition, vitamin C, which is water-soluble, is related to water content, an important factor in the retention of nutritional compounds. Thus, after mechanical operations, the decrease of water content and the oxidation reactions in fresh-cut mango experience significant losses of vitamin C.

Robles-Sánchez *et al.* (2009) reported 1.3 mg/100 g (fresh-wt) of vitamin E in fresh-cut mango, which is mainly present in the α-tocopherol form. Vitamin E can easily suffer lipid oxidation by light effects throughout storage (Robles-Sánchez *et al.* 2009). Generally, the incorporation of antioxidant compounds in the surface treatment of fresh-cut mango is used to prevent loss of hydrophilic and lipophilic vitamins.

8.5 Conclusions and Future Trends

Minimally processed mangoes undergo physical transformation when they are peeled, cut, and the seed is removed. Consequently, physiological cascade effects, which trigger deleterious reactions in the fruit matrix, can occur. The application of technologies such as physical treatments and edible coatings has been observed to improve quality in terms of microbiological, nutritional, and physicochemical parameters. In this sense, the increase in bioactive compounds concentration and firmness values in coated fresh-cut mangoes is remarkable.

Although high quality of fresh-cut products is obtained when appropriate preservation techniques are used, the consumption of fresh-cut mangoes among the consumers remains relatively low. Two of the main obstacles are production costs and transportation challenges. Therefore, packaging and careful handling during storage and transportation is key to assure marketable fresh-cut mangoes. Therefore, different hazards from processing to consumers should be addressed to maintain quality of minimally processed mango as fresh-like as possible. A combination of emerging technologies allows synergic or additive effects and, as a result, enhances the product's shelf life. Consequently, future research should focus on the development of other effective strategies, such as the application of combined techniques such as intense light pulses and/or edible coatings, to incorporate high quality fresh-cut mango in the worldwide market. Likewise, research on appropriate packaging materials and atmosphere gases composition for fresh-cut mango are also a concern. In addition, the emerging preservation methods for minimally processed products should be investigated for different mango varieties to potentially expand the fresh-cut mangoes consumption.

Acknowledgments

This work was supported by the University of Lleida (Spain) and financed by the Tecnológico de Monterrey, Monterrey, Mexico.

References

Campaniello, D., Bevilacqua, A., Sinigaglia, M. and Corbo, M.R. (2008) Chitosan: Antimicrobial activity and potential applications for preserving minimally processed strawberries. *Food Microbiology*, **25**, 92–1000.

Chien, P.-J., Sheu, F. and Yang, F.-H. (2007) Effects of edible chitosan coating on quality and shelf life of sliced mango fruit. *Journal of Food Engineering*, **78**, 225–229.

Chiumarelli, M. and Hubinger, M.D. (2012) Stability, solubility, mechanical and barrier properties of cassava starch – Carnauba wax edible coatings to preserve fresh-cut apples. *Food Hydrocolloids*, **28**, 59–67.

Chiumarelli, M., Pereira, L.M., Ferrari, C.C., Sarantopoulos, C.I. and Hubinger, M.D. (2010) Cassava starch coating and citric acid to preserve quality parameters of fresh-cut "Tommy Atkins" Mango. *Journal of Food Science*, **75**, E297–E304.

Chiumarelli, M., Ferrari, C.C., Sarantópoulos, C.I. and Hubinger, M.D. (2011) Fresh-cut Tommy Atkins mango pre-treated with citric acid and coated with cassava (*Manihot esculenta* Crantz) starch or sodium alginate. *Innovative Food Science and Emerging Technologies*, **12**, 381–387.

Djioua, T., Charles, F., Lopez-Lauri, F., Filgueiras, H., Coudret, A. *et al.* (2009) Improving the storage of minimally processed mangoes (*Mangifera indica* L.) by hot water treatments. *Postharvest Biology and Technology*, **52**, 221–226.

Djioua, T., Charles, F., Freire, M., Filgueiras, H., Ducamp-Collin, M.N. and Sallanon, H. (2010) Combined effects of postharvest heat treatment and chitosan coating on quality of fresh-cut mangoes (*Mangifera indica* L.). *International Journal of Food Science and Technology*, **45**, 849–855.

Gil, M.I., Aguayo, E. and Kader, A.A. (2006) Quality changes and nutrient retention in fresh-cut versus whole fruits during storage. *Journal of Agricultural and Food Chemistry*, **54**, 4284–4296.

González-Aguilar, G.A., Wang, C.Y. and Buta, J.G. (2000) Maintaining quality of fresh-cut mangoes using antibrowning agents and modified atmosphere packaging. *Journal of Agricultural and Food Chemistry*, **48**, 4204–4208.

González-Aguilar, G.A., Wang, C.Y., Buta, J.G. and Krizek, D.T. (2001) Use of UV-C irradiation to prevent decay and maintain postharvest quality of ripe "Tommy Atkins" mangoes. *International Journal of Food Science and Technology*, **36**, 767–773.

González-Aguilar, G.A., Villegas-Ochoa, M.A., Martínez-Téllez, M.A., Gardea, A.A. and Ayala-Zavala, J.F. (2007a) Improving antioxidant capacity of fresh-cut mangoes treated with UV-C. *Journal of Food Science*, **72**, S197–S202.

González-Aguilar, G.A., Zavaleta-Gatica, R. and Tiznado-Hernández, M.E. (2007b) Improving postharvest quality of mango "Haden" by UV-C treatment. *Postharvest Biology and Technology*, **45**, 108–116.

Gonzalez-Aguilar, G.A., Celis, J., Sotelo-Mundo, R.R., De La Rosa, L.A., Rodrigo-Garcia, J. and Alvarez-Parrilla, E. (2008) Physiological and biochemical changes of different fresh-cut mango cultivars stored at 5°C. *International Journal of Food Science and Technology*, **43**, 91–101.

González-Aguilar, G.A., Villa-Rodriguez, J.A., Ayala-Zavala, J.F. and Yahia, E.M. (2010) Improvement of the antioxidant status of tropical fruits as a secondary response to some postharvest treatments. *Trends in Food Science and Technology*, **21**, 475–482.

Hodges, D.M. and Toivonen, P.M.A. (2008) Quality of fresh-cut fruits and vegetables as affected by exposure to abiotic stress. *Postharvest Biology and Technology*, **48**, 155–162.

Martin-Belloso, O. and Soliva-Fortuny, R.S. (eds) (2010) *Advances in Fresh-Cut Fruits and Vegetables Processing*. CRC Press, Boca Raton, FL, 424 p.

Martínez-Ferrer, M., Harper, C., Pérez-Muntoz, F. and Chaparro, M. (2002) Modified atmosphere packaging of minimally processed mango and pineapple fruits. *Journal of Food Science*, **67**, 3365–3371.

Masibo, M. and He, Q. (2008) Major mango polyphenols and their potential significance to human health. *Comprehensive Reviews in Food Science and Food Safety*, **7**, 309–319.

Mercadante, A.Z. and Rodriguez-Amaya, D.B. (1998) Effects of ripening, cultivar differences, and processing on the carotenoid composition of mango. *Journal of Agricultural and Food Chemistry*, **46**, 128–130.

Moreno, M.A., Castell-Perez, M.E., Gomes, C., Da Silva, P.F., Kim, J. and Moreira, R.G. (2007) Optimizing electron beam irradiation of "Tommy Atkins" mangoes (*Mangifera Indica* L.). *Journal of Food Process Engineering*, **30**, 436–457.

Munafo Jr J.P., Didzbalis, J., Schnell, R.J., Schieberle, P. and Steinhaus, M. (2014) Characterization of the major aroma-active compounds in mango (*Mangifera indica* L.) cultivars Haden, White Alfonso, Praya Sowoy, Royal Special, and Malindi by application of a comparative aroma extract dilution analysis. *Journal of Agricultural and Food Chemistry*, **62**, 4544–4551.

Narayana, C.K., Rao, D.S. and Roy, S.K. (2012) Mango production, postharvest physiology and storage. In: *Tropical and Subtropical Fruits: Postharvest Physiology, Processing and Packaging* (ed. M. Siddiq), pp. 259–276. John Wiley & Sons, Oxford.

Ngarmsak, M., Delaquis, P., Toivonen, P., Ngarmsak, T., Ooraikul, B. and Mazza, G. (2006a) Antimicrobial activity of vanillin against spoilage microorganisms in stored fresh-cut mangoes. *Journal of Food Protection*, **69**, 1724–1727.

Ngarmsak, M., Delaquis, P., Toivonen, P., Ngarmsak, T., Ooraikul, B. and Mazza, G. (2006b) Microbiology of fresh-cut mangoes prepared from fruit sanitised in hot chlorinated water. *Food Science and Technology International*, **12**, 95–102.

Ohlsson, T. and Bengtsson, N. (2002) *Minimal Processing Technologies in the Food Industry*. Woodhead Publishing, Cambridge, 288 p.

Oms-Oliu, G., Martín-Belloso, O. and Soliva-Fortuny, R. (2008) Pulsed light treatments for food preservation: A review. *Food and Bioprocess Technology*, **3**, 13–23.

Oms-Oliu, G., Rojas-Graü, M.A., Gonzalez, L.A., Varela, P., Soliva-Fortuny, R. *et al.* (2010) Recent approaches using chemical treatments to preserve quality of fresh-cut fruit: A review. *Postharvest Biology and Technology*, **57**, 139–148.

Pierson, J.T., Monteith, G.R., Roberts-Thomson, S.J., Dietzgen, R.G., Gidley, M.J. and Shaw, P.N. (2014) Phytochemical extraction, characterisation and comparative distribution across four mango (Mangifera indica L.) fruit varieties. *Food Chemistry*, **149**, 253–63.

Plotto, A., Narciso, J.A., Rattanapanone, N. and Baldwin, E.A. (2010) Surface treatments and coatings to maintain fresh-cut mango quality in storage. *Journal of the Science of Food and Agriculture*, **90**, 2333–2341.

Rattanapanone, N., Lee, Y., Wu, T. and Watada, A.E. (2001) Quality and microbial changes of fresh-cut mango cubes held in controlled atmosphere. *HortScience*, **36**, 1091–1095.

Ribeiro, S.M.R., Barbosa, L.C.A., Queiroz, J.H., Knödler, M. and Schieber, A. (2008) Phenolic compounds and antioxidant capacity of Brazilian mango (*Mangifera indica* L.) varieties. *Food Chemistry*, **110**, 620–626.

Robles-Sánchez, R.M., Islas-Osuna, M.A., Astiazarán-García, H., Vázquez-Ortiz, F.A., Martín-Belloso, O. *et al.* (2009) Quality index, consumer acceptability, bioactive compounds, and antioxidant activity of fresh-cut "Ataulfo" mangoes (*Mangifera Indica* L.) as affected by low-temperature storage. *Journal of Food Science*, **74**, S126–S134.

Robles-Sánchez, M., Astiazarán-García, H., Martín-Belloso, O., Gorinstein, S., Alvarez-Parrilla, E. *et al.* (2011) Influence of whole and fresh-cut mango intake on plasma lipids and antioxidant capacity of healthy adults. *Food Research International*, **44**, 1386–1391.

Robles-Sánchez, R.M., Rojas-Graü, M.A., Odriozola-Serrano, I., González-Aguilar, G. and Martin-Belloso, O. (2013) Influence of alginate-based edible coating as carrier of antibrowning agents on bioactive compounds and antioxidant activity in fresh-cut Kent mangoes. *LWT - Food Science and Technology*, **50**, 240–246.

Rojas-Graü, M.A., Tapia, M.S., Rodríguez, F.J., Carmona, A.J. and Martin-Belloso, O. (2007) Alginate and gellan-based edible coatings as carriers of antibrowning agents applied on fresh-cut Fuji apples. *Food Hydrocolloids*, **21**, 118–127.

Rojas-Graü, M.A., Soliva-Fortuny, R. and Martín-Belloso, O. (2009) Edible coatings to incorporate active ingredients to fresh-cut fruits: A review. *Trends in Food Science and Technology*, **20**, 438–447.

Salinas-Hernández, R.M., González-Aguilar, G.A. and Tiznado-Hernández, M.E. (2013) Utilization of physicochemical variables developed from changes in sensory attributes and consumer acceptability to predict the shelf life of fresh-cut mango fruit. *Journal of Food Science and Technology*, **52**, 63–77.

Salinas-Roca, B., Soliva-Fortuny, R., Welti-Chanes, J., Martín-Belloso, O. (2014) Quality properties of coated mango slices treated by intense light pulses. *EFFoST 28th Conference and 7th International Conference on the Food Factory for the Future.* Uppsala, Sweden.

Salinas-Roca, B., Soliva-Fortuny, R., Welti-Chanes, J. and Martín-Belloso, O. (2016) Combined effect of pulsed light, edible coating and malic acid dipping to improve fresh-cut mango safety and quality. *Food Control*, **66**, 190–197.

Santos, J.G., Fernandes, F.A.N., de Siqueira Oliveira, L. and de Miranda, M.R.A. (2015) Influence of ultrasound on fresh-cut mango quality through evaluation of enzymatic and oxidative metabolism. *Food and Bioprocess Technology*, **8**, 1532–1542.

Sarkiyayi, S., Mohammed, M. and Yakubu, A. (2013) Comparative analysis of nutritional and anti-nutritional contents of some varieties of mango (*Mangifera indica*) in Kaduna Metropolis-Nigeria. *Research Journal of Applied Sciences, Engineering and Technology*, **5**, 387–391.

Siddiq, M. (ed.) (2012) *Tropical and Subtropical Fruits: Postharvest Physiology, Processing and Packaging.* John Wiley & Sons, Oxford, 648 p.

Siddiq, M., Liavaga, A. and Greiby, I. (2012) Peaches and nectarines. In: *Handbook of fruits and fruit processing* (eds N. Sinha, J. Sidhu, J. Barta, J. Wu and M. Cano), pp. 535–549. Blackwell Publishing Co: Ames, IA.

Siddiq, M., Sogi, D.S. and Dolan, K.D. (2013) Antioxidant properties, total phenolics, and quality of fresh-cut "Tommy Atkins" mangoes as affected by different pre-treatments. *LWT–Food Science and Technology*, **53**, 156–162.

Sinha, N., Hui, Y.H., Evranuz, E.Ö., Siddiq, M. and Ahmed, J. (2010) *Handbook of Vegetables and Vegetable Processing.* John Wiley & Sons. Ames, IA. 788 p.

Sinha, N., Sidhu, J., Barta, J., Wu, J. and Cano, M.P. (eds) (2012) *Handbook of Fruits and Fruit Processing.* John Wiley & Sons. Ames, IA, 712 p.

Skibsted, L.H., Risbo, J. and Andersen, M.L. (eds) (2010) *Chemical Deterioration and Physical Instability of Food and Beverages.* Woodhead Publishing, Cambridge, 824 p.

Sogi, D.S., Siddiq, M., Roidoung, S. and Dolan, K.D. (2012) Total phenolics, carotenoids, ascorbic acid, and antioxidant properties of fresh-cut mango (*Mangifera indica* L., cv. Tommy Atkins) as affected by infrared heat treatment. *Journal of Food Science*, **77**, C1197–C1202.

Sothornvit, R. and Rodsamran, P. (2010) Mango film coated for fresh-cut mango in modified atmosphere packaging. *International Journal of Food Science and Technology*, **45**, 1689–1695.

Souza, B.S., Durigan, J.F., Donadon, J.R., Teixeira, G.H.A. and Durigan, M.F.B. (2004) Respiratory and storage behavior of fresh cut "Tommy Atkins" mango. *Acta Horticulturae*, **682**, 1909–1916.

Souza, M.P., Vaz, A.F., Cerqueira, M.A., Texeira, J.A., Vicente, A.A. and Carneiro-da-Cunha, M.G. (2015) Effect of an edible nano-multilayer coating by electrostatic self-assembly on the shelf life of fresh-cut mangoes. *Food and Bioprocess Technology*, **8**, 647–654.

Watson, R. and Preedy, V. (eds) (2009) *Bioactive Foods in Promoting Health: Fruits and Vegetables.* Academic Press, London, 754 p.

9

Innovative Processing Technologies for Mango Products

Deepti Salvi, Ender Arserim and Mukund Karwe

Department of Food Science Rutgers, The State University of New Jersey, New Brunswick, New Jersey, USA

9.1 Introduction

Mango is a commercially important tropical fruit native to southern Asia, especially Eastern India, Burma, and the Andaman Islands (de la Cruz Medina and García 2002). Mango is cultivated mainly in Asia and Pacific, Latin America, the Caribbean, and Africa. India is the world's largest mango-producing nation, accounting for over 40% of the total global output (Kaushik *et al.* 2014).

In recent years, there has been a growing demand for mango in European and US markets (Jha *et al.* 2010). Processing mango into different shelf stable product provides a solution to the problem of shelf life, avoids spoilage related losses, and reduces cost of transportation and storage. Processed mango products are gaining in popularity for export in various forms such as mango pulp, juice, nectar, dried mango slices, fruit drink, jam, squash, etc. Conventional thermal processing, such as blanching, pasteurization, sterilization, and thermal drying, are the most common methods used to process the majority of mango products (Rawson *et al.* 2011). Due to induced chemical and physical changes during thermal processing, deterioration of nutritional and sensory properties of the products is common (Chen *et al.* 2007; Kim *et al.* 2009; Vasquez-Caicedo *et al.* 2007). These problems can be minimized by using mild or minimal-processing technologies. Moreover, consumers also prefer minimally processed food with fresh-like quality attributes.

High pressure processing (HPP), irradiation, pulsed electric field (PEF), ohmic heating, power ultrasound, ozone, ultraviolet-C (UV-C), oscillating magnetic fields (OMF), and cold atmospheric plasma (CAPP) are some of the novel processing technologies which have gained popularity due to their ability to achieve food safety without compromising nutrition and fresh-like sensory properties of the food. This chapter reviews innovative technologies for mango processing, and the use of novel processing technologies such as postharvest quarantine methods for insect/pest control in mangoes, and non-destructive methods for quality inspection of mango fruit.

Handbook of Mango Fruit: Production, Postharvest Science, Processing Technology and Nutrition,
First Edition. Edited by Muhammad Siddiq, Jeffrey K. Brecht and Jiwan S. Sidhu.

9.2 High Pressure Processing (HPP)

During HPP, the food experiences instantaneous and uniform pressure in the range of 100–900 MPa with or without application of heat, which results in microbial and enzymatic inactivation (Kaushik, *et al.* 2014; Liu *et al.* 2014b). Pressures above 200 MPa are adequate to inactivate vegetative bacteria, yeast, and molds, while pressures above 300 MPa can denature some of the enzymes (Terefe *et al.* 2014). For bacterial spores which are more pressure resistant, a process involving high pressure combined with moderate temperature (pressure assisted thermal sterilization (PATS) was approved by the US Food and Drug Administration (USFDA) in 2009 (Bermúdez-Aguirre and Barbosa-Cánovas 2011). Quality degrading enzymes, such as polygalacturonase (PG) and lipoxygenase (LOX), can be substantially inactivated by HPP, while pectin methylesterase (PME), polyphenol oxidase (PPO), and peroxidase (POD) are partially inactivated by HPP. Mild heat along with high pressure can be used to increase the efficiency of inactivation of these enzymes (Terefe *et al.* 2014).

HPP is being used commercially, though on a limited scale, for processing of fruit and vegetable juices, nectars, and purees and these products are gaining higher acceptance among consumers. HPP use has been reported for a number of mango products: nectars (Liu *et al.* 2013, 2014a,b), pulps (Ahmed *et al.* 2003, 2005; Liu *et al.* 2013, 2014c; Kaushik *et al.* 2014), puree (Lemmens *et al.* 2013), and pre-cut mango slices (Boynton *et al.* 2002). These studies mainly focused on the effects of HPP on microbial and enzyme inactivation, physical, nutritional, and sensory quality attributes.

9.2.1 Effect of HPP on Nutrients

Mango fruit is rich in carotenoids, flavonoids, phenolic compounds, and vitamins. Kaushik *et al.* (2014) reported that HPP-treated (100–600 MPa, 5–20 min) mango pulp retained 85% ascorbic acid, 92% total phenolic, and 90% *in-vitro* antioxidant capacity compared to untreated samples. This study did not compare their results with thermally processed puree. In another study, β-carotene isomerization in high pressure homogenized mango puree was compared with thermally processed puree (Lemmens *et al.* 2013). High pressure homogenization was carried out in the pressure range of 30–30 MPa. The study showed that high pressure homogenization produced negligible β-carotene isomerization, compared to both raw (untreated) and thermally processed mango puree. It is to be noted that if during processing β-carotene is converted to its cis-isomers, its health benefits are negatively affected. Liu *et al.* (2014a) reported that no significant changes in the antioxidant capacity and concentrations of L-ascorbic acid, total phenols, and total carotenoids of mango nectar were observed when HPP (600 MPa/1 min) and high temperature short time (HTST, 110 °C/8.6 s) treatments were compared.

9.2.2 Effect of HPP on Microbial Inactivation

Resistance to pressure inactivation of microorganisms, termed as baro-resistance, varies depending on the strain type as well as external factors, including processing conditions (pressure level, dwell time, processing temperature, number of pressure cycles, rate of compression and decompression, etc.), and properties of the surrounding media (i.e., pH, sugar content, water activity, and inherent microflora) (Kaushik *et al.*

Table 9.1 Estimated pressure sensitivity (z_p) and change in activation volume (ΔV) of natural microflora in high pressure processed mango pulp.

Group of microorganisms	Z_p value (MPa)	ΔV ($\times 10^{-6}$ m^3 mol^{-1})	R^2
Aerobic mesophiles	593.82 ± 12.65	−9.77 ± 0.32	0.963
Yeast and mold	630.52 ± 21.97	−9.20 ± 0.14	0.971
Coliforms	306.75 ± 10.20	−18.92 ± 0.24	0.998
Lactic acid bacteria	388.50 ± 15.32	−14.94 ± 1.02	0.965
Psychrotrophs	479.80 ± 11.71	−12.10 ± 0.80	0.979

Source: Kaushik *et al.* (2014)

2014). Although membrane damage is one suggested mechanism for microbial inactivation during HPP, many other mechanisms such as ATPase inactivation, macromolecular trans-conformations, inactivation of nucleic material, ionic dissociation, and pH changes in water, could be related to cell death (Karwe *et al.* 2015).

Microbial kinetics during HPP were studied by Kaushik *et al.* (2014) for destruction of natural microflora in mango pulp based on the dual mechanism of pulse pressure effect and isobaric inactivation. Pulse pressure effect or instantaneous pressure kill (pressure hold time of the order of 1 sec) was considered responsible for partial inactivation of microorganisms due to rapid change in pressure. Pulse pressure effect was calculated using the difference between the log of initial microbial count and the log of surviving cells after an HPP pulse (pressure hold time of 1 sec). First-order kinetics was used to calculate decimal reduction time (D_p value, min). The Eyring model is given below for calculating pressure sensitivity (z_p value, MPa) and activation volume (ΔV, m^3/mol) was used to understand the effect of pressure on inactivation of aerobic mesophiles, yeast and mold, coliforms, lactic acid bacteria, and psychrotrophs (Table 9.1).

$$\Delta V = -RT\left[\frac{\Delta(\ln K)}{\Delta P}\right] \tag{9.1}$$

where P = pressure (MPa), K = microbial death rate constant (min^{-1}), T = absolute temperature (303 K), R = gas constant (8.314×10^{-6} m^3 mol^{-1} MPa^{-1} K^{-1}), and ΔV = change in activation volume (m^3/mol).

Pressure treatment of 600 MPa for 5 min was enough to decrease the microbial counts of indigenous microflora to less than 10 CFU/g, achieving 4.6 log CFU/g reduction. Based on the z_p value, pressure sensitivity of coliforms was found to be the maximum and the yeast and the mold had to be the least sensitivity to pressure.

Liu *et al.* (2013) reported complete inactivation of aerobic bacteria (~10^5 CFU/ml), and of yeast and mold (~10^4 CFU/ml) in HPP-treated mango pulp at higher pressures above 400 MPa and hold times of more than 5 min. Boynton *et al.* (2002) found that HPP-treated pre-cut and peeled Tommy Atkins mango cubes showed lower microbial counts (total plate count: 10^2 CFU/ml–10^3 CFU/ml) during storage at 3°C compared to untreated slices.

HPP treatment (600 MPa for 1 min) of mango nectar (water-based beverage flavored by the mango pulp) resulted in comparable total aerobic bacterial and yeast and mold counts with HTST (110 °C/8.6 s) (Liu *et al.* 2014b). The total aerobic bacterial count was reduced from 6.81 log CFU/ml to 1.57 log CFU/ml and 1.11 log CFU/ml after HPP and HTST treatments respectively, and it did not change significantly during 16 weeks of storage at 4 °C and 25 °C. The yeast and mold counts reduced from 3.16 log CFU/ml to less than 1.00 log CFU/ml, and did not change during storage.

9.2.3 Effect of HPP on Enzymes

Active enzymes during storage are the main cause in deteriorating the quality of fruit and fruit products. HPP has shown to inactivate the many quality degrading enzymes at room temperature, while preserving nutritional and sensory qualities of the product (Terefe *et al.* 2014). In a study by Liu *et al.* (2013), the residual activity of enzyme PME was studied for HPP treatment of blanched and unblanched mango pulp and was compared with HTST treatment (Figure 9.1). The residual activity of PME was considered 100% in the control sample of unblanched mango pulp, and for the blanched mango pulp it was 15.2%. They found that increase in pressures (300–400 MPa) significantly activated PME for both blanched and unblanched mango pulp, and at pressure 500–600 MPa it was reduced. For HTST treatment, the PME activity was reduced and the PME was completely inactivated in unblanched and blanched HPP-treated mango pulp, respectively.

PPO enzyme, responsible for the biochemical reactions leading to discoloration and browning of plant tissue, was found to be relatively less pressure stable (Terefe *et al.* 2014). Guerrero-Beltrán *et al.* (2005) studied residual PPO activity in HPP-treated mango puree containing anti-browning agents (ascorbic acid) and found improved inhibition of PPO with no noticeable color difference between HPP-treated mango puree and control at the end of a 30-day storage period.

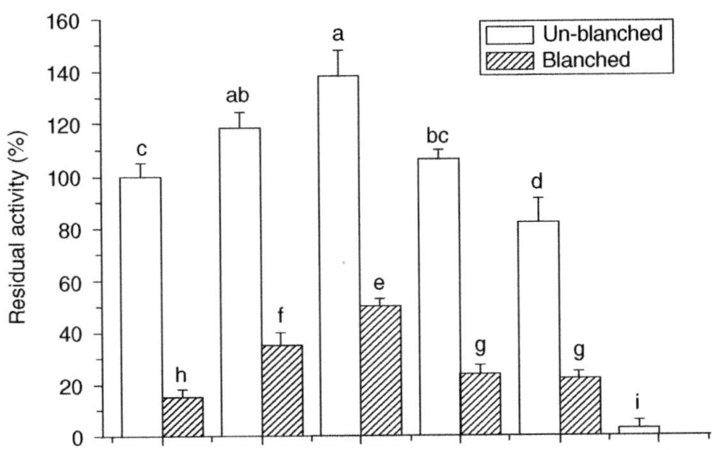

Figure 9.1 Effects of high hydrostatic pressure and high-temperature, short-time treatments on pectin methylesterase in unblanched and blanched mango pulp. *Source:* Liu *et al.* (2013).

9.2.4 Effect of HPP on pH, Acidity, and Soluble Solids

Liu *et al.* (2014) reported no change in pH and titratable acidity (TA) immediately after HPP treatment of mango nectars, but showed a decrease in pH and an increase in the TA after 16 weeks of storage at 4 °C and 25 °C. Total soluble solids (TSS) did not change significantly after HPP treatment compared to untreated mango nectar. Kaushik *et al.* (2014) reported a slight increase in TSS, decrease in pH, and increased TA of the HPP-treated mango pulp compared to untreated. Pressure levels affected these values, while dwell time did not show any impact on pH and TA.

9.2.5 Effect of HPP on Rheological Properties

It has been shown that HPP treatment affects the rheology of food products (Ahmed *et al.* 2003; 2005; Liu *et al.* 2014; Walker *et al.* 2006). Ahmed *et al.* (2005) studied effects of HPP (100–400 MPa, 15 and 30 min, at 20 °C) on the rheological properties of canned mango pulp from Alphonso and fresh Chousa mango pulp. Various rheological models (Newtonian, Power Law, Bingham, Casson, Herschel Bulkley) were tested using measured values of shear stress and shear rate. The Herschel Bulkley model (Equation 9.2) was found to be the best fit and was used for rheological characterization of the mango pulp, as given below:

$$\sigma = \sigma_0 + K(\gamma)^n \tag{9.2}$$

where σ is the shear stress (Pa), σ_0 is the yield stress (Pa), γ is the shear rate (s^{-1}), K is the consistency (Pa s^n), and n is the flow behavior index.

Pseudoplasticity with yield stress, which is the one the major quality characteristic of semi-solids, was reported also for both pulps. For Alphonso mango pulp, the shear stress values of HPP-treated mango pulp at any given shear rate were higher than untreated samples. Whereas, for Chousa mango pulp, the shear stress values of HPP-treated mango pulp at any given shear rate were lower than the control (no HPP treatment), as shown in Figure 9.2. This difference was potentially attributed to changes in the rheological properties due to thermal treatment of Alphonso mango pulp (canned) prior to HPP.

Similarly, Lui *et al.* (2013) reported significantly higher shear stress of unblanched and blanched HPP-treated mango pulps (from Southern China mangoes) at any shear rate compared to the control, as seen with the higher consistency coefficient values (K) reported in Table 9.2. These authors also reported pseudoplastic properties ($0 < n < 1$) for mango pulp with yield stress. No considerable effect of different pressure magnitudes (300–500 MPa) was observed for viscosity increase for both unblanched and blanched HPP-treated mango pulps. It was stated that HPP treatment increased the viscosity of unblanched and blanched HPP-treated mango pulp at lower pressure due to increased PME activity. In contrast, for unblanched HPP-treated mango pulp at higher pressure (600 MPa), protein or pectin coagulation and enhanced water binding capacity of pectin were the reasons reported for increased viscosity (Lui *et al.* 2013).

In another study by Liu *et al.* (2014b), HPP treatment (600 MPa, 1 min) of mango nectar (a water-based beverage flavored from mango pulp) did not show any change in the shear rate compared to the untreated sample, while an increase in viscosity was observed in HTST-treated mango nectar.

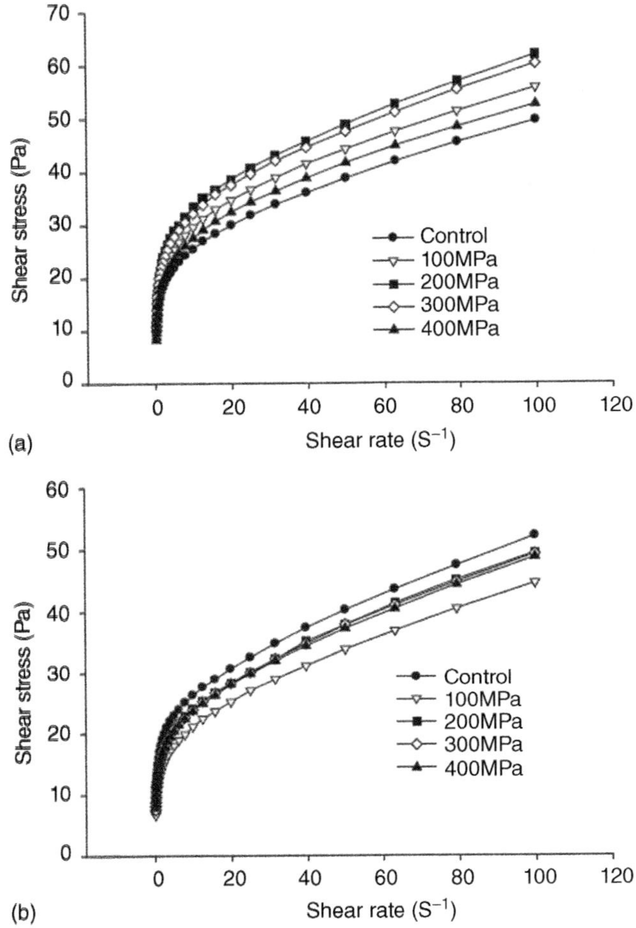

Figure 9.2 Effect of high-pressure on rheograms of Chousa (a) and Alphonso (b) mango pulps at 20 °C. *Source:* Ahmed *et al.* (2005).

9.2.6 Effect of HPP on Color and Sensory Properties

Kaushik *et al.* (2014) found significant differences in the color of mango pulp after HPP treatment compared to untreated pulp. The browning index (BI) was used as an indicator of enzymatic and non-enzymatic browning. The value of the BI was 249.9 and 223–258 for untreated and for HPP-treated pulps, respectively. Both yellowness index and BI showed a slight increase with single pulse at lower pressure (100–200 MPa) and decreased or remained constant at higher pressure levels (300–600 MPa) at dwell times of 5–20 min. However, in another study, Ahmed *et al.* (2005) reported no significant variation in color of thermally-processed and unprocessed mango pulp compared to HPP-processed pulp (100–400 MPa, 15 or 30 min, at 20 °C). Between Chousa and Alphonso varieties, Alphonso pulp showed initiation of browning for treatments at pressures ≥300 MPa for 30 min.

HPP treatment of fresh-cut mangoes (Tommy Atkins and Keitt cultivars, from Florida) at 300 MPa and 600 MPa at 1 min dwell time followed by storage at 3 °C,

Table 9.2 Estimated rheological parameters of mango pulp by the Herschel–Bulkley model.

	Yield stress	Consistency coefficient	Flow behavior index		
Treatment	σ_{0HB} (Pa)	K (Pa s^n)	n	R^2	SE
Unblanched mango pulp:					
Control	2.335 ± 0.134	3.817 ± 0.180	0.434 ± 0.023	0.992	1.029
300 MPa, 15 min	5.703 ± 0.132	4.946 ± 0.237	0.360 ± 0.220	0.992	1.519
400 MPa, 5 min	7.378 ± 0.284	4.076 ± 0.115	0.384 ± 0.019	0.995	0.982
500 MPa, 2.5 min	6.244 ± 0.245	5.095 ± 0.186	0.353 ± 0.024	0.991	1.297
600 MPa, 1 min	7.167 ± 0.164	4.575 ± 0.172	0.367 ± 0.022	0.992	1.386
High temp. short time	5.265 ± 0.185	7.477 ± 0.171	0.372 ± 0.022	0.992	2.097
Blanched mango pulp:					
Control	2.277 ± 0.192	4.161 ± 0.186	0.308 ± 0.023	0.993	1.18
300 MPa, 15 min	4.118 ± 0.088	6.101 ± 0.334	0.335 ± 0.020	0.994	1.439
400 MPa, 5 min	2.478 ± 0.202	6.495 ± 0.304	0.341 ± 0.020	0.994	1.541
500 MPa, 2.5 min	5.736 ± 0.181	5.451 ± 0.198	0.345 ± 0.022	0.993	1.489
600 MPa, 1 min	6.131 ± 0.155	5.107 ± 0.200	0.356 ± 0.024	0.992	1.713
High temp. short time	7.821 ± 0.391	7.964 ± 0.271	0.321 ± 0.020	0.994	2.097

Source: Lui *et al.* (2013).

resulted in a slight reduction in fresh mango flavor and an increase in sweetness as per sensory evaluation (Boynton *et al.* 2002). The firmness for the Tommy Atkins variety was lowered in HPP-treated slices, while it did not change in Keitt mangoes.

9.2.7 HPP as a Quarantine Method

Effective quarantine treatments ensure removal of invasive pests and insects from fruit by killing their eggs and larvae. Quarantine treatments are mandated by the US Department of Agriculture prior to importing into the USA. Fruit flies are one of the major pests associated with mango fruit. Candelario-Rodríguez *et al.* (2009) studied inactivation of the Mexican fruit fly (*Anastrepha ludens* Loew) using HPP treatment (25–150 MPa, 0–20 min, at 25 °C) in whole raw and ripe mangoes. HPP treatment at 150 MPa for 20 min was able to inactivate the eggs of Mexican fruit flies by 98%. By increasing the temperature in the HPP treatment, the inactivation was higher at lower pressure (Velazquez *et al.* 2010). In another study (Candelario-Rodríguez *et al.* 2014), the effect of HPP as a quarantine treatment for Mexican fruit flies and on postharvest physiology of Keitt mango was studied. It was found that high pressure treatments (50, 70, and 90 MPa, 9 min, at 46 °C) of whole mangoes were effective in inactivating the pest and the treatments did not inhibit fruit ripening or change the quality of mango compared to the untreated mangoes.

9.3 Pulsed Electric Field (PEF)

PEF is a non-thermal food processing method that inactivates microorganism and preserves food quality by using short electric pulses with high electrical field intensity (10–80 kV/cm) for a duration of microseconds to milliseconds (Mohamed & Eissa 2012). Food to be treated with PEF is placed in a treatment chamber (either static or continuous design), where two electrodes are connected which provide high voltage electrical pulses. Such high electrical field intensity causes temporary or permanent permeabilization of cell membranes to inactivate the bacteria. Since inactivation takes place at ambient or sub-lethal temperatures, nutritional quality and sensory qualities are preserved compared to traditional thermal processing (Mohamed & Eissa 2012). Research on PEF has successfully demonstrated pasteurization of liquid food products, for example, juices, milk, soups, liquid eggs, etc. (Mohamed & Eissa 2012; Vega-Mercado *et al.* 1997). Studies on the application of PEF for mango products are rather limited and are mainly based on combining PEF with other hurdle technologies.

9.3.1 Effect of PEF on Microbial Inactivation

The main inactivation mechanism of PEF is the rupturing of the cell membrane due to the electric field. Membrane rupture or poration occurs when applied potential is greater than natural cell membrane potential (~1 V) (Palgan *et al.* 2012). Depending on the intensity of the electric field applied, duration and number of pulses, this damage can be reversible or irreversible (Vega-Mercado *et al.* 1997). The cell membrane becomes permeable to small molecules such as water, causing the cell to swell and eventually membrane rupture triggering cell lysis (Vega-Mercado *et al.* 1997).

A study on processing of mango juice with a combination of mild heat (50 °C), antimicrobial compounds (0.1 mg/ml of nisin and lysozyme), and PEF (20 or 50 pulses of 87 kV/cm) showed a 4.4 log CFU/ml reduction in microbial count (Mittal and Griffiths 2005). They also reported that nisin and lysozyme by themselves did not change the microbial count. In addition, temperatures lower than 50 °C showed less inactivation.

Palgan *et al.* (2012) studied inactivation of *Listeria innocua* in a milk-based smoothie, which contained 20% (v/v) mango puree using combined treatments of PEF and manothermossonication (MTS). PEF alone (34 kV/cm electric field and 32 μs pulse width) achieved a maximum of 2.7 ± 0.9 log CFU/ml reduction in the population of *L. innocua*. Furthermore, when the PEF treatment was followed after MTS treatment, 5.6 log CFU/ml reduction was observed. Permeabilization of the bacterial cell was considered responsible for the synergistic effect after mechanical effects on the cell by MTS treatment. In another study, the synergistic effect of PEF combined with naturally occurring antimicrobial essential oil carvacrol in mango juice was studied (Ait-Ouazzou *et al.* 2013). Simultaneous application of PEF (30 kV/cm and 20–50 pulses, 7 min) with 1.3 mM of carvacrol was able to achieve 5 log CFU/ml reduction of *E. coli O157:H7*, suggesting PEF permeabilized cells were sensitive to the presence of the antimicrobial agent. A correlation (Equation 9.3) was developed to predict the number of pulses required at 30 kV/cm for a 99 % reduction in the *E. coli O157:H7* population:

$$log\ t = -1.1391C + 2.019 \tag{9.3}$$

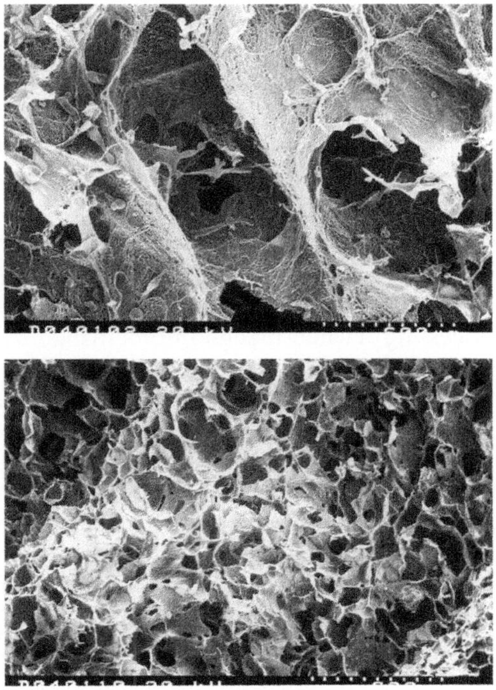

Figure 9.3 SEM-photographs of fresh mangoes ($Z_p = 0$) (*bottom*) and PEF pretreated mangoes ($Z_p = 0.54$) (*top*). *Source:* Tedjo *et al.* (2002).

where *t* is the number of pulses required at 30 kV/cm for a 99 % reduction in the *E. coli* O157:H7 population and C is the concentration (mM) of the antimicrobial essential oil carvacrol.

9.3.3 PEF as Pre-treatment for Osmotic Dehydration

Tedjo *et al.* (2002) applied PEF as a pre-treatment for mango slices followed by osmotic dehydration. Increase in the permeability index (Z_p) was observed with an increase in the field strength (0.8–2.9 kV/cm) at constant pulse numbers (50 at 1 Hz). The permeability index (Z_p) also increased with an increase in the number of pulses (10–100) at constant field strength (2.67 kV/cm). This indicates that the PEF parameters such as field strength and number of pulses affected cell disintegration (Figure 9.3). It was shown that the osmotically dehydrated mangoes pre-treated by PEF gained higher solids and lost more water than untreated mangoes. The texture of mango became softer when compared to the untreated mango and the color of the product was similar to the untreated one.

9.4 Ultraviolet Radiation (UV)

Ultraviolet radiation treatment involves exposing food to radiation with wavelengths in the range 100–400 nm. UV radiations are further divided into UV-A (315–400 nm),

UV-B (28–315 nm), and UV-C (100–280 nm). Due to its germicidal effect against bacteria, viruses, yeasts, and mold, UV-C radiation is most commonly used for inactivation of microorganisms on surfaces, in water, and on various food products (Keyser *et al.* 2008; Koutchma 2009). Generally, UV-C is used to process liquid food products like fruit juices. Due to its low penetration depth in turbid foods (300 mm in clear water, 1 mm in fruit juices, and 0.1 mm in milk), it is suggested to use the turbulent flow of the product or a thin film of liquid food to ensure uniform exposure to UV radiation (Guerrero-Beltrán and Barbosa-Cánovas 2006). UV radiation meets a 5-log CFU/ml reduction of a targeted pathogen in fruit juices and is approved as a non-thermal pasteurization technique by the USFDA (Santhirasegaram *et al.* 2015). Thus, UV radiation has great potential in processing of whole mango and cut mango, as well as liquid mango products such as juices and nectars.

9.4.1 Effect of UV-C Radiation on Microbial Inactivation

UV-C radiation inactivates microorganisms by damaging the nucleic acid and by creating pyrimidine dimers, which prevents the microorganisms from multiplying (Koutchma 2009). Guerrero-Beltrán and Barbosa-Cánovas (2006) studied inactivation of *Saccharomyces cerevisiae* in UV-C processed fresh mango nectar at five different flow rates (0.073–0.451 l/min) and at five different UV doses (75–450 kJ/m^2). Reduction of *S. cerevisiae* by UV-C followed first-order kinetics:

$$ln\left(N_t / N_0\right) = -kIt = -kF \tag{9.4}$$

where, N_t (CFU/ml) is the residual microbial load at a given time t (s), N_0 (CFU/ml) is the initial microbial load (CFU/ml), k is inactivation constant (m^2/kJ), I is intensity (kW/m^2), and F is the UV dosage (kJ/m^2). Based on these kinetics, decimal reduction dose for UV (kJ/m^2) to inactivate 90% of microorganisms was calculated:

$$\mathrm{D_{uv}} = -2.303 / \mathrm{k} \tag{9.5}$$

D_{uv} values ranged from 10.9–27.9 min (R^2 = 0.95) for total aerobic microorganisms and ranged from 11.8–26.0 min (R^2 = 0.962) for yeast. A maximum of 2.71 log CFU/ml and 2.94 log CFU/ml reduction was achieved in total plate count (TPC) and yeast, respectively. Addition of antimicrobial before UV treatment was suggested to increase shelf life of the mango nectar.

In a study by Keyser *et al.* (2008), application of UV-C dosage in mango juice (1,377 J/L) resulted in more than 10 CFU/ml aerobic plate count (APC) and yeast and mold count from initial low counts of 1.4 log CFU/ml and 2.8 log CFU/ml, respectively. Since mango juice was turbid and rich in fibers and suspended matter, it needed a higher dosage of UV-C. The same study reported lower amounts of UV-C dosage (459 J/L) in clear apple juice for higher inactivation (APC: 3.5 log CFU/ml, yeast and mold: 2.99 log CFU/ml).

Fresh mango juice from Thai mango cultivar Chokanan was treated with a UV-C dose of 3.525 J/m^2, for 15 and 60 min at 25 °C (Santhirasegaram *et al.* 2015). A maximum of 45% reduction in aerobic bacteria and 32% reduction in yeast and mold count from an initial count of 2.74 log CFU/ml and 2.42 log CFU/ml, respectively, was observed after

60 min. Yeast and mold were less susceptible to UV treatment due to differences in thickness of cell wall and size of microorganism. Due to high suspended solids, turbidity, and lack of turbulence for uniform exposure of UV radiation, low microbial inactivation values were observed.

9.4.2 Effect of UV-C Radiation on Enzymes

Guerrero-Beltrán and Barbosa-Cánovas (2006) used first-order kinetics to understand inactivation of PPO enzyme in mango nectar, wherein a higher inactivation of PPO was observed after a treatment of 75 kJ/m^2 for 5 min. Decimal reduction dose for UV (D_{uv} – kJ/m^2) to inactivate 90% of PPO ranged from 152–199 min.

9.4.3 Effect of UV-C Radiation on Color

Generally, carotenoids, responsible for the yellow to orange color, and other organic colorants become unstable and deteriorate when exposed to UV light, depending on its dosage (Guerrero-Beltrán and Barbosa-Cánovas 2006). They undergo photodegradation due to absorption of energy from UV and change in their chemical structure, as Guerrero-Beltrán and Barbosa-Cánovas (2006) reported that UV-C radiation (75–450 kJ/m^2) at flow rates of 0.073–0.451 l/min did not affect the bright yellow color of mango nectar significantly. Browning in the UV-treated mango nectar was higher at lower flow rates, hence the use of anti-browning agents was suggested. In a study by Santhirasegaram *et al.* (2015), fresh mango juice from Chokanan mangoes treated with UV-C radiation showed slightly noticeable difference in the color ($0.5 < \Delta E < 1.5$) as compared to well visible color change ($3.0 < \Delta E < 6.0$) in the thermally treated sample from the control.

9.4.4 Effect of UV-C Radiation on Shelf Life and Nutrients

Santhirasegaram *et al.* (2015) reported an increased shelf life of 5 weeks based on microbial count with UV treatment of 3.525 J/m^2 for 60 min compared to less than 1 week for fresh mango juice and lower shelf life compared to thermally-processed mango juice (beyond 5 weeks). The total carotenoid content in mango juice showed a slight increase with UV treatment (3.525 J/m^2 for 15 min and 30 min); which was attributed to inactivation of enzyme responsible for carotenoid loss (Santhirasegaram *et al.* 2015). A 12–15% reduction in ascorbic acid was observed, which was potentially related to the oxidation and the formation of free hydroxyl radicals by photochemical reaction. Due to activation of phenolic biosynthesis pathway by UV, significant increase (up to 31%) in total polyphenols was observed in UV-treated mango juice compared to control. Increase in carotenoids (6%) and flavonoids (3%) was also observed.

9.4.5 UV-C as Postharvest Treatment for Mangoes

Mangoes are susceptible to fungal attack and rapid physiological deterioration due to postharvest losses. UV treatment has been proposed to activate plant defenses against pathogens and reduce the rate of ripening (González-Aguilar *et al.* 2007a). Mangoes from the Haden cultivar (from Mexico) were treated with UV-C radiation at energy levels of 2.47–4.93 kJ/m^2 for a 5 min and 10 min duration. Enzymatic activity of enzymes related to the defense mechanism and postharvest shelf life, such as phenylalanine ammonia-lyase (PAL) and LOX, was studied to understand the effect of UV treatment.

The overall appearance of UV-treated mangoes was better and lower decay was observed than in untreated mangoes. Increased shelf life at 25 °C was seen with higher levels of total phenolics, total flavonoids, and increased enzymatic activity PAL and LOX.

UV-C radiation was used by González-Aguilar *et al.* (2001) to prevent decay and maintain postharvest quality of ripe Tommy Atkins mangoes at low temperature storage. UV-C irradiation (8,220 mW/m^2) for 10 min reduced decay and maintained firmness during storage at 5 °C or 20 °C, while maintaining sugar and organic acid levels similar to the control without any UV damage. UV treatment was recommended as an effective and rapid method for maintaining the postharvest quality of mangoes.

9.4.6 UV-C Application in Fresh-cut Mangoes

González-Aguilar *et al.* (2007b) used UV-C as a preservation technique for fresh-cut Tommy Atkins mangoes stored at 5 °C. Effects of the UV-C treatment time (0–10 min) on total phenolics, flavonoids, β-carotene, ascorbic acid contents, and antioxidant capacity were studied. Total phenolic content, flavonoid content, and antioxidant capacity were higher in UV-C-treated fresh-cut mangoes compared to the control, whereas UV-C-treated pre-cut mangoes showed decreased β-carotene and ascorbic acid content compared to the control. This decrease was related to the oxidation of ascorbic acid by UV and the oxidative stress of β-carotene due to UV.

George *et al.* (2015) investigated the effect on quality of fresh-cut Chokanan mangoes processed by UV-C treatment with a florescent lamp at 240 V, 50 Hz, and 10A (0, 15, 30, and 60 min) and by medium heat treatment (70 °C: 0, 5, 10, and 20 min) during storage at 4 ± 1 °C. No significant changes to the physicochemical attributes for both treatments were observed. Ascorbic acid content was preserved and antioxidants were enhanced by UV-C treatment compared to mild heat treatment. UV-C treated pre-cut mangoes were preferred by sensory panels over their mild heat treated counterparts. UV-C treatment was recommended as a preservation technique for better nutritional and sensory quality and for effective microbial inactivation of fresh-cut mangoes.

9.5 Ultrasound Treatment

Ultrasound processing or sonication is a technique where sound waves of frequencies above the hearing limit of the human ear (20–100 kHz) are applied to food. In liquid food, due to propagation high power at 20–100 kHz frequency, cavitation (formation of bubble and collapse) occurs, which generates localized spots of elevated pressure and temperature (Santhirasegaram *et al.* 2013). Authors reported that these hot spots provide energy that changes the product chemically or physically.

9.5.1 Ultrasound Processing for Mango Products

Sonication treatment for mango juice from Chokanan cultivar was carried out at 40 kHz and 130 W (for 15, 30, and 60 min at 25 °C) by Santhirasegaram *et al.* (2013). Though previous studies reported complete inactivation of microorganisms due to cavitation mechanism and formation of OH^- and H^+ free radicals during sonication, in this study no significant inactivation of microorganism was found at the studied frequency, power, and time. A maximum of 1-log CFU/ml reduction in coliforms was reported. Color, clarity, carotenoid content, and ascorbic acid content of ultrasound treated juice did not change significantly from freshly squeezed juice. This study suggested combination of hurdle technologies along with sonication for effective microbial inactivation.

Santos *et al.* (2015) studied ultrasound (25 kHz for 30 min, 55 W/L power density), a post-processing treatment of fresh-cut mangoes to preserve quality during refrigerated storage. Measurement of physicochemical and chemical variables (citric acid, pH, TSS, total sugar content, firmness) showed ultrasound affected negatively fresh-cut mangoes.

9.5.2 Ultrasonic as a Nondestructive Quality Technique

Ultrasonic waves are also used as a non-destructive technique for quality and maturity measurements of mango fruit (Mizrach *et al.* 1997; Mizrach 2000). Changes in reception of ultrasonic waves (50 kHz) passing through the peel and flesh were measured and related to fruit quality attributes such as firmness and physiological indices. Quantitative correlations between ultrasound parameters and sugar content, acidity, and firmness of mango fruits were obtained.

Valente *et al.* (2013) studied ultrasound (25 MHz) treatment and used ultrasound velocity and attenuation measurements to predict the biochemical compounds related to organoleptic quality of mangoes. Both velocity and attenuation showed variations based on fruit maturity and ripening staged in relation to TSS, fruit texture, and biochemical composition changes. They found that the prediction of sucrose content improved when the correlation was developed by combining soluble solids content and ultrasonic wave velocity.

9.6 Irradiation Treatment

The short shelf life of mango is a major concern and is responsible for great economic losses during its storage and transportation. Radiation processing has been shown to be a promising method to increase the shelf life of mango by delaying the ripening of fresh fruit. It is widely used as a shelf life extension method and as a quarantine treatment for mango fruit (Frylinck *et al.* 1987; Gregory *et al.* 1990; Youssef *et al.* 2002). It has been showed that irradiation minimizes the physical and chemical changes during storage, delays the ripening process, and extends the shelf life of mangoes (El-Samahy *et al.* 2000; Frylinck *et al.* 1987; Mahto and Das 2013; Sabato *et al.* 2009a). Other products like mango pulp have also shown increased shelf life due to microbial inactivation using irradiation treatment.

9.6.1 Effect of Irradiation on Texture of the Mango Fruit

Texture is the one of the major characteristics for identifying the degree of ripeness of mango. El-Samahy *et al.* (2000) found that firmness of green mature mangoes after irradiation (0.5–1.5 kGy) was proportional to the radiation dose. Immediately after irradiation exposure, the firmness decreased compared to the non-irradiated mangoes, which was related to the cell wall polysaccharides degradation and breakage of glycosides in pectin due to radiation. However, during storage, non-irradiated mangoes showed rapid decrease in firmness compared to irradiated mangoes (Figure 9.4). Overall, during

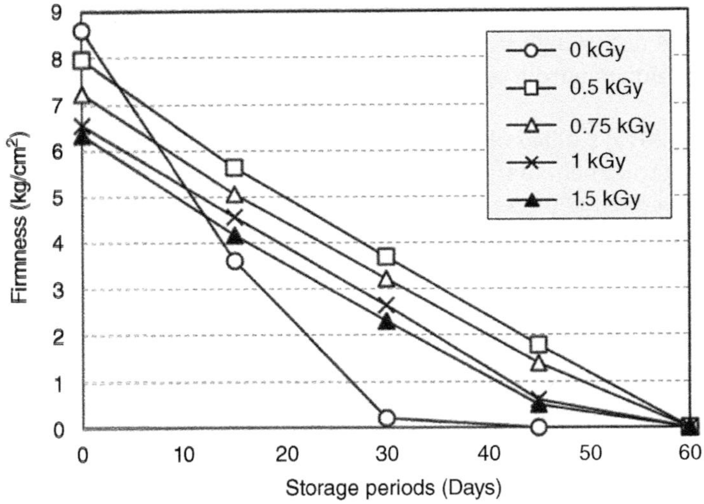

Figure 9.4 Effect of irradiation and storage on firmness of mango fruits. *Source:* Adapted from El-Samahy *et al.* (2000).

storage, firmness of both control and irradiated mangoes decreased (though at different rates), due to ripening related physiochemical changes.

Similarly, a recent study reported that the decrease in firmness with storage occurred faster in the control fruit compared to the irradiated fruit (0.3–10 kGy) for the Indian Dushehri and Fazli cultivar of mangoes (Mahto and Das 2013). Whereas, another study (Sabato *et al.* 2009a) reporting lower doses of irradiation (0.4 and 1.0 kGy) did not show any significant difference in firmness from control samples during 12 days of storage at 22 °C for Tommy Atkins mangoes.

9.6.2 Effect of Irradiation on Mango Pulp Rheology

Youssef *et al.* (2002) studied the effect of irradiation treatments of mango fruit (0.5–2.0 kGy) on the extracted pulp rheology. The Power-law model (Equation 9.6) was used to characterize the effect of irradiation on flow behavior of mango pulp:

$$\sigma = K\gamma^n \tag{9.6}$$

where σ = shear stress (Pa), γ =shear rate (1/s), K = consistency index (Pa.s^n), and n = power law index.

The pulp was found to be non-Newtonian, time-independent of pseudoplastic fluid. The *n* values (<1) increased with increasing irradiation doses and the consistency index K decreased with increasing irradiation, which was attributed to the degradation of carbohydrates.

9.6.3 Effect of Irradiation on Enzyme Activity

It has been reported that irradiation affects the enzyme activity. Activity of PAL and PPO significantly increased immediately after γ-irradiation treatment of mangoes

Table 9.3 Frequency distribution of the identified yeasts and their decimal radiation reduction doses in saline and mango pulp.

			D_{10}-values (kGy)		
Isolated yeasts	**Number of isolates**	**%**	**In saline**	**In mango pulp**	***P* value**
Candida tropicales	20	23.53	1.23	1.37	3.39E-05
Hanseniaspora uvarum	4	4.71	1.16	1.34	1.25E-05
Rhodoterula glutins	10	11.7	1.76	1.80	0.018
Saccharomyces cervisae	16	18.82	0.87	2.15	2.230E-09
Saccharomycodes ludwigii	6	7.06	1.75	2.23	4.454E-07
Zygosaccharomyces bailii	4	4.71	0.76	1.18	7.592E-07
Z. rouxii	25	29.41	1.19	1.65	4.163E-07

Source: Youssef *et al.* (2002).

(Frylinck *et al.* 1987). Youssef *et al.* (2002) found increased PPO activity with increase in γ-irradiation dose (0.5–2.0 kGy); this increased PPO activity was speculated to be due to an activation of a pro-enzyme or latent enzyme or due to conformational changes in the enzyme (Frylinck *et al.* 1987; Youssef *et al.* 2002).

9.6.4 Effect of Irradiation on Microbial Inactivation and Pests

Youssef *et al.* (2002) isolated six strains of yeast species from mango pulp to calculate the decimal radiation reduction dose (D_{10} value). Protective effect against γ-irradiation by yeast in mango pulp was seen with higher D_{10} values than in saline solution (Table 9.3). In general, γ-irradiation of mango pulp reduced significant amounts of total bacteria, lactic acid bacteria, yeast, and mold counts. The amount of log reduction increased with increasing irradiation dose (0.5–2.0 kGy).

A small dose of irradiation (74–101 Gy) was used as a successful quarantine treatment against Australian fruit flies (*Bactrocera tryoni* and *B. jarvisi*) on Kensington mangoes, and 99.997% mortality rate was obtained, which ensures quarantine security (Heather *et al.* 1991).

9.6.5 Effect of Irradiation on Nutrients

At low doses of γ-irradiation (up to 1 kGy), mango pulp and mango fruit did not show significant differences in total sugar and reducing sugar content (Cruz *et al.* 2012; Mahto and Das 2013; Youssef *et al.* 2002). Higher doses of irradiation (>1 kGy) increased total sugar and reducing sugar content (Mahto and Das 2013). Frylinck *et al.* (1987) reported that irradiation causes accumulation of flavanols during the storage period. However, Reyes and Cisneros-Zevallos (2007) pointed out that irradiation has no significant effect on the total phenolic content of mangoes.

Low doses of γ-irradiation (0.075, 0.3, and 0.6 Gy) did not show any significant change in carotene in mangoes (Gregory *et al.* 1990). Youssef *et al.* (2002) showed that total carotenoid of mango pulp was increased significantly by γ-irradiation of fruits

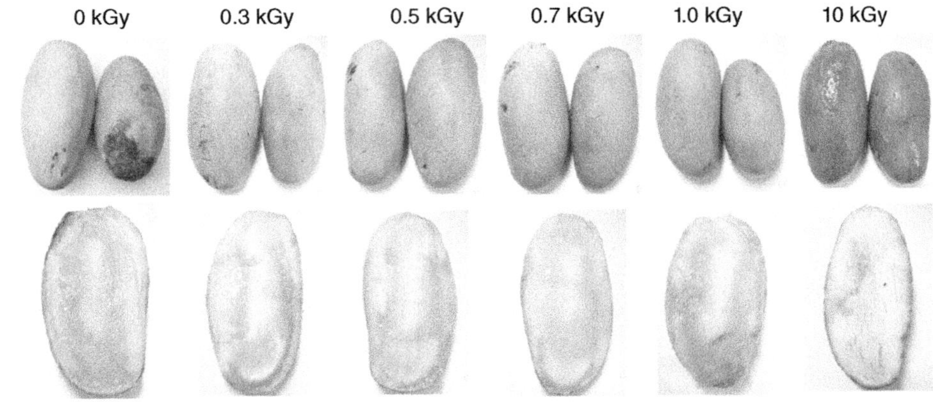

Figure 9.5 Effect of irradiation on appearance of the Indian "Dushehri" mango after 7 days of storage at 20 °C. *Source:* Mahto and Das (2013).

(0.5–2.0 kGy), which was explained by increased extractability of carotenoid due to irradiation. They also reported increased ascorbic acid degradation of mango pulp. Similarly, decreased ascorbic acid values were reported by Reyes and Cisneros-Zevallos (2007) for irradiated mango fruit.

9.6.6 Effect of Irradiation on Physical Appearance

No changes in physical appearance of γ-irradiated and non-irradiated mangoes were seen immediately after processing, that is, day 0 (Mahto and Das 2013). However, after storage for 7 days at 20 °C, noticeable visual differences of the skin and pulp were observed (Figure 9.5). The skin of the irradiated mangoes remained green throughout storage compared to the control and the pulp color remained lighter in irradiated mangoes. A lower rotting was also seen in the irradiated mangoes. However, a higher dose of irradiation resulted in brown patches on skin and pulp.

9.7 Machine Vision – Computer Vision System (CVS)

Machine vision or computer vision system (CVS) is a technology that provides objective assessment and has been used for grading and sorting the fruits based on their quality, physical characteristics, defects, and maturity. The CVS systems are also used to correlate the physical characteristics with biochemical quality indices. Artificial vision systems are used to acquire the image and various parameters, such as color, size, volume, shape, and texture are recorded based on image analysis with the help of computer software (Figure 9.6). Artificial vision systems may be based on broad spectrum of light, including visible light, ultraviolet (UV), and near-infrared (NIR) (Cubero *et al.* 2011). On the basis of the collected information, fruits are sorted by using computer algorithms. CVS, together with these algorithms, provide automated, non-destructive, and objective measurement, avoiding manual subjectivity.

Postharvest losses in mango can be reduced using machine vision or CVS to obtain correct information during harvesting, storage, transportation, and throughout the

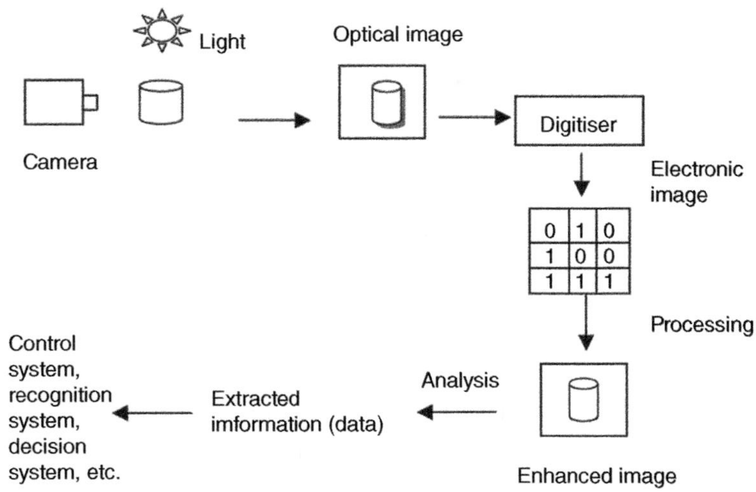

Figure 9.6 Basic principle of use of CVS. *Source:* Brosnan and Sun (2002).

process of ripening (Jha *et al.* 2010). These improved non-destructive technologies can be beneficial for determination of maturity, physical, and biochemical quality parameters, as well as internal disorders in mangoes.

9.7.1 CVS for Grading/Sorting of Mangoes

Ganiron (2014) used an image analysis algorithm to measure the size, roundness, area, and percent defect of a mangoes to sort them for export, for local market, or to reject. Based on the red, green, and blue (RGB) color data and size, the algorithm was trained to classify the fruits. The nearest neighbor technique with Euclidean distance was used where input pattern was compared to the closest match in the training set. The algorithm successfully identified defective areas in mangoes. Correct segmentation was performed irrespective of position of mangoes in the image. The algorithm was not able to differentiate between green and brown stems.

9.7.2 CVS for Measuring Maturity/Ripening

Ripening of mangoes can be understood based on various physical and biochemical parameters such as color, pigment content, firmness, TSS, and TA. A study was undertaken by Vélez-Rivera *et al.* (2014) using computer vision to relate optical measurement with physicochemical properties of mangoes to classify them based on their ripeness. Images of mangoes were taken and the values of the pixels were transformed into L*, a*, and b* coordinates of the CIELAB color space, and hue angle (H), saturation (S), and brightness (B) of the HSB color space. The L*, a*, and b* is an international standard for color measurement proposed by the Commission Internationale d'Eclairage (CIE), in which L* is the luminance component (0–100), and a* (from green to red) and b* (from blue to yellow) are the two chromatic components (from −20 to 120).

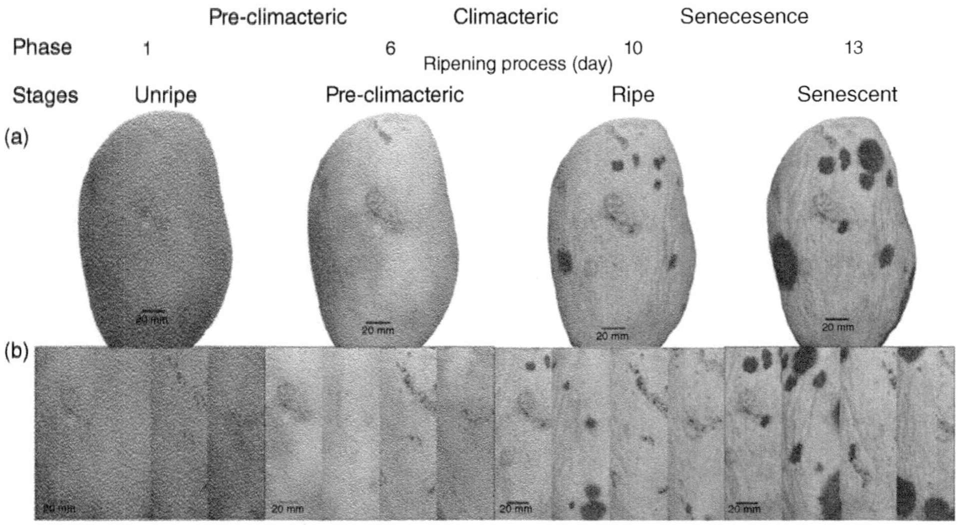

Figure 9.7 Phases and stages of the ripening (a) and stacked images of the mangoes at selected ripening stages (b). *Source:* Vélez-Rivera *et al.* (2014).

Physicochemical properties, such as firmness (F-Newton), TSS (TSS-°Brix), and TA (TA-g citric acid equivalent/100 g sample), were measured to obtain the ripening index (RPI):

$$\mathrm{RPI} = \ln\left(100 \times \mathrm{F} \times \mathrm{TA} / \mathrm{TSS}\right) \tag{9.7}$$

Based on data, from physiochemical analysis and significant variables from optical measurement, Principle component analysis (PCA) was used to classify the mangoes based on maturity as pre-climacteric, climacteric, and senescence (Figure 9.7). A total set of 117 samples with different RPI values was used to train the network and another set of 39 mangoes from a different season was used to validate the classification process with a 95% confidence interval.

In another study, by Kang *et al.* (2008), CIE L* a* b* color space was used to analyze images from mangoes at various levels of maturity. The L* a* b* values were converted to hue angle (Equation 9.8) and chroma (Equation 9.9), as:

$$\text{Hue angle} = \tan^{-1}\left(\frac{b^*}{a^*}\right) \tag{9.8}$$

$$\text{Chroma} = \sqrt{(a^*)^2 + (b^*)^2} \tag{9.9}$$

Changes in the color of captured images and hue angle can be clearly seen in Figure 9.8 of a mango at various stages of maturity during storage after harvest (0–34 days).

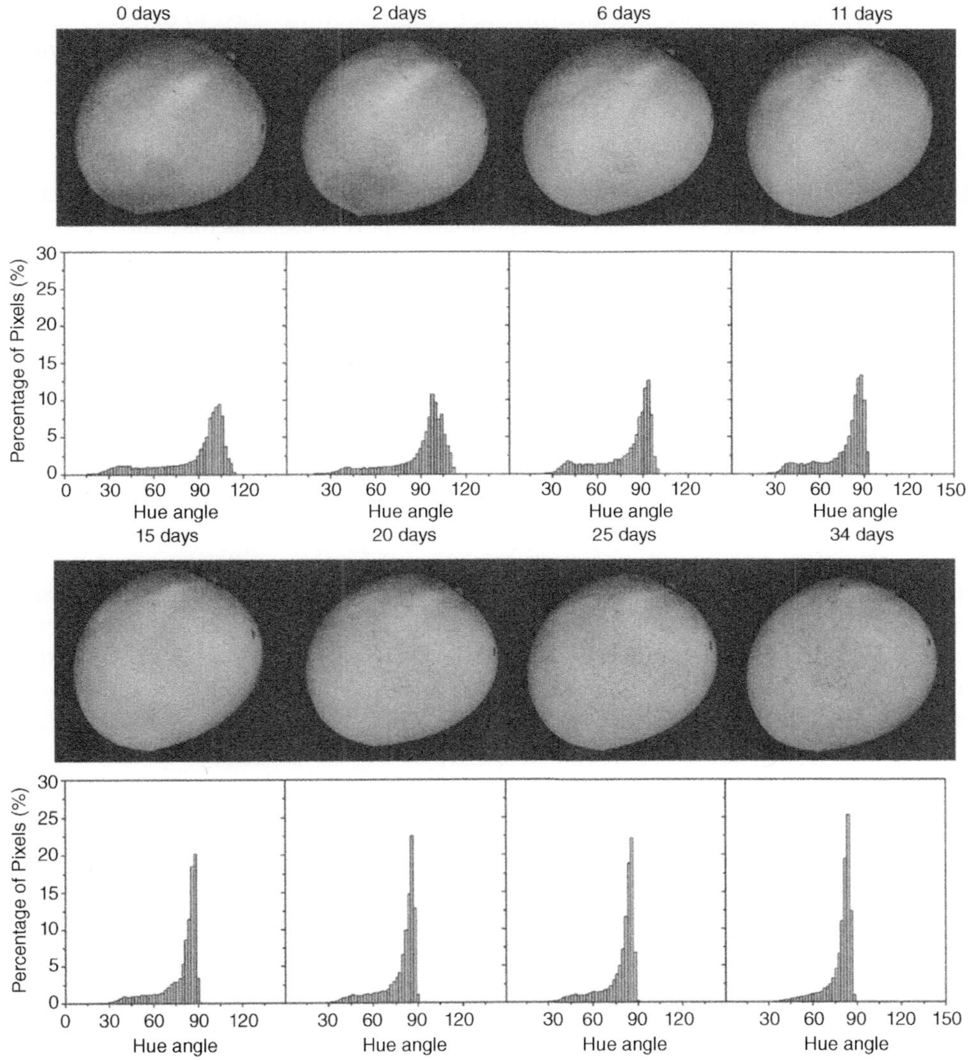

Figure 9.8 Color change of an individual mango stored at 17 °C, as described by images and hue angle histograms during various stages of maturity. Source: Kang *et al.* (2008)

Based on the fraction of pixels within a color range, data was used for maturity-based grading of mangoes.

9.7.3 CVS for Measuring Sweetness

The color of a mango changes from dark green to yellow/orange/red, as the fruit matures and its sugar content also increases. Bejo and Kamaruddin (2014) developed a correlation between the sweetness of the Chokanan mango and its mean pixel values in RGB color space, hue, saturation, and intensity. The correlation was used to predict sweetness at three different levels with 100% success rate for sweetness levels 1 and 2 and with 87% success rate for sweetness level 3.

9.7.4 CVS for Measuring Degree of Browning

Zheng and Lu (2012) developed an automatic sorting method for mangoes based on their browning degree, by using the least-squares support vector machine classifier as a function of fractal analysis and CIELAB color parameters. Classification accuracy based on the degree of browning was 100% when the classifier model was based on both fractal analysis and CIELAB color parameters, and an accuracy of 88.89 and 85.19% was achieved when the model was based on CIELAB color parameters and on fractal analysis, individually.

9.7.5 CVS for Measuring Mass/Weight

CVS can also be used to predict the mass of irregularly shaped mangoes with potential applications to measure *in-situ* growth of the fruits. Spreer and Müller (2011) measured length, maximum width, and maximum thickness of Chokanan mangoes and developed an equation to calculate the mass of the mangoes based on this application (R^2 = 0.97). In another study by Teoh and Syaifudin (2007), a relationship between mango pixels from images and mango weights was developed (R^2 = 0.98). When the linear model for prediction of mango weight was evaluated for a set of 50 mangoes, the mean absolute percentage error was 3.76%.

9.8 Other Innovative Processing Technologies

9.8.1 Cold Atmospheric Plasma Processing (CAPP)

Perni *et al.* (2008) studied decontamination of a pathogenic bacteria *Escherichia coli* and spoilage microorganisms *S. cerevisae, Pantoea agglomerans,* and *Gluconacetobacter liquefaciens* from the surface of mango skin using CAPP. CAPP is an emerging technology where non-thermal ionized gases/plasma containing mixture of electrons, ions, and free radicals are used to inactivate the microorganisms on surfaces at atmospheric pressure. Application of CAPP, generated using a helium and oxygen mixture with an AC voltage of 16 kV and an excitation frequency of 30 kHz, resulted in an approximate 4-log CFU/cm^2 reduction (from initial 10^7 CFU/cm^2 concentration) in *E. coli*. Spoilage organisms *P. agglomerans* and *G. liquefaciens* were inactivated at least by 3 log CFU/cm^2 (from initial 10^6 CFU/cm^2 concentration) based on the limitation of the recovery method (detection limit of 10^3 CFU/cm^2) after 2.5 sec plasma treatment. *S. cerevisae* needed 10 sec plasma exposure to achieve similar inactivation. This study for the first time showed potential application of CAPP in decontaminating fresh-cut fruits.

9.8.2 Ozone Pre-Treatment to Increase Shelf Life

Ozone is very effective in removing ethylene from fruits through a chemical reaction during the postharvest storage to increase the shelf life by physiological and biochemical changes. Ozone has also been used to reduce microbial population and diseases in the fruits. In a study by Tran *et al.* (2013), mango fruits were fumigated with ozone gas at concentrations of 2 μL/L for 20 min and 10 μL/L for 10 min, and the quality of the fruits was compared with untreated fruit. Ozone 10 μL/L for 10 min was more effective during initial storage duration (up to 3 days) in suppressing the respiration rate, inhibiting the color change in fruit skin and pulp, and decreasing the weight loss of mango fruit.

9.9 Future Research Trends

The researchers have studied processing of mango and its products using various innovative technologies such as HPP, PEF, UV-C, ultrasound, CAPP, ozone treatment, irradiation, and CVSs. HPP is one of the most extensively investigated processing methods by the researchers for various mango products and has shown great potential to increase the shelf life by inactivating the microorganisms and enzymes while retaining nutritional and sensory qualities. More studies focusing on shelf life and sensory qualities are required to establish HPP as an acceptable method on an industrial scale. HPP products due to their raw fresh-like characteristic have high commercial value and HPP products from exotic tropical fruit like mango have great profit-making potential.

PEF and ultrasound processing are less explored for mango produces. These technologies were not very efficient in inactivating the microorganisms. Future research on optimizing these technologies to ensure effective microbial inactivation by combining with other hurdle technologies is needed. UV-C is a well-established method for processing clear liquids and can be used for processing mango juices and nectars. UV-C has low penetration depth in turbid mango juice, hence researchers should consider using turbulent and thin film flow conditions to ensure uniform exposure and more effective processing.

Most of the research on mango products is focused on processing purees, juices, nectars, etc. Novel technologies, such as CAPP and UV-C, have shown the possibility to increase the shelf life of fresh-cut mango products, but more research is needed to substantiate the findings. There is an increasing demand for fresh-cut fruits due to convenience and high nutritive value. Innovative technologies such as HPP along with CAPP and UV-C can also be explored for increasing shelf life of these fresh-cut mango products.

Though irradiation has being used for a long time, its effect on extension of shelf life, quality, and safety of mango varies based on cultivar. More studies are required to understand these effects and irradiation tolerance on different cultivar. Although machine vision is a well-established commercial technique for detection of quality and sorting the mangoes, more research on reliable prediction of quality is required. Innovative processing technologies provide a promising solution in mango processing to create products with high microbial safety and fresh-like product characteristic with high nutritional value.

References

Ahmed, J., Ramaswamy, H.S., Alli, I. and Ngadi, M. (2003) Effect of high pressure on rheological characteristics of liquid egg. *LWT-Food Science and Technology*, **36**, 517–524.

Ahmed, J., Ramaswamy, H.S. and Hiremath, N. (2005) The effect of high pressure treatment on rheological characteristics and color of mango pulp. *International Journal of Food Science and Technology*, **40**, 885–895.

Ait-Ouazzou, A., Espina, L., García-Gonzalo, D. and Pagán, R. (2013) Synergistic combination of physical treatments and carvacrol for *Escherichia coli O157*: H7 inactivation in apple, mango, orange, and tomato juices. *Food Control*, **32**, 159–167.

Bejo, S.K. and Kamaruddin, S. (2014) Determination of Chokanan mango sweetness (*Mangifera indica*) using non-destructive image processing technique. *Australian Journal of Crop Science*, **8**, 475–480.

Bermúdez-Aguirre, D. and Barbosa-Cánovas, G.V. (2011) An update on high hydrostatic pressure, from the laboratory to industrial applications. *Food Engineering Reviews*, **3**, 44–61.

Boynton, B.B., Sims, C.A., Sargent, S., Balaban, M.O. and Marshall, M.R. (2002) Quality and stability of pre-cut mangoes and carambolas subjected to high-pressure processing. *Journal of Food Science*, **67**, 409–415.

Brosnan, T. and Sun, D.W. (2002) Inspection and grading of agricultural and food products by computer vision systems: A review. *Computers and Electronics in Agriculture*, **36**, 193–213.

Candelario-Rodríguez, H.E., Hurtado-González, M., Morales-Castro, J., Velazquez, G., Ramírez, J.A. *et al.* (2009) Efficacy of high hydrostatic pressure as a quarantine treatment to improve the quality of mango fruits infested by the Mexican fruit fly *Anastrepha ludens*. *CyTA Journal of Food*, **7**, 135–142.

Chen, J.P., Tai, C.Y. and Chen, B.H. (2007) Effects of different drying treatments on the stability of carotenoids in Taiwanese mango (*Mangifera indica* L.). *Food Chemistry*, **100**, 1005–1010.

Cubero, S., Aleixos, N., Moltó, E., Gómez-Sanchis, J. and Blasco, J. (2011) Advances in machine vision applications for automatic inspection and quality evaluation of fruits and vegetables. *Food and Bioprocess Technology*, **4**, 487–504.

Cruz, J.N., Soares, C.A., Fabbri, A.D.T., Cordenunsi, B.R. and Sabato, S.F. (2012) Effect of quarantine treatments on the carbohydrate and organic acid content of mangoes (cv. Tommy Atkins). *Radiation Physics and Chemistry*, **81**, 1059–1063.

de la Cruz Medina, J. and Garcia, H.S. (2002) Mango: Postharvest operations. In: *InPho Post-Harvest Compendium* (D. Mejia and B Lewis). AGSI/FAO [Online]. Available: http://www.fao.org/fileadmin/user_upload/inpho/docs/Post_Harvest_Compendium_-_Mango.pdf

El-Samahy, S.K., Youssef, B.M., Askar, A.A. and Swailam, H.M.M. (2000) Microbiological and chemical properties of irradiated mango. *Journal of Food Safety*, **20**(3), 139–156.

FAO (2015) Crops Production and Trade Statistics [Online]. Available: http://faostat3.fao.org/download/Q/QC/E [26 January 2016].

Frylinck, L., Dubery, I.A. and Schabort, J.C. (1987) Biochemical changes involved in stress response and ripening behaviour of γ-irradiated mango fruit. *Phytochemistry*, **26**, 684–686.

Ganiron Jr, T. U. (2014) Size Properties of mangoes using image analysis. *International Journal of Bio-Science and Bio-Technology*, **6**, 31–42.

George, D.S., Razali, Z., Santhirasegaram, V. and Somasundram, C. (2015) Effects of ultraviolet light (UV-C) and heat treatment on the quality of fresh-cut Chokanan mango and Josephine pineapple. *Journal of Food Science*, **80**, 426–34.

González-Aguilar, G.A., Wang, C.Y., Buta, J.G. and Krizek, D.T. (2001) Use of UV-C irradiation to prevent decay and maintain postharvest quality of ripe "Tommy Atkins" mangoes. *International Journal of Food Science and Technology*, **36**, 767–773.

González-Aguilar, G.A., Zavaleta-Gatica, R. and Tiznado-Hernández, M.E. (2007a) Improving postharvest quality of mango "Haden" by UV-C treatment. *Postharvest Biology and Technology*, **45**, 108–116.

González-Aguilar, G.A., Villegas-Ochoa, M.A., Martinez-Tellez, M.A., Gardea, A.A. and Ayala-Zavala, J.F. (2007b) Improving antioxidant capacity of fresh-cut mangoes treated with UV-C, *Journal of Food Science*, **72**, 197–202.

Gregory, E., Mitchell, G.E., Mclauchlan, R.L., Beattie, T.R., Banos, C., and Gillen, A.A. (1990) Effect of gamma irradiation on the carotene content of mangoes and red capsicums. *Journal of Food Science*, **55**, 1185–1186.

Guerrero-Beltran, J.A. (2005) High hydrostatic pressure processing of mango puree containing anti-browning agents. *Food Science and Technology International*, **11**, 261–67.

Guerrero-Beltran, J.A. and Barbosa-Cánovas, G.V. (2006) Inactivation of *Saccharomyces cerevisiae* and polyphenoloxidase in mango nectar treated with UV light. *Journal of Food Protection*, **69**, 362–368.

Heather, N.W., Corcoran, R.J. and Banos, C. (1991) Disinfestation of mangoes with gamma irradiation against two Australian fruit flies (Diptera: Tephritidae). *Journal of Economic Entomology*, **84**, 1304–1307.

Jha, S.N., Narsaiah, K., Sharma, A.D., Singh, M., Bansal, S. and Kumar, R. (2010) Quality parameters of mango and potential of non-destructive techniques for their measurement: A review. *Journal of Food Science and Technology*, **47**, 1–14.

Kang, S.P., East, A.R. and Trujillo, F.J. (2008) Color vision system evaluation of bicolor fruit: A case study with "B74" mango. *Postharvest Biology and Technology*, **49**, 77–85.

Karwe, M.V., Maldonado, J. and Mahadevan, S. (2015) High pressure processing: Current status. In: *Conventional and Advanced Food Processing Technologies* (ed. S. Bhattacharya), pp. 595–616. John Wiley & Sons, UK.

Kaushik, N., Kaur, B.P., Rao, P.S. and Mishra, H.N. (2014) Effect of high pressure processing on color, biochemical and microbiological characteristics of mango pulp (*Mangifera indica* cv. Amrapali). *Innovative Food Science and Emerging Technologies*, **22**, 40–50.

Keyser, M., Műller, I.A., Cilliers, F.P., Nel, W. and Gouws, P.A. (2008) Ultraviolet radiation as a non-thermal treatment for the inactivation of microorganisms in fruit juice. *Innovative Food Science and Emerging Technologies*,**9**, 348–354.

Kim, Y., Lounds-Singleton, A.J. and Talcott, S.T. (2009) Antioxidant phytochemical and quality changes associated with hot water immersion treatment of mangoes (*Mangifera indica* L.). *Food Chemistry*, **115**, 989–993.

Koutchma, T. (2009) Advances in ultraviolet light technology for non-thermal processing of liquid foods. *Food and Bioprocess Technology*, **2**, 138–155.

Lemmens, L., Tchuenche, E. S., Van Loey, A.M. and Hendrickx, M.E. (2013) Beta-carotene isomerisation in mango puree as influenced by thermal processing and high-pressure homogenisation. *European Food Research and Technology*, **236**, 155–163.

Liu, F., Wang, Y., Bi, X., Guo, X., Fu, S. and Liao, X. (2013) Comparison of microbial inactivation and rheological characteristics of mango pulp after high hydrostatic pressure treatment and high temperature short time treatment. *Food and Bioprocess Technology*, **6**, 2675–2684.

Liu, F., Wang, Y., Li, R., Bi, X. and Liao, X. (2014a) Effects of high hydrostatic pressure and high temperature short time on antioxidant activity, antioxidant compounds and color of mango nectars. *Innovative Food Science and Emerging Technologies*, **21**, 35–43.

Liu, F., Li, R., Wang, Y., Bi, X. and Liao, X. (2014b) Effects of high hydrostatic pressure and high-temperature short-time on mango nectars: changes in microorganisms, acid invertase, 5-hydroxymethylfurfural, sugars, viscosity, and cloud. *Innovative Food Science and Emerging Technologies*, **22**, 22–30.

Liu, F., Grauwet, T., Kebede, B.T., Van Loey, A., Liao, X. and Hendrickx, M. (2014c) Comparing the effects of high hydrostatic pressure and thermal processing on blanched and unblanched mango (*Mangifera indica* L.) nectar: Using headspace fingerprinting as an untargeted approach. *Food and Bioprocess Technology*, **7**, 3000–3011.

Mahto, R. and Das, M. (2013) Effect of gamma irradiation on the physico-chemical and visual properties of mango (*Mangifera indica* L.), cv. "Dushehri" and "Fazli" stored at 20 °C. *Postharvest Biology and Technology*, **86**, 447–455.

Mittal, G.S. and Griffiths, M.W. (2005) Pulsed Electric Field processing of liquid foods and beverages. In: *Emerging Technologies for Food Processing* (ed. Da-Wen Sun), pp. 99–131. Elsevier Academic Press, San Diego, CA.

Mizrach, A. (2000) Determination of avocado and mango fruit properties by ultrasonic technique. *Ultrasonics*, **38**, 717–722.

Mizrach, A., Flitsanov, U. and Fuchs, Y. (1997) An ultrasonic nondestructive method for measuring maturity of mango fruit. *Transactions of the ASAE*, **40**, 1107–1111.

Mohamed, M.E., Ayman, H. and Eissa, A. (2012) Pulsed electric fields for food processing technology. In: *Structure and Function of Food Engineering* (ed. A.A. Eissa). pp. 275–306. INTECH, Open Access Publisher.

Palgan, I., Muñoz, A., Noci, F., Whyte, P., Morgan, D.J. *et al.* (2012) Effectiveness of combined Pulsed Electric Field (PEF) and Manothermosonication (MTS) for the control of *Listeria innocua* in a smoothie type beverage. *Food Control*, **25**, 621–625.

Perni, S., Liu, D.W., Shama, G. and Kong, M.G. (2008) Cold atmospheric plasma decontamination of the pericarps of fruit. *Journal of Food Protection*, **71**, 302–308.

Rawson, A., Patras, A., Tiwari, B.K., Noci, F., Koutchma, T. and Brunton, N. (2011) Effect of thermal and non-thermal processing technologies on the bioactive content of exotic fruits and their products: Review of recent advances. *Food Research International*, **44**, 1875–1887.

Reyes, L.F. and Cisneros-Zevallos, L. (2007) Electron-beam ionizing radiation stress effects on mango fruit (*Mangifera indica* L.) antioxidant constituents before and during postharvest storage. *Journal of Agricultural and Food Chemistry*, **55**, 6132–6139.

Sabato, S. F., da Silva, J. M., da Cruz, J.N., Salmieri, S., Rela, P.R. and Lacroix, M. (2009a) Study of physical–chemical and sensorial properties of irradiated Tommy Atkins mangoes (*Mangifera indica* L.) in an international consignment. *Food Control*, **20**, 284–288.

Santhirasegaram, V., Razali, Z. and Somasundram, C. (2013) Effects of thermal treatment and sonication on quality attributes of Chokanan mango (*Mangifera indica* L.) juice. *Ultrasonics Sonochemistry*, **20**, 1276–1282.

Santhirasegaram, V., Razali, Z., George, D.S. and Somasundram, C. (2015) Comparison of UV-C treatment and thermal pasteurization on quality of Chokanan mango (*Mangifera indica* L.) juice. *Food and Bioproducts Processing*, **94**, 313–321.

Santos, J.G., Fernandes, F.A.N., de Siqueira Oliveira, L. and de Miranda, M.R.A. (2015) Influence of ultrasound on fresh-cut mango quality through evaluation of enzymatic and oxidative metabolism. *Food and Bioprocess Technology*, **8**, 1532–1542.

Spreer, W. and Müller, J. (2011) Estimating the mass of mango fruit (*Mangifera indica* cv. Chokanan) from its geometric dimensions by optical measurement. *Computers and Electronics in Agriculture*, **75**, 125–131.

Tedjo, W., Taiwo, K.A., Eshtiaghi, M.N. and Knorr, D. (2002) Comparison of pretreatment methods on water and solid diffusion kinetics of osmotically dehydrated mangoes. *Journal of Food Engineering*, **53**, 133–142.

Teoh, C.C. and Syaifudin, A.M. (2007) Image processing and analysis techniques for estimating weight of Chokanan mangoes. *Journal of Tropical Agriculture and Food Science*, **35**, 183.

Terefe, N.S., Buckow, R. and Versteeg, C. (2014) Quality-related enzymes in fruit and vegetable products: effects of novel food processing technologies. Part I: High-pressure processing, *Critical Review in Food Science Nutrition*, **54**, 24–63.

Tran, T.T.L., Aimla-or, S., Srilaong, V., Jitareerat, P., Wongs-Aree, C. and Uthairatanakij, A. (2013) Fumigation with ozone to extend the storage life of mango fruit cv. Nam Dok Mai No. 4. *Agricultural Science Journal*, **44**, 663–672.

Valente, M., Prades, A. and Laux, D. (2013) Potential use of physical measurements including ultrasound for a better mango fruit quality characterization. *Journal of Food Engineering*, **116**, 57–64.

Vásquez-Caicedo, A. L., Schilling, S., Carle, R. and Neidhart, S. (2007) Effects of thermal processing and fruit matrix on β-carotene stability and enzyme inactivation during transformation of mangoes into purée and nectar. *Food Chemistry*, **102**, 1172–1186.

Vega-Mercado, H., Martin-Belloso, O., Qin, B.L., Chang, F.J., Góngora-Nieto, M.M. *et al.* (1997) Non-thermal food preservation: Pulsed electric fields. *Trends in Food Science and Technology*, **8**, 151–157.

Velazquez, G., Candelario, H.E., Ramírez, J.A., Montoya, P., Loera-Gallardo, J. and Vázquez, M. (2010) An improved quarantine method for mangoes against the Mexican fruit fly based on high-pressure processing combined with heat. *Foodborne Pathogens and Disease*, **7**, 493–498.

Vélez-Rivera, N., Blasco, J., Chanona-Pérez, J., Calderón-Domínguez, G., de Jesús Perea-Flores, M. *et al.* (2014) Computer vision system applied to classification of "manila" mangoes during ripening process. *Food and Bioprocess Technology*, **7**, 1183–1194.

Walker, M.K., Farkas, D.F., Loveridge, V. and Meunier-Goddik, L. (2006) Fruit yogurt processed with high pressure. *International Journal of Food Science and Technology*, **41**, 464–467.

Youssef, B.M., Asker, A.A., El-Samahy, S.K. and Swailam, H.M. (2002) Combined effect of steaming and gamma irradiation on the quality of mango pulp stored at refrigerated temperature. *Food Research International*, **35**, 1–13.

Zheng, H. and Lu, H. (2012) A least-squares support vector machine (LS-SVM) based on fractal analysis and CIELab parameters for the detection of browning degree on mango (*Mangifera indica* L.). *Computers and Electronics in Agriculture*, **83**, 47–51.

10

Mango Processing and Processed Products

Muhammad Siddiq[1], *Dalbir S. Sogi*[2] *and Sunisa Roidoung*[3]

[1] *Department of Food Science and Human Nutrition, Michigan State University, East Lansing, Michigan, USA*
[2] *Department of Food Science and Technology, Guru Nanak Dev University, Amritsar, Punjab, India*
[3] *Department of Food Technology & Nutrition, Maha Sarakham University, Maha Sarakham, Thailand*

10.1 Introduction

Mango is one of the major tropical fruit and considered to be among one of the most consumed fresh fruits in the world. In recent years, mango consumption has grown significantly outside the major mango producing and consuming countries, for example, in North America and the European Union. About 75% of the total import is by the USA and Europe. Table 10.1 shows the volume and value of US imports of fresh mangoes and mango products for selected years. The 2010 to 2015 increase in volume and value for different mango products was 21.8–289.0% and 61.5–266.9%, respectively. The US Department of Agriculture has predicted a 7% growth of mango imports into the US.

Mango is a unique fruit in terms of the diverse products processed from it, especially from its immature to fully-ripe stages. The most common processed mango products are puree/pulp, nectar, juice, juice concentrate, and dried/dehydrated mangoes. Besides these common products, there are a number of traditional products which are processed commercially in major mango producing countries, such as pickles, sweet or sour chutney (a tomato ketchup type sauce), *amchoor* or dried powder, mango leather, and a variety of soft drinks and beverages (Siddiq *et al.* 2012). In addition, during the mango production season, freshly made mango shakes and smoothies are popular in South Asian countries. This chapter covers processing and quality aspects of mango products, and their nutrition and health benefits. The food safety aspects and the use of innovative processing technologies are also covered in this chapter.

10.2 Processed Mango Products

Mangoes are used for preparing a number of flavorful and nutritious products. Currently, chemically preserved, canned, dried, and frozen forms of mango pulp and jams, jellies, canned slices, dehydrated pulp (mango leather), frozen chunks and slices,

Handbook of Mango Fruit: Production, Postharvest Science, Processing Technology and Nutrition,
First Edition. Edited by Muhammad Siddiq, Jeffrey K. Brecht and Jiwan S. Sidhu.

Table 10.1 US imports of fresh mangoes and processed mango products for the years 2010 and 2015.

	Volume (million metric tons)		Value (million US $)		2010-15 Increase (%)	
Mango product	**2010**	**2015**	**2010**	**2015**	**Volume**	**Value**
Fresh	320.5	390.6	248.4	401.1	21.8	61.5
Frozen	29.3	63.3	38.6	131.2	115.7	240.2
Dried	5.9	8.8	43.3	82.1	50.0	89.9
Juice	2.2	8.5	2.3	8.5	289.0	266.9
Others	45.9	88.1	66.1	125.8	92.0	90.2

Source: Adapted from USDA-ERS (2016).

traditional pickles, and chutneys, are commonly processed from mangoes (Occena-Po 2006). The bulk of immature mangoes that are often blown down by heavy winds are best utilized to produce mango powder (*amchoor*). Owing to its higher moisture content and relatively shorter shelf life, the postharvest losses of mango are excessive in India and Pakistan, where less than 10% of mangoes are processed (Siddiq *et al.* 2012). Other mango products of significance include sauces and jelly made of half-ripe or green mangoes, and peeled and sliced as fillings for pies. Many types of chutneys are manufactured by using green mangoes with added sugar, salt, and various spices (Morton 1987).

Detailed process flow outlines for the production of various mango products are shown in Figure 10.1 (except for the fresh-cut mangoes that are covered in Chapter 8, the rest of the products shown in this figure are discussed in this chapter). Siddiq *et al.* (2012) noted that most of the preparation steps, for example selection of fruit, grading/sorting, washing, peeling, and de-stoning, are generally similar for most of the processed mango products.

10.2.1 Mango Puree and Pulp

Generally, terms "pulp" and "puree" are interchangeably used, although the puree is more "refined" since it has less fibrous content and smooth consistency. These two products serve as a base for a variety of processed mango products, e.g., nectar, beverages, jam, jelly, and leather. Both pulp and puree are primarily preserved by chemical preservatives, canning, or aseptic processing/packaging, and occasionally by freezing. Concentrated mango pulp holds a great commercial significance and is bulk-packaged for local as well as export markets.

A typical mango puree production outline is shown in Figure 10.1. Harvesting mangoes at the full-ripe stage is recommended for the production of puree. Peeling is optional, and washed fruits are mashed and passed through a pulper. Mango pits (stones), peels, and residual parts are removed by successive pressing of the pulp through different size screens. The final refining of the puree is achieved by passing it through a 30-mesh (~0.6 mm) screen. For a finer puree, the use of a finishing screen (pore size <0.5 mm) further removes fibrous material, thereby giving a smooth consistency to the product (Occena-Po 2006). Mango puree is typically sweetened to 42 °Brix

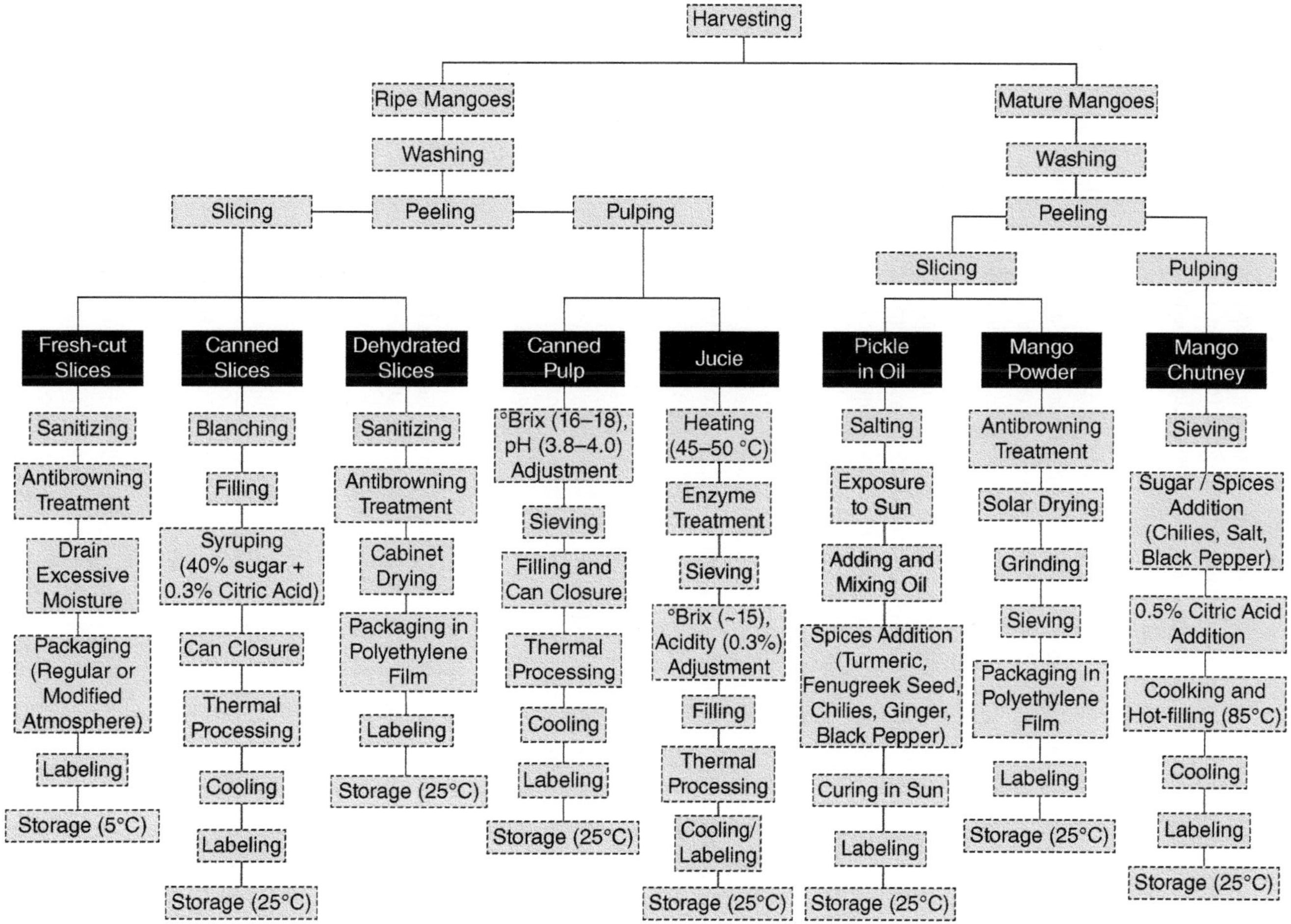

Figure 10.1 Processing steps for the production of different mango products.

and citric acid is added to acidify the product and lower its pH, which helps to retain and improve color, flavor, and overall quality. For canning, mango puree is hot-filled into cans at 85 °C and thermally processed at 100 °C for 45 min (Narain *et al.* 1998).

Kermani *et al.* (2015) studied the feasibility using mango endogenous pectinases to improve the viscosity of mango puree. Pectin in the mango puree contained relatively high amounts of galactose and considerable β-galactosidase activity. These researchers further noted that creating processing conditions allowing endogenous β-galactosidase activity did not change the consistency of mango puree significantly. Sakhale *et al.* (2016) reported that the pectinase and gelatin each at 0.5% concentration significantly improved the yield, soluble solids, and ascorbic acid content of the pulp. The sensory quality was improved significantly, both by pectinase alone and also with pectinase and gelatin combined. Kaushik *et al.* (2015) assessed the enzyme inactivation in mango pulp during high pressure processing (HPP), with pH and total soluble solids (TSS) as additional hurdles. Response surface and contour models predicted the kinetic behavior of mango pulp enzymes adequately, as indicated by the small error between predicted and experimental data. These researchers concluded that the exploration of additional hurdles such as pH, TSS, and temperature for enzyme inactivation during HPP processing of fruits is useful from an industrial point of view, as these parameters play a key role in the preservation process design.

The physicochemical attributes of mango fruit, puree, and pulp are given in Table 10.2. Guerrero-Beltrán *et al.* (2005) reported that high hydrostatic pressure (HHP) processing with added ascorbic acid improves the color stability of mango puree during refrigerated storage; however, the addition of cysteine was shown to have a negative effect on puree color during 48-day storage. Guerrero-Beltrán *et al.* (2006) also studied the effect of ascorbic acid on the microbiological quality of HPP-processed mango puree. Mango puree containing 500 ppm ascorbic acid, standardized at pH 3.5 (with phosphoric acid), and inoculated with *Saccharomyces cerevisiae*, was processed at 207, 345, 483, and 552 MPa. The HHP-processed (552 MPa, 5 min) mango puree was stored at 3 °C for 1 month. No microbial growth, as shown by less than 10 CFU/g, was observed in HHP-processed mango puree.

Mango pulp can be effectively preserved with the help of chemical preservatives, such as potassium metabisulfite and sodium benzoate at suitable concentrations. Both of these preservatives have antimicrobial effect, but potassium metabisulfite is more

Table 10.2 Proximate and physicochemical composition of raw mangoes, mango puree, and pulp (%, fresh-wt basis).

Composition	Mango fruit[1]	Puree[1]	Pulp[2]
Water	72.1–85.5	67–86	81–83
Protein	0.30–5.42	0.3–1.5	0.5–0.6
Crude fiber	0.30–2.38	0.05–0.6	0.3–0.4
Ash	0.29–1.13	0.13–0.61	0.4
Total sugars	10.5–18.5	12.28–17.54	15.4–16.9
Titratable acidity, as citric acid	0.327	0.1–1.1	0.4–0.5

Source: [1]Occena-Po (2006), [2]Anon. (2016).

effective than sodium benzoate. Slight negative effects of these preservatives have been observed on the sensory characteristics of juices prepared from chemically preserved mango pulp (Akhtar *et al.* 2010). Ahmed *et al.* (2005) evaluated the rheological properties and color of mango pulp processed under HPP; 100–400 MPa for 15 or 30 min at 20 °C. The consistency index (K) of fresh pulp increased with increasing pressure, from 100 to 200 MPa, while a steady decrease was observed for canned pulp. The HPP treatment was shown to have no detrimental effects on pulp color.

10.2.2 Mango Juice and Juice Concentrate

Mango juice can be prepared either from pre-processed puree (a common practice commercially) or from fresh mangoes, as shown in Figure 10.1. Depending on the final TSS of the juice (~12–15 °Brix), the puree is mixed with water at a predetermined ratio (Siddiq *et al.* 2012). Occena-Po (2006) reported that mango juice is used directly as single-strength, blended with other juices, sold as juice blends, or used in fruit smoothies/shakes. As compared to the pulp, the chemical composition of canned mango juice is characterized by a lower content of ascorbic acid, free amino nitrogen, reducing sugars, and carotenoids (Table 10.3), while the pulp is characterized by lower TSS, higher pH, lower total sugars, and lighter color (Ragab and El-Nemr 1990).

The enzyme treatment of puree is a critical step in mango juice production. In addition to the combination and rate of enzymes used, treatment time and temperature also play a significant role in the final consistency of the juice. To maximize juice yield with optimum viscosity and cloud stability, the use of pectinase and cellulase enzymes is common to break down pectin and cellulose. Enzyme treated puree is filtered through a fine screen, typically 0.4 mm or US sieve size 5. To adjust the desired acidity level, citric and/or ascorbic acid is added at the 0.4–0.5% level (Occena-Po 2006). Added organic acids also help to achieve an acceptable sugar–acid ratio for optimum flavor. For processing, mango juice is heated to 85 °C using heat exchangers and filled into 16 oz or 64 oz cans. After sealing, cans are processed at 100 °C for 20–30 min, followed by cooling to under 40 °C temperature. After drying the outer surface of cans with air jets, cans are labeled and palletized for shipment to the retail markets.

Table 10.3 Physico-chemical properties of mango pulp and canned mango juice (fresh-wt basis).

Composition/Property	Mango Pulp[1]	Mango Pulp[2]	Canned Juice[1]
Total soluble solids (°Brix)	12.1	14.2–14.5	18.5
pH value	4.65	4.0–4.4	4.23
Acidity (% citric acid equiv.)	–	0.25–0.32	–
Ascorbic acid (mg/100 g)	15.56	–	5.6
Total sugars (%)	8.59	12.6–13.1	17.41
Reducing sugars (%)	1.47	1.9–2.5	1.13
Non-reducing sugars (%)	7.12	10.5–10.7	16.28
Carotenoids (mg/100 g)	3.95	–	1.4

Source: [1]Ragab & El-Nemr (1990); [2] Singh *et al.* (2000).

Vaillant *et al.* (2001) reported that the enzymes treatment applied at 0.14 g per liter at 27–30 °C for 30 min produced the highest juice yield. These researchers also proposed a microfiltration process for producing clarified juices and concentrates from pulp-rich juices, such as mango (Figure 10.2). The appropriate enzyme treatment of diluted fruit purees and pulps can increase the efficiency of the membrane filtration process. Incubation of enzymes-added macerate at 45 °C for 6 hours was shown to be optimum for maximum juice yield (Chauhan *et al.* 2001). A 30 min liquefaction with 1.1 ml enzyme mixture/kg mango pulp was reported to be optimal for the production of clear mango juice following subsequent centrifugation (Singh *et al.* 2000). The mixture consisted of commercial pectinolytic and cellulytic enzymes, such as Pectinex, Celluclast, and Rapidase.

Mango juice concentrate can be produced from mango juice or puree as a base or starting material. When the concentrate is to be prepared from puree, treatment with enzymes to break down pectin and cellulose, followed by centrifugation or fine-mesh filtration, is necessary. Vacuum evaporation is the most widely-used process for the concentration of mango juice. Depending on the intended end use, juice concentrates of variable TSS contents, ranging from 28 to 60 °Brix are prepared (Siddiq *et al.* 2012; Wu *et al.* 1993). Concentration of fruit juices by evaporation or osmosis is greatly facilitated by removal of pulp and pectin by ultrafiltration or cross-flow microfiltration (Johnson *et al.* 1996). Wrolstad *et al.* (1993) reported that concentration of fruit juices by membrane technology is an energy efficient alternative to traditional processes.

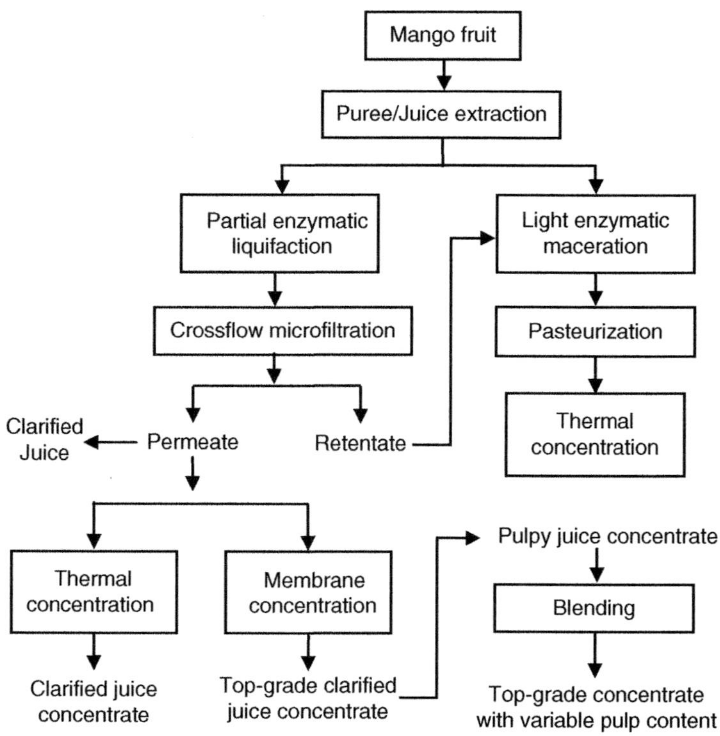

Figure 10.2 Proposed flowchart for producing clarified juice and juice concentrate from pulp juices on an industrial scale. *Source:* Vaillant *et al.* (2001).

Table 10.4 Effects of UV-C treatment and thermal pasteurization on clarity, total carotenoids, and ascorbic acid content of Chokanan mango juice.

Treatment	Clarity, transmittance (at 660 nm)	Total carotenoids (μg/100 ml)	Ascorbic acid (mg/100 ml)
Control	25.50 ± 0.22	82.03 ± 1.29	8.91 ± 0.51
UV-C, 15 min	25.40 ± 0.08	87.10 ± 1.14	7.85 ± 0.24
UV-C, 30 min	25.17 ± 0.15	84.97 ± 1.35	7.55 ± 0.30
UV-C, 60 min	24.25 ± 0.10	80.16 ± 1.80	6.87 ± 0.12
Pasteurization	7.30 ± 0.14	48.92 ± 1.32	3.10 ± 0.62

Source: Adapted from Santhirasegaram *et al.* (2015).

A number of non-traditional technologies have been explored for the processing of mango juice. Santhirasegaram *et al.* (2015) compared the effect of UV-C light treatment (15, 30, and 60 min, at 25 °C) and thermal pasteurization (90 °C, 1 min) of Chokanan mango juice on microbial inactivation, physicochemical properties, antioxidant activities, and other quality attributes. Results showed that a significant increase in extractability of carotenoids (6%), polyphenols (31%), and flavonoids (3%) was observed in juice exposed to UV-C light for 15 and 30 min, when compared to the control (selected data shown in Table 10.4). Thermal pasteurization and UV-C treatment exhibited significant reduction in microbial load and prolonged shelf life of juice. The UV-C treatment also resulted in the enhancement of antioxidant activity.

Hiremath and Ramaswamy (2012) investigated the high-pressure (HP) destruction kinetics of three common spoilage microorganisms, *Zygosaccharomyces bailii, Pichia membranaefaciens,* and *Leuconostoc mesenteroides,* and two pathogenic microorganisms, *Escherichia coli* O157:H7 and *Listeria monocytogenes* Scott A, in mango juice. Processing conditions used were 250–550 MPa and 0–60 min hold time at 20–25 °C. *L. mesenteroides* was found to be the most resistant spoilage bacteria, yet its complete destruction was achieved by a 5-min treatment at 400 MPa. The pathogenic *E. coli* was the most resistant of all the bacterial strains tested. A six-log reduction in *E. coli O157:H7* was observed at 400 MPa when treated for 10 min, while at 500 and 550 MPa, there were no survivors after the 1-min treatment.

The ultra-high pressure processing (UHP; 500 MPa, 25 °C, 20 min) for the retention of volatile compounds of mango juice was studied by Zhang *et al.* (2008). It was shown that UHP treatment led to increases in the contents of linalool, furfural, cis-3-hexen-1-ol, and hexanol by 23.4, 79.1, 93.8, and 128.6%, respectively. Moreover, 2-hexenal (0.28%) and octadecanal (0.24%) were only found in UHP-treated mango juice. Sensory analysis showed that UHP processing preserved the flavor and characteristic aroma of mango fruit in the juice.

10.2.3 Mango Nectar

Mango nectar, a ready-to-drink beverage, is prepared by diluting single-strength mango pulp with the addition of sugar, water, and preservatives to make a product that has a certain percentage of mango pulp/puree. Mango nectar can be made either from mangoes directly or from canned, aseptically packaged, or frozen puree (Siddiq *et al.* 2012). The use of honey as a natural sweetener for mango nectar has also been reported (Lakhanpal and Vaidya 2015). According to the guidelines in the *Codex Alimentarius*, the percentage of pure juice product in the beverage must be noted on the label, e.g., a statement like "this nectar contains at least 30% pure mango pulp" must be put on the label (FSA 2007). In the US and European markets, regulations state that mango nectar must have at least 25% pure mango pulp. Occena-Po (2006) noted that the major difference between mango juice and nectar is that the latter contains lower fruit content.

For making nectar from mango puree, 20–30 °Brix puree is mixed with water and other ingredients (sugar, citric acid, vitamin C), and carboxymethylcellulose is added as a stabilizer. After thoroughly mixing the ingredients, the nectar is filtered through a finishing screen of appropriate size. Nectar viscosity plays a critical role in the selection of process parameters, such as time, temperature, initial fill temperature, etc. Mango nectars are available with 20–33% pulp content, and the commercially packed nectar typically contains more than 25% mango pulp (Occena-Po 2006). Wu *et al.* (1993) reported that nectar can be processed using one of the three methods, where nectar is:

- vacuum-sealed in cans and then processed at 100 °C for 3 min in an agitating retort;
- pasteurized at 80 °C, using a plate heat exchanger, then filled into cans, sealed, and processed at 80 °C for 10 min; or
- pasteurized at 95 °C for 1 min, using a plate heat exchanger, and then aseptically packaged in plastic-lined bulk drums.

In recent years, a number of innovative technologies have been investigated for the processing of mango nectar to assess the effect on microbial and other quality attributes. Liu *et al.* (2014) processed mango nectar using HHP (600 MPa, 1 min) and high-temperature short-time (HTST; 110 °C, 8.6 sec) treatments and studied their effect on natural microorganisms and other physico-chemical properties. The results of microbial counts are shown in Table 10.5, which were not different statistically between HPP and HTST treatments. Similarly, there were no differences in the viscosity of HPP- and HTST-treated nectars, although viscosity decreased gradually during 16 weeks of storage at 4 and 25 °C. Tribst *et al.* (2011) evaluated the effect of high-pressure homogenization (HPH) with heat shock on *Aspergillus niger*, vitamin C, and color of mango nectar, which was processed at 200 MPa followed by heat shock at 60–85 °C for 10–20 min. The results showed that *A. niger* can be inactivated by applying HPH with heat shock, particularly as an option to commercial pasteurization. Kumar *et al.* (2015) studied the feasibility of pulsed electric fields (PEF) processing for mango nectar at different pulse frequencies (70–120 Hz) and pulse widths (15–24 μsec). Mango nectar exhibited high retention of carotene content at higher frequencies and pulse widths as compared to lower levels. Maximum inactivation of native microflora, retention of carotenoids content, and best overall sensory acceptability were obtained when PEF treatments were set up to 38.0 kV/cm for 24 μsec using bipolar pulses at 120 Hz.

10.2.4 Aseptically-packaged Mango Beverages and Squash

Since the early 1980s, aseptically-packaged mango beverages have been popular among consumers due to convenience and shelf-stable "brick" type packaging (trade name

Table 10.5 Microbial counts in HPP and HTST processed mango nectar during 16 weeks of storage at 4 °C and 25 °C ($\log_{10}$CFU/ml).

	4 °C		25 °C	
Storage (weeks)	**HTST[1]**	**HHP[2]**	**HTST**	**HHP**
Total aerobic bacteria (6.81 ± 0.03[3]):				
0	1.11 ± 0.05	1.57 ± 0.04	1.11 ± 0.05	1.57 ± 0.24
1	1.01 ± 0.10	1.70 ± 0.10	1.02 ± 0.01	1.42 ± 0.09
2	1.13 ± 0.15	1.66 ± 0.08	1.14 ± 0.07	1.20 ± 0.16
4	1.04 ± 0.02	1.63 ± 0.13	1.08 ± 0.13	1.66 ± 0.14
8	1.09 ± 0.04	1.53 ± 0.04	1.14 ± 0.11	1.26 ± 0.13
12	1.07 ± 0.05	1.49 ± 0.15	1.11 ± 0.06	1.01 ± 0.13
16	1.03 ± 0.07	1.53 ± 0.13	1.13 ± 0.08	1.18 ± 0.11
Yeast and mold (3.16 ± 0.13[3]):				
0	<1.00	<1.00	<1.00	<1.00
1–16	<1.00	<1.00	<1.00	<1.00

[1] High temperature short time treatment.
[2] High hydrostatic pressure treatment.
[3] Microbial count in untreated or control nectar.
Source: Liu *et al.* (2014).

Tetra Pak®). The process of aseptically-packaged mango beverages is the same as for many other similar drinks, where a sterile product is filled into sterile packages and hermetically sealed in a sterile environment. The formulation with respect to mango pulp varies widely in aseptically-packaged mango beverages and some are marketed as blends containing pulp from other fruits (Occena-Po 2006).

For mango squash, a concentrated mango drink, juicy varieties of mangoes are preferred. Mango nectar and squash are similar in composition, except for the presence of a preservative – 0.1% sodium benzoate or 350 ppm SO_2 in squash (Occena-Po 2006). Mango squash formulation consists of blending 68 °Brix sugar syrup with mango puree, citric acid, and preservative (Narain *et al.* 1998). The contents are heated to approximately 90 °C and hot-filled into pre-sterilized glass bottles. Bottles are closed with screw-type caps or cork stoppers; if cork stoppers are used, the bottles are inverted in hot-wax to put a coating on the cork and bottle neck to make the closure air tight. Mango squash is diluted about 3 times with water before serving in the chilled form (Siddiq *et al.* 2012).

Hussain *et al.* (2005) studied the effect of cultivar and storage on quality of mango squash from Fajri, Desi, Ting, Chonsa, and Totapari mangoes, by mixing pulp with sugar and water (1:1:1) and 0.06% potassium metabisulfite added as a preservative. Results of this study showed that the sensory score for color and flavor decreased significantly during storage. Soluble solids, reducing sugars, and pH increased significantly during storage, while contents of ascorbic acid and non-reducing sugars and acidity decreased.

Bhatt *et al.* (2015) investigated the feasibility of producing ready-to-serve (RTS) beverages and squash from naturally dropped unripe mangoes, at three stages of fruit weight (11.82 ± 0.84, 33.50 ± 1.63, and 65.27 ± 5.86 g). Their results showed that fruit from all three stages can be utilized successfully in preparation of refreshing and nutritious RTS beverages and squash, with overall acceptability being the highest where these products were prepared from third-stage (highest weight/fruit) dropped mangoes.

10.2.5 Dried Mango Products

Dried mango products (slices, pieces, powders, and flakes) are prepared from ripe mangoes, and drying is done using a variety of methods (hot-air cabinet drying, vacuum-drying, spray drying, freeze drying, foam-mat drying, and osmotic dehydration). Dried mango products are intended either for direct consumption (dried slices, dices/cubes, mango chips) or for use in other food formulations (mango leather and powder) (Siddiq *et al.* 2012). Corzo and Alvarez (2014) suggested that to obtain a better understanding of the industrial process applications, it is important to model and quantify the kinetics and process parameters, especially the effects of air temperature and velocity on color change kinetics of mango slices at different maturity stages during air drying.

10.2.5.1 Mango Slices/Dices and Chips

Among dried mango products, dehydrated slices have a significant share of the market. Dehydrated mango slices have become a popular snack item outside the markets of major mango producers, for example in North America and the EU. A typical process outline for the production of dehydrated mango slices is shown in Figure 10.1. The processing steps for dehydrated slices, dices/cubes, and chips are similar; the only difference is the shape and size of the product. After washing and peeling/pitting, mangoes are sliced using sanitized stainless steel knives. The cut slices are subjected to selected pretreatments, such as dips in approximately 40° Brix sugar, 0.03% SO_2, 0.2% ascorbic acid, and 0.5–1% citric acid, used to preserve product color and improve product stability (Siddiq *et al.* 2012). The pretreated slices are generally dried in regular or forced-air cabinet dryers at a temperature of 60–65 °C. The primary process factors that can influence color and water activity of slices are air temperature and drying time (Pott *et al.* 2005). The dried mango slices have a shelf life of one year, and slices keep good sensory and physicochemical quality without any sign of visible browning. The final moisture content of the dried mangoes is critical for maintaining the optimum shelf life.

Dissa *et al.* (2009) developed a prototype indirect solar dryer, as shown in Figure 10.3. The solar energy collector panel converts radiant energy received on its surface into heat. The air crossing the porous matrix of its absorber then receives a portion of this energy by convective heat exchange. The resulting heated air circulates in the unit for drying mango slices laid out in a single layer on mesh trays. Using this solar dryer, 8-mm thick mango slices can be dried in 3 days, with the final water content corresponding to 0.6 water activity, which is suitable for optimum shelf life (Dissa *et al.* 2009). In the developing countries, mango slices are also sun-dried or by using solar dryers, the main reasons for the use of these indigenous methods being the high energy requirements and equipment costs for the commercial equipment.

Chen *et al.* (2007) reported that drying method and pretreatments were shown to have a significant effect on total carotenoid content of dried mangoes (Figure 10.4). Soaking in sodium bisulfate ($NaHSO_3$) prior to freeze-drying gave the highest yield of

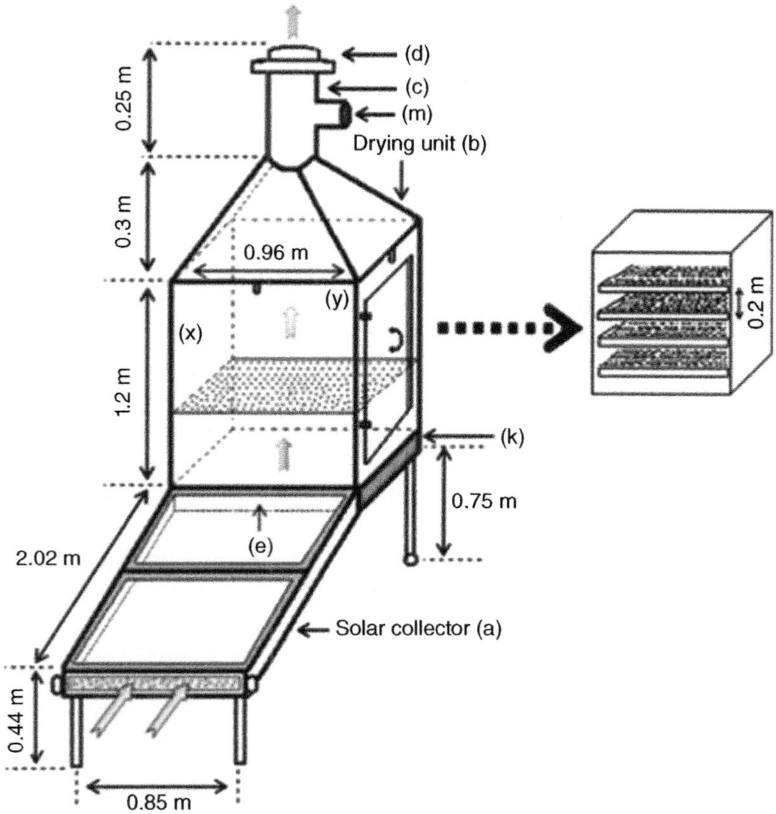

Figure 10.3 Prototype of indirect solar dryer. Labeled parts are: (a) solar collector; (b) drying unit; (c) chimney; (d) air extractor; (e) air entrance; (k) drying unit bottom side; (x) drying unit small side; (y) drying unit large side; (m) air recycling pipe. *Source:* Dissa *et al.* (2009).

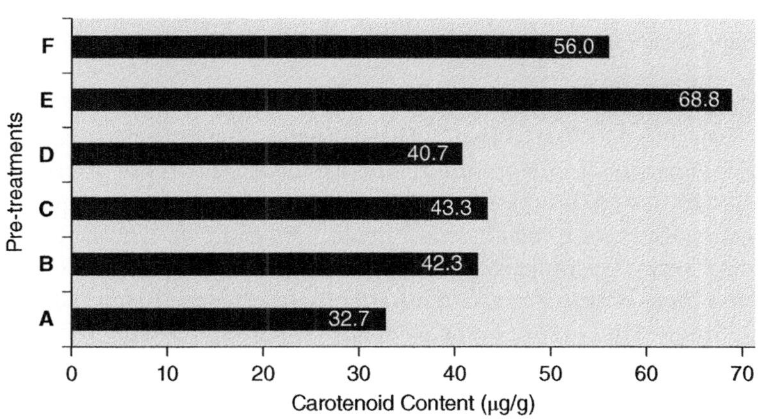

Figure 10.4 Effect of different drying methods on total carotenoid content of mango. Mango pieces were: (A) dried under hot-air at 60 C for 20 hours (B) soaked in 1% sodium hydrogen sulfite solution for 30 min prior to treatment-A; (C) soaked in 1% ascorbic acid solution for 30 min prior to treatment-A; (D) freeze-dried (−53 C, 0.06 torr) for 32 hours; (E) soaked in 1% sodium hydrogen sulfite solution for 30 min prior to treatment-D; and (F) soaked in 1% ascorbic acid solution for 30 min prior to treatment-D. *Source:* adapted from Chen *et al.* (2007).

total carotenoids (68.81 μg/g), followed by soaking in ascorbic acid (55.98 μg/g) before freeze-drying, soaking in ascorbic acid (43 μg/g) or $NaHSO_3$ (42 μg/g) before hot-air drying, freeze-drying (41 μg/g), and hot-air drying (33 μg/g) without soaking.

Besides air drying, osmotic dehydration is another method being used increasingly to produce dehydrated mango products (Giraldo *et al.* 2003). Osmotic dehydration involves dipping mango slices in 70 °Brix sugar syrup, draining, dipping in sulfite solution followed by a second draining, and then drying in an air convection oven. Vacuum dryers can also be used (at 2–4 mm Hg, 60–66 °C) to achieve the desired moisture content, typically below 10% (Nagai *et al.* 2015; Narain *et al.* 1998). In general, the use of highly concentrated osmotic solutions and the high level of osmo-dehydration induces loss of volatiles. On the other hand, more diluted solutions and shorter treatment times (lower osmo-dehydration level) can give rise to the enhancement of volatile production (Torres *et al.* 2007).

In recent years, some innovative methods have been used for drying mangoes, with the objectives of improving process efficiency and/or for better retention of nutrients. Pu and Sun (2016) used microwave-vacuum drying for mango slices. These authors reported that the moisture content can be predicted accurately using NIR hyperspectral imaging. It was shown that moisture content uniformity is affected by different shapes of slices – square, rectangle, regular triangle, and round. De Medeiros *et al.* (2016) explored a dual-stage technique as a pretreatment for studying mango drying kinetics and its effect on the finished product quality. These pretreatments consisted of two stages in which high-calorie sugars are partially removed from fruit pieces in the first stage and, in the second stage, low-calorie sugar is incorporated into the fruit to maintain its sweetness. Sensory evaluation showed pretreated samples having good consumer acceptance. Guiamba *et al.* (2015) evaluated infrared (IR) dry blanching with conventional water blanching prior to hot-air drying of mango to inactivate polyphenol oxidase and ascorbic acid oxidase enzymes, and assessed the effect on color change and retention of vitamin C and beta-carotene. The results showed that IR blanching of mango prior to drying can improve the retention of vitamin A and C, but not the retention of carotenoids and vitamin E, which demonstrated that this effect was not dependent on the water- or lipid-solubility of these vitamins.

10.2.5.2 Mango Chips and Flakes

Nunes and Moreira (2009) developed high quality mango chips using vacuum frying. Tommy Atkins mango slices were pretreated with maltodextrin at concentrations of 40–65%, and dehydrated osmotically (at 22 and 40 °C solution temperature, and 45–70 min). Pretreated slices were vacuum fried at 120, 130, and 138 °C and different quality attributes were evaluated. Mango chips under atmospheric frying had 32% less carotenoid retention than those under vacuum frying. Consumer taste panel showed that all dehydrated mango chips were of acceptable quality (scores >5, on a 9-point scale).

Drum-drying technology can be used to produce mango flakes. This technique has been used for drying pulps or purees from fruits and vegetables. Drum-drying of mango puree is an efficient, economical process for producing dried mango powder and flakes. The production of drum-dried mango flakes involves blending mango pulp with small amounts of wheat flour and sugar. The homogenized mixture is dried using an atmospheric double-drum dryer. Flakes can be milled and screened into mango powder or used directly, depending on the intended final use (Wu *et al.* 1993).

10.2.5.3 Mango Leather

Fruit leather is made by drying a very thin layer of fruit puree to obtain a product with a chewy texture similar to soft leather (Andress and Harrison 2006). The process involves passing the puree through a mechanical finisher with a stainless steel sieve (30–50 mesh, 0.595–0.297 mm) and soluble solids adjusted to 25–35 °Brix. Potassium metabisulfite may be added as a preservative for extended shelf life. The puree is spread evenly in a thin layer of 1 cm on glycine-coated metal trays and dried in a forced-air dryer (50–80 °C, 2–20 hours) to a final moisture content of 15–20% (Wu *et al.* 1993).

Gujral *et al.* (2013) studied the effect of adding sucrose (0, 5, 10%), pectin (0, 1, 2%), and maltodextrin (0, 2.5, 5%) on the moisture diffusion from pulp during the making of mango leather in a cabinet dryer. The drying rate of mango leather was reduced by the addition of sucrose, pectin, and maltodextrin. The drying rate constant (k) of 0.116 liter/hour was the lowest in the mango pulp containing 10, 2, and 5% of sucrose, pectin, and maltodextrin, respectively. Singh and Singh (2003) reported that the addition of hydrocolloids (guar gum or pectin) reduces redness and yellowness of dried mango leather.

In developing countries, sun drying is also employed for preparing mango leather from ripe fruit pulp; however, the sun-dried product becomes discolored and the process is unhygienic and lengthy. In India, mango leather production involves manual extraction of mango puree and straining through bamboo sieves to remove the fibrous material. The puree is spread on date palm leaf mats and dried for about 40 days. Upon drying, another layer is spread over the first, and the process is repeated until a desired thickness of 2–3 cm is obtained. Mango leather thus prepared is sliced and packaged in cellophane, polyethylene, or waxed paper (Narain *et al.* 1998).

10.2.5.4 Mango Powder

Dried mango powder can be produced by drying mango puree to about 3% moisture using one of the following methods: spray drying, freeze drying, foam-mat drying, puff drying, vacuum drying, or drum drying (Occena-Po 2006). Mango powder stores well for up to 6 months in polyester or metallized polyester pouches at 27–32 °C. It has been observed that extended storage of mango powder in polyester packaging beyond 6 months increases its total sugar and reduces ascorbic acid concents (Hymavathi and Khader 2005a). Most appropriate packaging material suggested for the mango powders is high density polyethylene (HDPE) film with aluminum lamination (Hymavathi and Khader 2005b). Dry mango powders can be reconstituted into juice or used as a base for the preparation of products such as beverages, baby foods, baked good, and snacks.

Sogi *et al* (2015) dehydrated Tommy Atkins mango cubes to produce powder using different methods: hot-air/cabinet, vacuum, infra-red, and freeze drying. As shown in Table 10.6, mango powders had 5.86–10.61% moisture, 6.22–7.86% acidity, and pH 3.65–3.83. Water and oil absorption capacities of mango powder were 2.54–3.14 and 1.69–2.75 g/g, respectively. Solubility of cabinet-dried powder was the highest. Color values indicated dominance of yellowness. Mango powder contained a high quantity of phenolics (936.2–1,725.2 mg GAE/100 g), ascorbic acid (97–225 mg/100 g), and total carotenoids (3.3–5.2 mg/100 g); freeze-dried powder had the highest antioxidant capacity than samples from the other three drying techniques. The antioxidant capacity, determined as the ORAC value, ranged from 408–651 μmol TE/100 g (Figure 10.5). The mango powder can be utilized for enhancing flavor and added fiber and

Table 10.6 Effect of drying techniques on the physico-chemical properties of mango powder (dry-wt basis except moisture).

Properties	FD[1]	CD[2]	VD[3]	IRD[4]
Moisture (%)	9.83	7.24	5.86	10.61
pH	3.65	3.72	3.83	3.63
Titratable acidity (%)	7.86	6.74	6.22	7.52
Solubility (%)	77.60	78.69	74.17	66.80
Water Absorption Index (g/g)	2.76	2.87	2.54	3.14
Oil Absorption Index (g/g)	2.75	1.93	1.69	1.95
Hunter color *L* value	88.93	73.07	54.33	49.37
Hunter color *a* value	−0.77	9.53	11.40	9.33
Hunter color *b* value	35.67	32.30	18.70	14.07

[1] Freeze drying.
[2] Cabinet drying.
[3] Vacuum drying.
[4] Infrared drying.
Source: Sogi *et al.* (2015).

antioxidant-rich phytochemicals for preparing healthy and nutritious food products, especially snacks. Different methods of producing mango powder are discussed below.

10.2.5.4.1 Spray Drying

Spray drying is the most commonly used commercial process for drying fruit purees. Mango puree, liquid glucose (5%), and tri-calcium phosphate (0.5%, as an anti-caking agent) are mixed. The mixture is then heated at 50 °C for 30 min, homogenized at 2,800 rpm, pumped to the atomizer (at 2,800 rpm), and spray dried at 165 °C inlet temperature. The powder is cooled to about 4 °C before transferring to a dehumidified room for packaging (Wu *et al.* 1993). Spray-dried mango powder has a nice color, but lacks characteristic mango flavor. The addition of sugar, milk solids, glyceryl monostearate, and alginate to the puree improves the flavor and overall quality of the mango powder (Nanjundaswamy 1997).

10.2.5.4.2 Foam-mat Drying

Mango powder produced by foam-mat drying has a better nutritional quality and enhanced shelf life. This method involves the incorporation of foaming agents into liquid foods (e.g. mango puree) with subsequent whipping to form a stiff foam. The foam is then spread on a tray and dried in an air stream. In general, drying of foamed material is faster than the non-foamed ones (Wilson *et al.* 2014). Hardy and Jideani (2016) reported that the foam-mat drying process is suitable for heat sensitive, viscous, and sticky products, which cannot be efficiently dried using other methods. Foaming of juice or pulp makes the drying medium porous and increases penetration of heat up to the center of film layers (Venkataraman 1996). Mango pulp was foam-mat dried by Wilson *et al.* (2012) using 0, 3, 5, 7, and 9% egg white as the foaming agent, followed by air drying at 65, 75, and 85 °C. It was observed that there was significant effect of drying

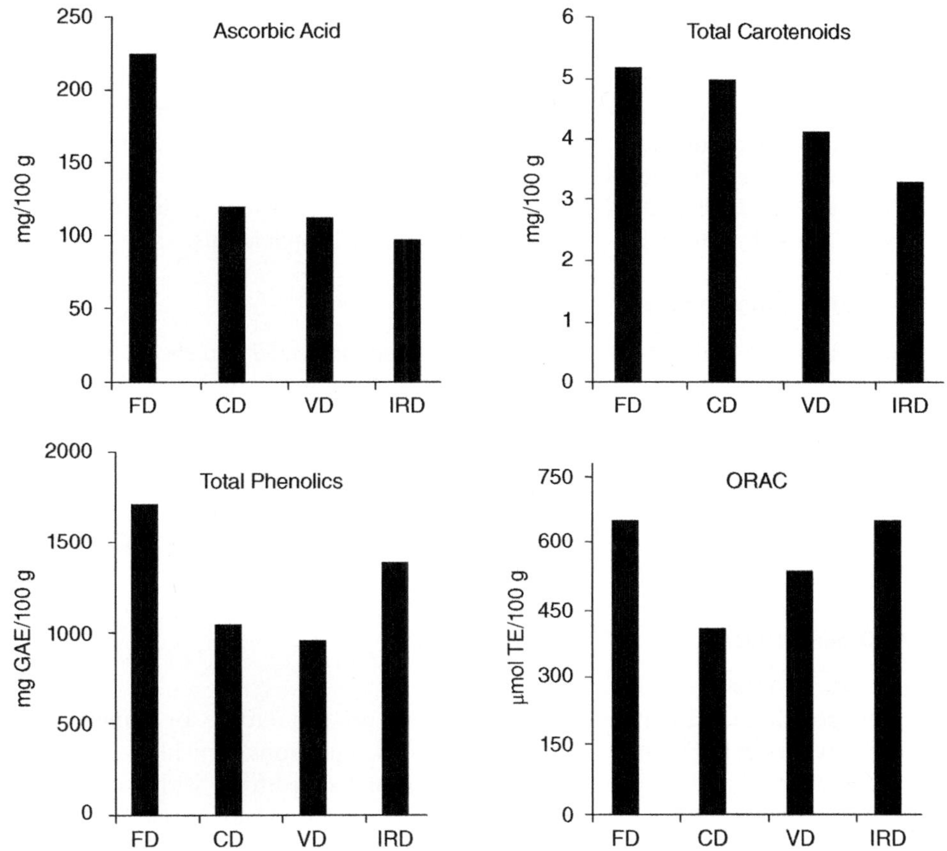

Figure 10.5 Effect of different drying methods on ascorbic acid, total carotenoids, total phenolics, and oxygen reducing antioxidant capacity or ORAC of dried mango powder, dry-wt basis (FD, freeze drying; CD, cabinet drying; VD, vacuum drying; IRD, infrared drying; GAE, gallic acid equiv.; TE, Trolox equiv.). *Source:* Sogi *et al.* (2015).

temperatures and egg white concentration on carotenoid retention. A foam-mat using 3% egg white at 65 °C air drying temperature was found to be the best combination.

10.2.5.4.3 Freeze Drying

Freeze drying is done by direct immersion of puree in liquid nitrogen (at −192 °C) with the shortest drying time of 16 hours compared with 18 hour freezing at −40 °C and 22.5 hour at −65 °C. Since the sugar content of the mango powder is high enough, no extra sugar is added before freeze drying. Freeze-dried powder can be used for the preparation of a variety of beverages, ice creams, jellies, cakes, and candies as flavoring and coloring agents and the process have the capability to produce mango powder with minimum microbial load (Kadam *et al.* 2010). Caparino *et al.* (2013) reported that freeze-dried mango powders exhibited greater surface area and porosity in comparison to Refractance Window dried mango powders. Nanjundaswamy (1997) recommended that the shelf life of freeze-dried mango powder can be increased by sugar addition.

10.2.5.4.4 Cabinet Drying

Mango pulp is poured in thin layers in aluminum trays and subjected to drying in a cabinet dryer using approximately 60 °C hot air for 6–8 hours. The control of drying temperature is very critical, since very high temperatures cause case hardening, thus hindering the outflow of water (Azeredo *et al.* 2006). The process of dehydration is terminated when the dried pulp is easily peeled from the waxy or plastic surface trays in the form of a sheet (Gujral and Khanna 2002). The dried pulp can be ground to prepare the powder or used as mango leather for a variety of food applications.

10.2.6 Frozen Mango Products

Generally, fully ripe mangoes are frozen, either whole or peeled and sliced, and packaged in sugar and quick frozen in moisture-proof packaging. In addition, diced mangoes, soaked in sweetened or unsweetened lime juice to prevent discoloration, are quick-frozen (FAO 1995). Mango puree is another product that is commonly frozen. The puree is blanched in a heat exchanger at 90–93 °C for 1 min, rapidly cooled to 0 °C, filled into polyethylene-lined containers, and frozen at −23.5 °C (Narain *et al.* 1998). Wu *et al.* (1993) reported that for sweetened frozen puree, sugar is added to a desired level and the puree is frozen in a blast freezer.

10.2.7 Other Products

10.2.7.1 Mango Pickles

Green mangoes are used to make pickles. A general process outline for mango pickle production is shown in Figure 10.1. There are two classifications of pickles – salt pickles and oil pickles. They are processed from whole and sliced fruit, with and without stones. The basic recipe for the preparation mango pickle in oil consists of: mango pieces (250 g), salt (60 g), oil (30 g), chilli powder (20 g), mustard powder (30 g), turmeric powder (2–4 g), and fenugreek seeds (2–4 g). The ingredients are mixed together and filled into wide-mouthed bottles. Three days later, the contents are thoroughly mixed and refilled into the bottles. Extra oil is added to form a 1–2 cm layer over the pickles (FAO 1995).

10.2.7.2 Mango Chutney

Mango chutney, a spicy relish popular in the Indian sub-continent, is prepared from peeled, sliced, or grated unripe or semi-ripe fruit by cooking the shredded fruit with salt over a medium heat for 5–7 min, mixed, and then sugar, spices, and vinegar are added. Cooking over mild heat continues until the product resembles a thick puree, then adding the remaining ingredients and simmering for another 5 min. Thus, prepared chutney is hot filled (~85 °C) into sterilized jars. A typical mango chutney recipe consists of: prepared mango flesh (1.35 kg), water (2000 ml), sugar (450 g), vinegar (600 ml), salt, ginger, garlic, and chilli powders (75 g each), cinnamon, and raisins (125 g each) (FAO 1995). A general process outline for mango chutney production is shown in Figure 10.1.

10.2.7.3 Mango Jam and Jelly

Mango jams and jellies are semi-solid gels, which are made using the same general process. Both of these products are made from fruit pulp, with added sugar, pectin, calcium chloride, and citric acid. Jelly is a clear or translucent fruit spread made from sweetened fruit juice and set using naturally occurring pectin. Jelly can be made from sweet, savory, or spicy ingredients. It is made by a process similar to that used for making jam, with the additional step of filtering out the fruit pulp after the initial heating. It is important not to attempt to force the straining process, e.g., by squeezing the mass of fruit in the muslin, or the clarity of the resulting jelly will be compromised.

Basu and Shivhare (2013) investigated the production of low-calorie mango jam by substituting sucrose with sorbitol. Yield stress of the jam increased with TSS content during the gelation process and decreased with increasing temperature. Hardness of mango jam decreased with increasing sorbitol concentration, due to weaker junction zones of pectin molecules. Microstructure of sorbitol jam was found to be composed of network regions with large pores as well as dense, compact regions with small pores.

10.3 Utilization of Mango Processing Waste

Wu *et al.* (1993) reported that, depending on the mango cultivar and fruit size, the edible pulp makes up 33–85% of the fresh fruit, while the peel and the kernel amount to 7–24% and 9–40%, respectively. Larrauri *et al.* (1996) reported that the by-products of industrial mango processing may amount to 35–60% of the total fruit weight. Mango peel and kernel are rich in many bioactive compounds, such as phenolics, antioxidants, dietary fiber, vitamins, minerals, and oil (Ajila *et al.* 2010; Ashoush and Gadallah 2011). Traditionally, animal feed has been the usual application of these wastes; however, in recent years, the value-added utilization of mango-processing wastes is being explored on an increasing scale (Chapter 14 covers the value-added processing and utilization of mango waste in detail).

10.4 Nutritional Composition and Health Benefits

The USDA data on the nutritional profile of raw mangoes, mango nectar, and sweetened-dried mangoes is presented in Table 10.7. Mangoes are an excellent source of vitamins C and A, both being important in human health. Vitamin C, with its antioxidant properties, promotes healthy immune function and collagen formation. Mango fruit is an excellent source of flavonoids, beta-carotene, and beta-cryptoxanthin. Consumption of natural fruits rich in carotenes is known to protect the body from lung and oral cavity cancers (Anon. 2015). According to NMB (2016), a 100 g edible potion of raw mango provides 765 μg or 85% of recommended daily levels of vitamin A, which is important for vision and bone growth. Mangoes are a good source of dietary fiber, too.

All parts of the mango fruit and tree potentially have some medicinal uses. Mango has been recommended for many traditional remedies for a number of disorders. Some of the traditional remedies for which mango or mango parts/products are used, especially in Southeast Asia, are indigestion, blood disorders, skin disorders, gum inflammation, dysentery and diarrhea, weight loss diets, and bilious disorders. Mangoes are suggested to be good for the kidneys, digestive system, and the skin (Siddiq *et al.* 2012). It must be noted that the scientific evidence for some of the health benefits is rather inconclusive and needs further research.

Table 10.7 Composition and nutritional profile of mango fruit, canned nectar, and sweetened dried mango (per 100 g).

	Raw Mango	Mango Nectar	Dried Mango
Proximate:			
Water (g)	83.46	86.63	16.6
Energy (kcal/kJ)	60/250	51/212	319/1366
Protein (g)	0.82	0.11	2.45
Total lipid/fat (g)	0.38	0.06	1.18
Ash (g)	0.36	0.08	1.19
Carbohydrate[1] (g)	14.98	13.12	78.58
Fiber, total dietary (g)	1.6	0.3	2.4
Sugars, total (g)	13.66	12.45	66.27
Minerals:			
Calcium (mg)	11	17	0
Iron (mg)	0.16	0.36	0.23
Magnesium (mg)	10	3	20
Phosphorus (mg)	14	2	50
Potassium (mg)	168	24	279
Sodium (mg)	1	5	162
Zinc (mg)	0.09	0.02	0.3
Vitamins:			
Vitamin C[2] (mg)	36.4	15.2	42.3
Folate, total (μg)	43	7	68
Choline, total (mg)	7.6	1.5	23.7
β-Carotene (μg)	640	402	786
β-Cryptoxanthin (μg)	10	26	29
Vitamin A (IU)	1082	692	1343
Lycopene (μg)	3	0	0
Lutein + zeaxanthin (μg)	23	0	52
Vitamin E[3] (mg)	0.9	0.21	4.02
Vitamin K[4] (μg)	4.2	0.8	13.2

[1] by difference.
[2] as total ascorbic acid.
[3] as α-tocopherol.
[4] as phylloquinone.
Source: USDA (2016).

10.5 Summary

Mango is one of the most popular, nutritionally rich fruits with unique flavor, aroma, taste, and health-promoting qualities, making it a choice ingredient in new functional foods. In recent years, the growing popularity of mangoes in North American and

European markets offers more opportunities for trade expansion of this nutritious fruit and its processed products. Mangoes are processed into a variety of value-added products, for example, nectar, puree, juice, juice concentrate, dried slices and pieces, pickles, leather, and powder. In addition, the traditional mango processing in different countries offer a wide range of flavorful products (sweet, sour, and spicy) in different forms and choices. Beyond commercially processed products, mango and its various forms are being used increasingly in culinary applications, as fruit salad, mango-flavored ice cream, other desserts, and relish. Mango processing waste (peel, seed/kernel), which contains high amounts of many nutrients and antioxidant compounds, present future opportunities for their value-added utilization in functional foods.

References

Ahmed, J., Ramaswamy, H.S. and Hiremath, N. (2005) The effect of high pressure treatment on rheological characteristics and color of mango pulp. *International Journal of Food Science and Technology*, **40**, 885–895.

Ajila, C.M., Aalami, M., Leelavathi, K. and Prasada Rao, U.J.S. (2010) Mango peel powder: A potential source of antioxidant and dietary fiber in macaroni preparations. *Innovative Food Science and Emerging Technologies*, **11**, 219–224.

Akhtar, S., Riaz, M., Ahmad, A. and Nisar, A. (2010) Physico-chemical, microbiological and sensory stability of chemically preserved mango pulp. *Pakistan Journal of Botany*, **42**, 853–862.

Andress, E.L. and Harrison, J.A. (2006) *So Easy to Preserve*, 4th edition. University of Georgia Cooperative Extension Service, Athens, GA. 374 p.

Anon. (2015) Mango Fruit Nutrition Facts [Online]. Available: http://www.nutrition-and-you.com/mango-fruit.html [18 September 2015].

Anon. (2016) Mango Pulp [Online]. Available: http://www.agafruit.com/products.htm [25 May 2016].

Ashoush, I.S. and Gadallah, M.G.E. (2011) Utilization of mango peels and seed kernels powders as sources of phytochemicals in biscuit. *World Journal of Dairy and Food Science*, **6**, 35–42.

Azeredo, H.M.C., Brito, E.S., Moreira, G.E.G., Faria, V.L. and Bruno, L.M. (2006) Effect of drying and storage time on the physico-chemical properties of mango leathers. *International Journal of Food Science and Technology*, **41**. 635–638.

Basu, S. and Shivhare, U.S. (2013) Rheological, textural, microstructural, and sensory properties of sorbitol-substituted mango jam. *Food and Bioprocess Technology*, **6**, 1401–1413.

Bhatt, A., Chopra, C.S. and Sharma, S.K. (2015) Screening of naturally dropped unripe mango fruits for making beverages. *Journal of Hill Agriculture*, **6**, 264–268.

Caparino, O.A., Sablani, S.S., Tang, J., Syamaladevi, R.M. and Nindo, C.I. (2013) Water sorption, glass transition, and microstructures of refractance window–and freeze-dried mango (Philippine "Carabao" Var.) powder. *Drying Technology*, **31**, 1969–1978.

Chauhan, S.K., Tyagi, S.M. and Singh, D. (2001) Pectinolytic liquefaction of apricot, plum and mango pulps for juice extraction. *International Journal of Food Properties*, **4**, 103–109.

Chen, J.P., Tai, C.Y. and Chen, B.H. (2007) Effects of different drying treatments on the stability of carotenoids in Taiwanese mango (*Mangifera indica* L.). *Food Chemistry*, **100**, 1005–1010.

Corzo, O. and Álvarez, C. (2014) Color change kinetics of mango at different maturity stages during air drying. *Journal of Food Processing and Preservation*, **38**, 508–517.

de Medeiros, R.A.B., Barros, Z.M.P., de Carvalho, C.B.O., Neta, E.G.F., Maciel, M.I.S. and Azoubel, P.M. (2016) Influence of dual-stage sugar substitution pretreatment on drying kinetics and quality parameters of mango. *LWT-Food Science and Technology*, **67**, 167–173.

Dissa, A.O., Bathiebo, J., Kam, S., Savadogo, P.W., Desmorieux, H. and Koulidiati, J. (2009) Modelling and experimental validation of thin layer indirect solar drying of mango slices. *Renewable Energy*, **34**, 1000–1008.

FAO (1995) Fruit and Vegetable Processing (Chapter 8) [Online]. Available: http://www.fao.org/docrep/v5030e/V5030E0p.htm [6 June 2016].

FSA [Food Standard Agency] (2007) Fruit Juices and Fruit Nectar Regulations 2003 [Online]. Available: http://www.food.gov.uk/sites/default/files/multimedia/pdfs/fruitjuiceguidefeb07.pdf [26 July 2015].

Giraldo, G., Talens, P., Fito, P. and Chiralt, A. (2003) Influence of sucrose solution concentration on kinetics and yield during osmotic dehydration of mango. *International Journal of Food Engineering*, **58**, 33–43.

Guerrero-Beltrán, J.A., Swanson, B.G. and Barbosa-Cánovas, G.V. (2005) Inhibition of polyphenoloxidase in mango puree with 4-hexylresorcinol, cysteine and ascorbic acid. *LWT-Food Science and Technology*, **38**, 625–630.

Guerrero-Beltrán, J.A., Barbosa-Cánovas, G.V., Moraga-Ballesteros, G.E.M.M.A., Moraga-Ballesteros, M.J. and Swanson, B.G. (2006) Effect of pH and ascorbic acid on high hydrostatic pressure-processed mango puree. *Journal of Food Processing and Preservation*, **30**, 582–596.

Guiamba, I.R., Svanberg, U. and Ahrné, L. (2015) Effect of infrared blanching on enzyme activity and retention of β-carotene and vitamin C in dried mango. *Journal of Food Science*, **80**, E1235–E1242.

Gujral, S.H. and Khanna, G. (2002) Effect of skim milk powder, soy protein concentrate and sucrose on the dehydration behavior, texture, color and acceptability of mango leather. *International Journal of Food Engineering*, **55**, 343–348.

Gujral, H.S., Oberoi, D.P.S., Singh, R. and Gera, M. (2013) Moisture diffusivity during drying of pineapple and mango leather as affected by sucrose, pectin, and maltodextrin. *International Journal of Food Properties*, **16**, 359–368.

Hardy, Z. and Jideani, V.A. (2016) Foam-mat drying technology: a review. *Critical Reviews in Food Science and Nutrition*, doi:10.1080/10408398.2015.1020359.

Hiremath, N.D. and Ramaswamy, H.S. (2012) High-pressure destruction kinetics of spoilage and pathogenic microorganisms in mango juice. *Journal of Food Processing and Preservation*, **36**, 113–125.

Hussain, I., Gilani, S.N., Khan, M.R., Khan, M.T. and Shakir, I. (2005) Varietal suitability and storage stability of mango squash. *International Journal of Agriculture and Biology*, **7**, 1038–1039.

Hymavathi, T.V. and Khader, V. (2005a) Carotene, ascorbic acid and sugar content of vacuum dehydrated ripe mango powders stored in flexible packaging material. *Journal of Food Composition and Analysis*, **18**, 181–192.

Hymavathi, T.V. and Khader, V. (2005b) Storage stability of the vacuum dehydrated ripe mango-mix powders. *Journal of Food Science and Technology*, **41**, 521–524.

Johnson, J.R., Braddock, R.J. and Chen, C.S. (1996) Flavor losses in orange juice during ultra-filtration and subsequent evaporation. *Journal of Food Science*, **61**, 540–543.

Kadam, D.M., Wilson, R.A. and Kaur, S. (2010) Determination of biochemical properties of foam-mat dried mango powder. *International Journal of Food Science and Technology*, **45**, 1626–1632.

Kaushik, N., Nadella, T. and Rao, P.S. (2015) Impact of pH and total soluble solids on enzyme inactivation kinetics during high pressure processing of mango (*Mangifera indica*) pulp. *Journal of Food Science*, **80**, E2459–E2470.

Kermani, Z.J., Shpigelman, A., Houben, K., Geuzendam, B., Van Loey, A.M. and Hendrickx, M.E. (2015) Study of mango endogenous pectinases as a tool to engineer mango purée consistency. *Food Chemistry*, **172**, 272–282.

Kumar, R., Bawa, A.S., Kathiravan, T. and Nadanasabapathi, S. (2015) Optimization of pulsed electric field parameters for mango nectar processing using response surface methodology. *International Food Research Journal*, **22**, 1353–1360.

Lakhanpal, P. and Vaidya, D. (2015) Development and evaluation of honey based mango nectar. *Journal of Food Science and Technology*, **52**, 1730–1735.

Larrauri, J.A., Rupérez, P., Borroto, P. and Saura-Calixto, F. (1996) Mango peels as a new tropical fibre: Preparation and characterization. *LWT–Food Science and Technology*, **29**, 729–733.

Liu, F., Li, R., Wang, Y., Bi, X. and Liao, X. (2014) Effects of high hydrostatic pressure and high-temperature short-time on mango nectars: changes in microorganisms, acid invertase, 5-hydroxymethylfurfural, sugars, viscosity, and cloud. *Innovative Food Science and Emerging Technologies*, **22**, 22–30.

Morton J. (ed.) (1987) Mango: In: *Fruits of Warm Climates*, pp. 221–239. Julia F. Morton, Miami, FL.

Nagai, L.Y., Santos, A.B., Faria, F.A., Boscolo, M. and Mauro, M.A. (2015) Osmotic dehydration of mango with ascorbic acid impregnation: Influence of process variables. *Journal of Food Processing and Preservation*, **39**, 384–393.

Nanjundaswamy, A.M. (1997) Processing. In: *The Mango – Botany, Production and Uses* (ed. R.E. Litz), pp. 509–539. CAB International, Cambridge, MA.

Narain, N., Bora, P.S., Narain, R. and Shaw, P.E. (1998) Mango. In: *Tropical and Subtropical Fruits* (eds P.E. Shaw, H.T. Chan and S. Nagy), pp. 1–77. AgScience, Inc., Auburndale, FL.

NMB [National Mango Board]. (2016) Mango Nutrition [Online]. Available: http://www.mango.org/nutrition-professionals [6 June 2016].

Nunes, Y. and Moreira, R.G. (2009) Effect of osmotic dehydration and vacuum-frying parameters to produce high-quality mango chips. *Journal of Food Science*, **74**, 355–362.

Occena-Po, L.G. (2006) Banana, mango, and passion fruit. In: *Handbook of Fruits and Fruit Processing* (ed. Y.H. Hui), pp. 635–650. Blackwell Publishing, Ames, IA.

Pott, I., Neidhart, S., Muehlbauer, W. and Carle, R. (2005) Quality improvement of non-sulfphited mango slices by drying at high temperatures. *Innovative Food Science and Emerging Technologies*, **6**, 412–419.

Pu, Y.Y. and Sun, D.W. (2015) Prediction of moisture content uniformity of microwave-vacuum dried mangoes as affected by different shapes using NIR hyperspectral imaging. *Innovative Food Science and Emerging Technologies*, **33**, 348–356.

Ragab, M. and El-Nemr, S.E. (1990) Chemical composition and volatile constituents of canned mango juice. *Nahrung*, **34**, 591–599.

Sakhale, B.K., Pawar, V.N. and Gaikwad, S.S. (2016) Studies on effect of enzymatic liquefaction on quality characteristics of Kesar mango pulp. *International Food Research Journal*, **23**, 860–865.

Santhirasegaram, V., Razali, Z., George, D.S. and Somasundram, C. (2015) Comparison of UV-C treatment and thermal pasteurization on quality of Chokanan mango (*Mangifera indica* L.) juice. *Food and Bioproducts Processing*, **94**, 313–321.

Siddiq, M., Akhtar, S. and Siddiq, R. (2012) Mango processing, products and nutrition. In: *Tropical and Subtropical Fruit Processing and Packaging* (ed. M. Siddiq), pp. 277–297. John Wiley & Sons, Ames, IA.

Singh, N.G.I., Dhuique-Mayer, C. and Lozano, Y. (2000) Physicochemical changes during enzymatic liquefaction of mango pulp. *Journal of Food Processing and Preservation*, **24**, 73–85.

Singh, G.H. and Singh, B.S. (2003) Effect of hydrocolloids on the dehydration kinetics, color, and texture of mango leather. *International Journal of Food Properties*, **6**, 269–279.

Sogi, D.S., Siddiq, M. and Dolan, K.D. (2015) Total phenolics, carotenoids and antioxidant properties of Tommy Atkins mango cubes as affected by drying techniques. *LWT-Food Science and Technology*, **62**, 564–568.

Torres, J.D., Talens, P., Carot, J.M., Chiralt, A. and Escriche, I. (2007) Volatile profile of mango (*Mangifera indica* L.), as affected by osmotic dehydration. *Food Chemistry*, **101**, 219–228.

Tribst, A.A.L., Franchi, M.A., de Massaguer, P.R. and Cristianini, M. (2011) Quality of mango nectar processed by high-pressure homogenization with optimized heat treatment. *Journal of Food Science*, **76**, M106–M110.

USDA [United States Department of Agriculture]. (2016) National Nutrient Database for Standard Reference, Release 28 [Online]. Available: https://ndb.nal.usda.gov/ndb/search [26 May 2016).

USDA-ERS [United States Department of Agriculture – Economic Research Service] (2016) Fruit and Tree Nut Data [Online]. Available: http://www.ers.usda.gov/data-products/fruit-and-tree-nut-data [23 May 2016]

Vaillant, F., Jeanton, E., Dornier, M., O'Brien, G.M., Reynes, M. and Decloux, M. (2001) Concentration of passion fruit juice on an industrial pilot scale using osmotic evaporation. *Journal of Food Engineering*, **47**, 195–202.

Venkataraman, P.S. (1996) Success and failures in mango pulp processing industries in Chitoor district of A.P. *Indian Food Packer*, **51**, 57–58.

Wilson, R.A., Kadam, D.M., Chadha, S. and Sharma, M. (2012) Foam mat drying characteristics of mango pulp. *International Journal of Food Science and Nutrition Engineering*, **2**, 63–69.

Wilson, R.A., Kadam, D.M., Chadha, S., Grewal, M.K. and Sharma, M. (2014) Evaluation of physical and chemical properties of foam-mat dried mango (*Mangifera indica*) powder during storage. *Journal of Food Processing and Preservation*, **38**, 1866–1874.

Wrolstad, R.E., McDaniel, M.R., Durst, R.W., Micheals, N., Lampi, K.A. and Beaudry, E.G. (1993) Composition and sensory characterization of red raspberry juice concentrated by direct-osmosis or evaporation. *Journal of Food Science*, **58**, 633–637.

Wu, J.S., Chen, H. and Fang, T. (1993) Mango juice. In: *Fruit Juice Processing Technology* (eds S. Nagy, C.S. Chen and P.E. Shaw), pp. 620–655. AgScience, Inc., Auburndale, FL.

Zhang, J., Wang, J., Zhang, W. and Wang, H. (2008) Effect of ultra-high pressure treatment on flavor components in mango juice. *Journal of the Chinese Institute of Food Science and Technology*, **8**, 118–122.

11

Composition and Nutritional Properties of Mangoes

Tasleem A. Zafar and Jiwan S. Sidhu

Department of Food Science and Nutrition, College of Life Sciences, Kuwait University, Safat, Kuwait

11.1 Introduction

Mango (*Mangifera indica* L.), family *Anacardiaceae*, is a fruit plant grown widely in the tropical and subtropical countries of the world. Mango fruit fits very well into the healthy diet for humans, as consumers are increasingly gearing toward the healthy and sustainable food consumption (Friel *et al.* 2014). The mango, in fresh products or a variety of processed products, offers a nutrient-dense and satiety choice for a healthy diet. Moreover, mango fruit is rich in a variety of bioactive compounds, for example polyphenols, which, according to Lima *et al.* (2014), act as antioxidants and have been shown to exert anti-inflammatory, anti-carcinogenic, and anti-aging properties. Mango is a fruit that has been utilized at every stage of growth, starting from very raw immature to fully ripe. Raw fruit is utilized for processing into pickle, chutney, mango sauce, raw mango powder (*amchoor*), and green mango drink (*panna*), whereas the fully-ripe fruit is used for making pulp, squash, nectar, drinks, mango leather (*Amb Papad*), mango puree, mango fruit bars, frozen and canned mango slices, and jam (Singh 1990). This chapter reviews the chemical and nutritional profiles of mango, including the effects of processing on the nutritional properties, and mango based functional foods research and development.

11.2 Composition and Nutrition

Mango is a nutrient-rich fruit containing carbohydrates, dietary fiber (pectin), vitamin C, vitamin A (from β-carotene), and many other phytochemicals for maintaining normal health. Table11.1 shows composition of raw mangoes, per 100 g and per serving size, that is, one cup or 165 g (USDA 2016). It should be noted that some differences can be expected in the composition of mango fruit and its processed products from different regions, due to variable climatic and soil conditions, agricultural practices, postharvest handling, and processing methods. The peel/skin of fully ripe mangoes is known to be rich in many nutrients such as terpenoids, but it is not consumed because of its unpleasant taste. The essential oils isolated from the mango peel oil are

Handbook of Mango Fruit: Production, Postharvest Science, Processing Technology and Nutrition, First Edition. Edited by Muhammad Siddiq, Jeffrey K. Brecht and Jiwan S. Sidhu.

Table 11.1 Proximate, mineral, and vitamins composition of raw mangoes.

Composition	Unit	Raw (100 g)	Raw (1 cup, 165 g)
Proximate			
Water	g	83.46	137.71
Energy	Kcal (KJ)	60 (250)	99 (412)
Protein	g	0.82	1.35
Total lipid (fat)	g	0.38	0.63
Ash	g	0.36	0.59
Carbohydrate, by difference	g	14.98	24.72
Fiber, total dietary	g	1.6	2.6
Sugars, total	g	13.66	22.54
Minerals			
Calcium	mg	11	18
Iron	mg	0.16	0.26
Magnesium	mg	10	16
Phosphorus	mg	14	23
Potassium	mg	168	277
Sodium	mg	1	2
Zinc	mg	0.09	0.15
Copper	mg	0.111	0.183
Manganese	mg	0.063	0.104
Selenium	μg	0.6	1
Vitamins			
Vitamin C, total ascorbic acid	mg	36.4	60.1
Thiamin	mg	0.028	0.046
Riboflavin	mg	0.038	0.063
Niacin	mg	0.669	1.104
Pantothenic acid	mg	0.197	0.325
Vitamin B-6	mg	0.119	0.196
Folate, total	μg	43	71
Carotene, beta	μg	640	1056
Carotene, alpha	μg	9	15
Cryptoxanthin, beta	μg	10	16
Vitamin A	IU	1082	1785
Lycopene	μg	3	5
Lutein + zeaxanthin	μg	23	38
Vitamin E (alpha-tocopherol)	mg	0.9	1.48
Vitamin K (phylloquinone)	μg	4.2	6.9

Source: USDA (2016).

Table 11.2 Chemical composition of mango pulp and seeds.

		African bush mango	
Parameter	**Indian ripe mango Pulp***	**Pulp****	**Seeds****
Moisture (%)	78.9–82.8	80.0	3.36
Energy (Kcal)	62.1–63.7	–[1]	–
Crude protein (%)	0.36–0.40	1.09	7.76
Crude Fat (%)	0.30–0.53	1.06	65.46
Carbohydrates (%)	16.20–17.18	16.7	10.93
Crude fiber (%)	0.85–1.06	0.4	10.23
Ash (%)	0.34–0.52	0.8	2.26
Calcium (mg/100g)	6.1–12.8	262.3	–
β-carotene (mg/100g)	0.14–1.192	–	–
Thiamine (mg/100g)	0.02–0.07	–	–
Riboflavin (mg/100g)	0.03–0.07	–	–
Niacin (mg/100g)	0.03–0.7	–	–
Ascorbic acid (mg/100g)	7.8–172.0	66.7	–

[1] Not determined.
Source: *Gopalan *et al.* (1977) and Tharanathan *et al.* (2006); **Onimawo *et al.* (2003).

rich in monoterpenes (83.2%). The dominant compounds in mango peel oil have been reported to yield mainly 2-3-carene (58.2%) and α-pinene (13%), while 70.3% of the mango leaf oil is sesquiterpenes (Dzamic *et al.* 2010). However, in raw mangoes, which are usually pickled-in-oil (as *Achaar*), mango fruit peel is also consumed. In mango pickle, sometimes the mango kernel (stone or pit) is also consumed by some. The general chemical composition of mango pulp and mango pits has been reported by many workers. The variations in the chemical composition of pulp and pits obtained from mango fruits grown in different parts of the world are summarized in Table 11.2 (Gopalan *et al.* 1977; Onimawo *et al.* 2003; Tharanathan *et al.* 2006).

11.2.1 Carbohydrates, Pectin, and Titratable Acidity

As the mango fruit matures, there is a significant increase in the dry matter (DM) and starch contents; however, no significant differences in total sugars, reducing sugars, fructose, or sucrose contents have been reported (Saranwong *et al.* 2004). Sucrose is the major sugar (nearly 50%) in mango fruit, while fructose is about 33%, and the remaining sugar is glucose. It is shown that the DM and starch contents in mango fruit at the time of harvest have a strong correlation with the eating quality. The higher the starch content in mango fruit, the higher will be the level of sugar synthesis during the ripening stage, thus leading to better eating quality. During ripening of mango fruit, the total soluble solids (TSS) content has been reported to increase from 9.1% to 17.3%, while the titratable acidity decreased from 0.6% to 0.2% (Padda *et al.* 2011). Citric acid and malic acids are reported to be the major organic acids responsible for titratable acidity, probably due to their utilization as a substrate for the respiration activities of the mango fruit (Medicott and Thomson 1985). Mahto and Das (2013) reported that the rate of increase of reducing and total sugars was faster in the non-irradiated control mango fruits than in the irradiated ones. However, during faster utilization of sugars during respiration activity, both the reducing and total sugars showed a decrease after peaking.

Mango fruit varieties and cultivars are reported to differ widely in their sugar, organic acids, ascorbic acid, and TSS content (Ali *et al.* 1992; Liu *et al.* 2013; Rodriguez-Pleguezuelo, *et al.* 2012; Vijayanand *et al.* 2015). Among the three cultivars grown in Karnataka, India, Mallika had higher TSS content than the Sindura and Totapuri cultivars (Vijayanand *et al.* 2015), though mango nectar prepared from Sindura was found to be highly acceptable. Liu *et al.* (2013) studied various mango cultivars being grown in southern China, among which the Keitt mango cultivar showed significantly higher content of *p*-hydroxybenzoic acid, succinic acid, catechin, sucrose, total sugars, TSS, and polyphenol oxidase (PPO) activity, but had the lowest amounts of total phenolics, ascorbic acid, titratable acidity, citric acid, and oxalic acid. The mango fruit cultivars grown in the Mediterranean region of Spain have been studied by Rodriguez-Pleguezuelo *et al.* (2012). Gleen, Palmer, and Osteen cultivars gave the highest pulp yield (about 85%), with Osteen having the lowest seed weight (4.2%) in the fruit. The Valencia Pride cultivar had the highest acidity (0.22%), while Lippens contained the highest TSS (20%). Physicochemical characteristics of some of the important cultivars grown in Pakistan are shown in Table 11.3 (Ali *et al.* 1992).

Thanaraj *et al.* (2009) reported that in Sri Lankan mangoes, fructose was the predominant sugar in both peel (56.2–106 mg/g dry wt.) and pulp (67.4–141 mg/g, dry wt), followed by glucose and sucrose. The starch content increased with maturity and was higher in the pulp (26.0–55.0%, dry wt.) than in the peel (18.2–38.9%, dry wt.) at the full maturity stage of the mango fruits. Mango kernel (average on dry weight basis) is reported to have 58% starch, 2.9% reducing sugars, 5.7% proteins, 0.8% pectin, 9.3% fat, and 1.1% tannins (Garg and Tandon 1997). Kaur *et al.* (2004) reported the physicochemical, thermal, morphological, and rheological characteristics of starches prepared

Table 11.3 Chemical characteristics of various cultivars/varieties of mango grown in Pakistan.

Cultivar	TSS, %	Total Invert sugars, %	Acidity,%	Sugar/acid ratio	Pulp/stone ratio
Pohilot	22.81	14.60	0.52	28.09	43.90
Shah Pasand	21.89	15.80	0.30	53.10	73.66
Baganpalli	23.20	14.50	0.49	28.59	47.40
Sanglakhi	21.14	14.10	0.80	17.63	26.43
Tamuria	23.20	14.80	0.20	69.90	116.00
Chaunsa	24.90	16.40	0.28	58.54	89.22
Malda	17.34	17.20	0.31	53.80	64.08
Ghulab Khas	27.50	14.30	0.29	49.36	94.96
Swarnareeka	23.25	14.10	0.46	30.72	50.67

Source: Adapted from Ali *et al.* (1992).

from the kernels of five important mango cultivars (Chausa, Totapuri, Kuppi, Langra, and Dassehri).

Kermani *et al.* (2015) studied the functional properties of citric acid extracted mango peel pectin for use in the food industry. Although the pectin had comparable molecular weight to that of apple pectin, it did not form a strong gel, even when the concentrations of pectin and sugar were increased under acidic conditions. They hypothesized that the MPP had neutral sugar branches, which caused steric hindrance to the gel network formation leading to a thickening effect. They suggested the use of MPP in stabilizing acidified milk drink-based functional foods. Partial or full substitution of mango pulp with mango peel for the formation of jam has been reported by Damiani *et al.* (2009). They found no significant difference in the vitamin C and TSS in the jam prepared using mango peel instead of pulp. However, as the level of peel substitution was increased, they observed an increase in ash, insoluble fiber, and sucrose contents, but a decrease in reducing sugar content in the jam. It was concluded that a partial or total replacement of pulp with mango peel in the formulation of jam would improve the nutritional and functional value of this product.

11.2.2 Antioxidant Compounds in Mango

Mango fruit is known to be rich in ascorbic acid (vitamin C), polyphenolic compounds, and carotenoids (Garcia-Magana *et al.* 2013). The antioxidant composition varies widely among cultivars, as shown in Figure 11.1 for vitamin C content. Carotenoids content varies widely among cultivars, affecting their peel and pulp color, and also provide health benefits to the consumer. The nutritional importance of mango fruit is mainly due to its α-carotene and β-carotene, both of which are precursors of vitamin A. Hymavathi and Khader (2005) reported the β-carotene content in Mulgoa and Alphonso cultivars to be 800 μg, and 13,000 μg/g, respectively. The concentration of β-carotene may vary from 1.7 μg/g to about 18 μg/g in different varieties of mango fruit (Rodriguez-Amaya *et al.* 2008).

A number of workers have tried to estimate the β-carotene content of mango fruits using rapid and non-destructive determination methods. Bicanic *et al.* (2010) have determined the β-carotene contents in mango homogenates by measuring the amplitude

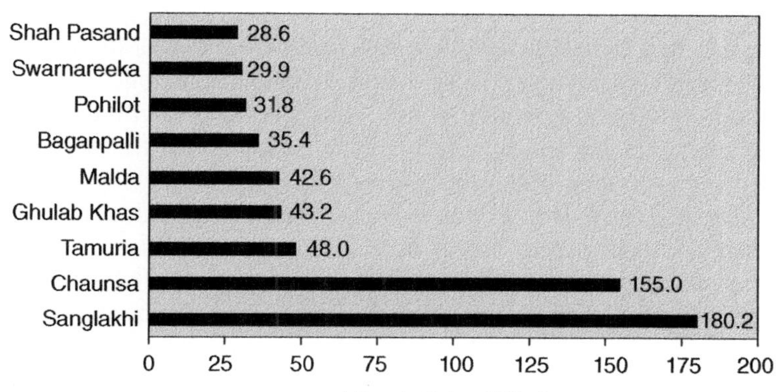

Figure 11.1 Vitamin C content of different mango cultivars grown in Pakistan, *Source:* Adapted from Ali *et al.* (1992).

of optothermal signals, chromaticity indices, and the intensities of Raman peaks. The three methods used were laser-based optothermal window (OW), the resonance Raman spectroscopy, and the tristimulus colorimetry. They opined that among these methods, the OW technique holds promise for a rapid screening/quantitative assessment of β-carotene in mango fruit, mainly because of the compact instrument, low cost, and the ease of operation. Rungpichayapichet *et al.* (2015) compared the non-destructive method (NIR spectroscopy) with the colorimetric method for the measurement of β-carotene content in mango. Long-wave NIR gave better prediction ability than the short-wave. They recommended NIR spectroscopy for the non-destructive and reliable measurement of β-carotene content in mango fruit.

11.2.3 Polyphenolics and Dietary Fiber

Mango fruit is considered to be an excellent source of many phytochemicals, including phenolics, carotenoids, and ascorbic acid (Masibo and He 2008; Nunez-Selles *et al.* 2015; Sogi *et al.* 2012). Mango stone and peels are an important source of many bioactive phenolic compounds. Dorta *et al.* (2014) identified the phenolic compounds in stones and peels of three mango cultivars (Keitt, Sensation, and Gomera 3) from Spain, by using high-performance liquid chromatography coupled to electrospray ionization and quadrupole time-of-flight mass spectrometry (HPLC-ESI-QTOF-MS) technique and multivariate analysis for use in functional food product development. They identified 30 different compounds belonging to 5 families: such as gallates and gallotannins; flavonoids, mainly quercetin derivatives; ellagic acid and derivatives; xanthones, primarily mangiferin; and benzophenones and derivatives, such as maclurin derivatives. The extraction method had a greater effect on the phenolics than the variety or the mango by-product analyzed. Polyphenolics are oxidized by the PPO enzyme, resulting in dark coloration of the fruit when bruised during handling and transportation, which lowers its acceptability by the consumer. Cheema and Sommerhalter (2015) investigated the PPO activity in the Ataulfo mango, their results showed that the PPO activity increased with ripeness and was always higher in the peel compared with the pulp.

The mechanism of anticancer activity of the ethanolic extract of mango peel (EEMP) and their constituents has been studied by Kim *et al.* (2012). Using liquid chromatography-electrospray ionization tandem mass spectroscopy and GCMS, they found the mango peel to contain quercetin 3-O-galactoside, mangiferin gallate, isomangiferin gallate, quercetin 3-O-arabinopyranoside, and mangiferin along with unsaturated fatty acids, such as oleic acid, linoleic acid, and ethyl linoleate. As the EEMP induced death of human cervical carcinoma HeLa cells through apoptosis, they suggested that these compounds may help to prevent cervical cancer, and may be useful for the treatment of other malignancies. Pandit *et al.* (2009) analyzed 27 mango cultivars for their aroma volatiles to explain the diversity in the flavor, their results concluded that mango flavor is dominated qualitatively and quantitatively by terpene hydrocarbons.

Mango peel is rich in polyphenolic compounds, especially in unripe versus ripe. Distribution of various polyphenolic compounds in mango peel, pulp, and kernel is shown in Table 11.4. The amount of polyphenolics and total flavonoid contents decreased as the mango fruit reached the fully ripe maturity stage (Abdul Aziz *et al.* 2011a). This study further showed that the total carotenoid and total anthocyanin

Table 11.4 Phenolic compounds in unripe and ripe mango pulp and peel flours (mg/g, dry-basis).

	Unripe (green) mango		Ripe mango	
Component	**Peel flour**	**Pulp flour**	**Peel flour**	**Pulp flour**
Total phenolics	102.41	54.36	70.20	14.57
Total carotenoids	96.91	56.46	160.64	126.77
Total anthocyanins	215.74	172.55	425.02	252.25
Total flavonoids	33.00	17.07	29.24	5.43
Antioxidant activity[1]				
FRAP method	70.53	16.42	65.92	15.30
DPPH method	54.23	12.18	43.30	9.82

Source: Adapted from Abdul Aziz *et al.* (2011a).
[1] Trolox equivalent.

contents were lower in both peel and pulp from unripe fruit, but increased as the fruit reached full ripe stage of maturity. The flavonoids and phenolic compounds have been reported in the extracts of mango peel and kernel (Table 11.5).

The polyphenolics and carotenoids present in ripe and unripe mango peel have been shown to offer protection against the hydrogen peroxide-induced oxidative damage in rat erythrocytes (Ajila and Prasada Rao 2008). The hemolysis of rat erythrocytes by hydrogen peroxide was inhibited by mango peel extract in a dose-dependent manner.

After extracting juice, the remaining by-products contain both the soluble dietary fiber (SDF) as well as insoluble dietary fiber (IDF), in addition to many colorants, antioxidants, and other bioactive compounds with positive health benefits. Ajila and Prasada Rao (2013) studied the mango peel as a source of dietary fiber from two locally grown cultivars (Raspuri and Badami). The dietary fiber content of these cultivars ranged from 40.6% to 72.5% (dry basis). Galactose, glucose, and arabinose were the major sugars associated with both the SDF and IDF. The bound polyphenolic and flavonoid contents ranged from 8.1 to 29.5 and 0.10 to 0.39 mg/g, respectively, and were found to be higher in the ripe peel than the unripe peel. The bound phenolics in peels were gallic acid, protocatechuric acid, and syringic acids, whereas the major flavonoids were kaempferol and quercetin.

11.3 Changes during Fruit Ripening

Mango fruit is harvested when fully mature, because being a climacteric fruit it ripens during transportation and storage. The management and maintenance of fruit quality during the export chain, including external and internal appearance, flavor, texture, and nutritional value is extremely important for the international mango trade (Sivakumar *et al.* 2011). Fruit quality is known to depend on the concentration of many metabolic compounds, such as sugars, organic acids, amino acids, pigments, vitamins, and aromatic volatiles produced during the fruit ripening process (Kienzle *et al.* 2014).

Table 11.5 Flavonoids and phenolic compounds in mango peel and kernel extracts (g/kg).

Compound	Peel extract*	Kernel extract**
Flavonoids		
Rutin	1.70[1] (0.01[2])	–[3]
Kaempferol	1.22 (0.26)	–
Catechin	0.39 (nd[4])	–
Genistein	1.41 (nd)	–
Phenolic acids		
Gallic acid	5.02 (3.50)	20.7
Protocatechuic acid	1.58 (1.01)	–
Gentisic acid	4.75 (5.27)	–
Coumarin	–	6.0
Caffeic acid	–	12.6
Vanillin	–	7.7
Mangiferin	–	20.2
Ferulic acid	–	4.2
Cinnamic acid	–	10.4

[1] Free.
[2] Bound.
[3] Not determined.
[4] Not detected.
Source: *Gondi *et al.* (2015); **Abdalla *et al.* (2007).

11.3.1 Aroma Volatiles Production

During postharvest handling and storage, mango fruit produces ethylene gas, a plant hormone that plays an important role in the ripening process, accompanied by fruit softening, change in skin color of fruit, and the production of a variety of aroma volatiles (Willis *et al.* 1998). Mango aroma is one of the most important attributes affecting the consumer acceptability of this fruit. Mango fruit aroma consists of an array of chemical compounds, such as monoterpenes, sesquiterpenes, aldehydes, ketones, alcohols, esters, and lactones (Lalel *et al.* 2003a; Singh and Singh 2012). The composition and production of aroma compounds has been shown to be influenced by the cultivar (Pandit *et al.* 2009), stage of ripening (Lalel *et al.* 2003b), growing location, and application of various treatments to the harvested fruit (Singh and Dwivedi 2008). The vapor heat treatment resulted in a significant lag for the emission of aroma volatiles than the fruit ripening. This may adversely affect the consumer acceptance of such vapor heat treated fruits (Singh and Saini 2014).

11.3.2 Retention of Antioxidant Compounds

Singh and Dwivedi (2008) reported the effect of the application of ethrel and 1-methylcyclopropene (1-MCP) on the antioxidant levels in mango fruit during ripening.

The 1-MCP treatment led to decreased levels of hydrogen peroxide and lipid oxidation in the mango fruits. Changes in the antioxidant phytochemicals in the mature-green mangoes with hot water immersion treatment (46 °C, 75 min) and modified atmosphere (3% oxygen and 97% nitrogen gas; 3% oxygen, 10% carbon dioxide and 87% nitrogen gas) were evaluated after storage for 2 weeks at 10 °C and subsequent ripening in air at 25 °C (Kim *et al.* 2007). The results showed that gallic acid and phenolics decreased throughout the fruit ripening period. However, the overall decline in polyphenolic compounds was inhibited by the modified atmosphere treatments due to delayed ripening.

Mango fruits immersed in hot water (46.1 °C, for 70–110 min) were evaluated for changes in polyphenolics, antioxidant capacity, and fruit quality (Kim *et al.* 2009). Free gallic acid and four gallotannins were tentatively identified as the major polyphenolics by HPLC. Although antioxidant capacity remained unchanged as a result of hot water treatments for prolonged periods, two major polyphenolics (gallic acid and gallotannins) and the soluble phenolics decreased. During storage for 4 days at 25°C, the total soluble phenolics and antioxidant capacity were decreased for all heat treatments, regardless of the duration of treatment times.

11.4 Changes during Fruit Processing

11.4.1 Pulp, Nectar, Juice, and Dried Mangoes

Mango is a tropical fruit that is highly valued, not only for its flavor and eating quality, but also as a good source of antioxidants such as carotenoids, ascorbic acid, and phenolic compounds. Since mango is a seasonal crop and highly perishable fruit, it needs to be processed into various shelf-life stable products, such as mango pulp, puree, nectar, juice, beverages, jam, pickles, chutney (dip), and dried products (slices, cubes, and powder). Though mangoes are processed into many different products, most of the compositional data on the effect of processing is mostly limited to pulp, juice, nectar, and dried mangoes, whereas the evaluation of physical properties and processing characteristics are widely reported in the literature. Since processing is known to affect the physico-chemical, nutritional, and sensory quality of processed mango products, it is important to design and optimize processes to preserve these qualities. Table 11.6 shows selected composition of dried mangoes and mango nectar as compared to raw mangoes (USDA 2016).

Rajab and El-Nemr (1990) evaluated the chemical composition of mango pulp and canned mango juice. Whole mangoes were blanched in boiling water for 3 min and mechanically crushed to obtain pulp. Juice was obtained by filtering the pulp and diluting it with water. Then the juice was preheated through a plate heat exchanger at 80 °C and hot-filled into cans, and thermally processed at 100 °C for 30 min. Mango juice had a significantly lower content of ascorbic acid and carotenoids as compared to pulp (Figure 11.2), since both ascorbic acid and carotenoids are heat sensitive. Mango pulp had higher total TSS, which were due mainly to added sucrose to enhance flavor.

Kadam *et al.* (2010) investigated the effect of drying air temperature (65, 75, and 85 °C) on the chemical properties of foam-mat dried mango juice powder. Chemical

Table 11.6 Effect of processing on the nutritional profile of mango products compared to raw mangoes (per 100 g).

Composition	Unit	Raw	Dried, sweetened	Nectar
Water	g	83.46	16.6	86.63
Energy	Kcal (KJ)	60 (250)	319 (1366)	51 (213)
Protein	g	0.82	2.45	0.11
Total lipid/fat	g	0.38	1.18	0.06
Carbohydrate	g	14.98	78.58	13.12
Total dietary fiber	g	1.6	2.4	0.3
Total sugars	g	13.66	66.27	12.45
Magnesium	mg	10	20	3
Phosphorus	mg	14	50	2
Potassium	mg	168	279	24
Sodium	mg	1	162	5
Vitamin C[1]	mg	36.4	42.3	15.2
Total folate	µg	43	68	7
Beta-carotene	µg	640	786	nr[2]
Vitamin A	IU	1082	1343	692

[1] As total ascorbic acid.
[2] Not reported.
Source: USDA (2016).

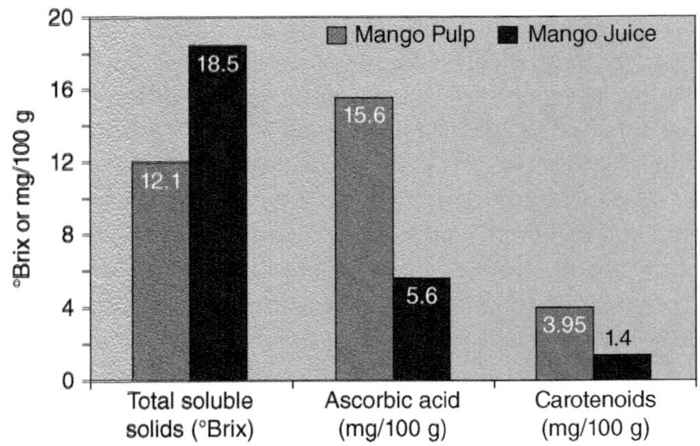

Figure 11.2 Total soluble solids, ascorbic acid and carotenoids content of mango pulp and canned mango juice (*juice sweetened to 18.5°Brix with 50% sucrose solution for flavor enhancement). *Source:* Ragab and El-Nemr (1990).

properties such as total sugars, ascorbic acid, total carotenes, minerals, total acid, pH, TSS, and microbial load (fungal and bacterial) of foam-mat dried mango powder were determined. Generally, all chemical properties showed a decreasing trend with an increase in drying air temperature. The decrease in ascorbic acid content was most pronounced (5.60, 4.05, and 3.20 mg/100 g in 65, 75, and 85 °C dried mango powder). Total acids and total carotenoids also decreased significantly with the corresponding increase in drying air temperature.

Siddiq and Sogi (2012, unpublished data) studied the effect of different drying methods (freeze, cabinet, vacuum, and infra-red drying) on the antioxidant quality of dried mango cubes. It was shown that in comparison to freeze-dried cubes' ascorbic acid content (225 mg/100 g, db), degradation levels of 47, 51, and 57% were observed in cabinet, vacuum, and infra-red dried mango cubes. However, total carotenoids were relatively stable and showed a degradation of 4–37% as compared to freeze-dried.

11.4.2 Fresh-cut Mangoes

Minimally-processed products, such as fresh-cut fruits, are one of the latest growing segments of the food industry. Robles-Sanchez *et al* (2009) investigated the influence of dipping fresh-cut Kent mango slices in ascorbic acid, citric acid, and calcium chloride solutions on the retention of bioactive compounds and antioxidant activity. All these dipping treatments resulted in the better retention of ascorbic acid and antioxidant activity compared with the controls. In a later study, Robles-Sanchez *et al.* (2013) applied alginate-based coating to prevent browning in fresh-cut Kent mango cubes and to enhance the bioactive compounds and antioxidant activity. The results showed that the combination of alginate and anti-browning agent not only preserved the color, but also increased the antioxidant potential of fresh-cut Kent mango cubes.

Siddiq *et al.* (2013) studied the effect of selected pre-treatments and storage time on total phenolics, antioxidant properties of mango cubes treated with ascorbic acid + citric acid + $CaCl_2$ (AA–CA–Ca), sodium acid sulfate (SAS), and 5-min infrared heat (IR-5). The phenolics content of the untreated cubes was 21.16 mg GAE/100 g on day 0, whereas treated samples had higher contents of 25.98–70.82 mg GAE/100 g. Similarly, higher antioxidant activity was generally observed with all treatments, with this effect most pronounced in AA–CA–Ca treated samples. Total phenolics and antioxidant activities were found to be stable during 12 days of storage at 4 °C.

Heat treatment given to any fresh-cut fruit cubes is known to result in the reduction of nutritional quality. Consequently, non-thermal technologies are being explored as alternatives. Charles *et al.* (2013) investigated the impact of pulsed light treatment on the physical and nutritional quality of fresh-cut Kent mango cubes. Pulsed light treatment maintained the firmness, color, phenolics, ascorbic acid, and carotenoid content of fresh-cut mangoes. Vega-Vega *et al.* (2013) used ethanolic extract of Haden mango kernels to treat fresh-cut mango cubes. The fresh-cut cubes retained the highest content of total phenolics (7.4 times), flavonoids (3.1 times), and antioxidant capacity done by DPPH, TEAC, and ORAC methods (2.9, 2.3, and 2.8 times, respectively), when compared with the control samples. This treatment also reduced the mesophilic plate count by 80% and total mold count by 97%.

11.4.3 High Hydrostatic Pressure (HHP) Processing

HHP, carried out at significantly lower temperatures than traditional thermal processing, has a lower negative impact on various quality attributes, for example, β-carotene, is the major micronutrient responsible for the bright yellow and orange color of mango pulp and a major provider of health benefits (Masibo and He 2009). Lemmens *et al.* (2013) investigated the effect of HHP homogenization and thermal processing on the isomerization of β-carotene to *cis*-isomers, which offer lower health benefits. Their results showed a negligible conversion to *cis*-isomers of β-carotene during high-pressure homogenization of mango puree, but during thermal processing they observed an increasing intensity of *cis*-isomer formation with increasing treatment intensity.

Liu *et al.* (2014) reported the effect of HHP and high temperature short time (HTST) treatments on the antioxidant activity and antioxidant compounds of mango nectars during a 16-week storage at 4 and 25 °C. Both the HHP and HTST treatments showed insignificant changes in antioxidant capacity (FRAP assay), L-ascorbic acid, sodium erythrobate, total phenolics, and total carotenoids. Kaushik *et al.* (2014) studied the effect of HHP (100–600, 1–20 min, at 30 °C) on the color, biochemical, and microbial load of mango pulp. The HHP-treated pulp samples were shown to retain 85, 92, and 90% of their original ascorbic acid, total phenolics, and antioxidant activity, respectively.

11.5 Health Benefits and Medicinal Uses

In recent years, enough evidence has accumulated to suggest that consuming carotenoids and antioxidant rich foods may provide a variety of health benefits. Boon *et al.* (2010) reviewed various factors, such as heat, light, oxygen, pH, transition metals, or interactions with radical species, which may influence the chemical stability of carotenoids in foods. Mastication is known to affect the bioavailability of carotenoids from fruit and vegetable tissues. Smaller particle size of plant tissues results in improved bioavailability of β-carotene and xanthophylls in mango pulp (Low *et al.* 2015). An extensive review on the functional properties of carotenoids in human health and chemistry of carotenoids, their absorption, transport, bioavailability, metabolism, and their mode of action has been published by Perera and Yen (2007).

Mango leaves, bark, and fruit (pulp, peel, kernel) are a rich source of many bioactive compounds such as proteins, vitamin A, vitamin C, carotenoids, phenolic compounds (mangiferin), dietary fiber, carbohydrates, and minerals, offering many medicinal, nutritional, and industrial benefits, and these have been discussed in a recent review by Rymbai *et al.* (2013). These antioxidant compounds are known to protect against diseases related to oxidative stress. They suggested using all mango tree parts in functional foods to improve nutrition and health. Much evidence has supported the involvement of reactive oxygen species (ROS) in many pathological conditions such as ischemia-reperfusion injury, inflammatory and degenerative diseases, certain types of cancer, DNA damage, and aging. Garrido *et al.* (2008) suggested that mangiferin (a phenolic compound) may be responsible for the antioxidant activity of the mango tree extract. Mangiferin, present in highest amounts in leaves, bark, then peel and pulp, has also been reported to be the major bioactive compound in mango (Masibo and He 2009).

Kim *et al.* (2010) investigated the cytoprotective effect of mango pulp and peel on the oxidative damage induced by hydrogen peroxide in a human hepatoma cell line. The mango peel extract exhibited stronger free radical scavenging ability than mango pulp extract, regardless of the stage of maturity. The mango peel extract also exhibited significant antiproliferative effect on all tested cancer cell lines in a dose-dependent manner, when compared with mango pulp extract. Their results suggested that the antiproliferative activity of mango peel and pulp extracts correlated well with their phenolic and flavonoid contents.

Gondi *et al.* (2015) investigated the anti-diabetic effect of mango peel in streptozotocin-induced diabetic rats. Diabetic rats, when fed mango peel powder (MPP) at 5 and 10% levels in their basic diet, had their antioxidant enzyme activities increased and lipid peroxidation in plasma, kidney, and liver, when compared to the untreated diabetic rats.

11.6 Functional Foods from Mango

Research on phytochemicals for incorporation into functional foods has increased in recent years. Fruits and vegetables are the richest sources of many of these bioactive compounds, such as polyphenolics, carotenoids, dietary fiber, vitamins, and minerals (Paz *et al.* 2015; Reis *et al.* 2015). Some bioactive compounds (polyphenolics) have been valued not only for their antioxidant properties but also for their antimicrobial activities. Mango tree bark, leaves, fruit pulp, and kernel are valued for their antioxidant activities. In addition, mango pulp is being investigated for the preparation of biodegradable materials for food packaging as a renewable natural resource (Reis *et al.* 2015). Medium chain and behenic acid has been incorporated into the mango fat triglycerides (TG) by Bebarta *et al.* (2013). These modified fats are known to have lower calorie content than the short chain fatty acid TG and could be used in developing functional foods with lower calorie contents. The use of value-added ingredients from mango (dietary fiber, peel, and kernel powder) have been explored in cereal-based functional foods – bread, cookies, soft dough, biscuits, pasta, and macaroni (Abdul Aziz *et al.* 2011b; Ajila *et al.* 2008, 2010; Vergara-Valencia *et al.* 2007) and dairy-based functional foods – yogurt, yogurt powder, and milk beverages (Karunasawat and Anprung 2010; Kaur and Kumar 2008; Narayana and Gupta 2013).

11.7 Chemical Composition of Mango By-products

Interestingly, in the case of ripe mango fruit, only the pulp is eaten, but the peel and kernel (stone) end up as waste products. During the processing of mango fruit into juice and other products, a large amount of waste (~28–43% of the total mango fruit) is generated mainly from the peel and kernels. Peel and pulp of raw fruit are being used extensively for preparing popular condiments such as mango pickle and chutney. Mango kernel has been reported to be rich in fat, protein, minerals, dietary fiber, and many bioactive compounds which, if consumed, may provide health benefits. Mango peel is also a rich source of dietary fiber, starch, pectin, minerals, and phytochemicals (Kumar *et al.* 2012; Sun and Chen 2012). The mango peel powder showed a high antioxidant capacity as determined by DPPH assays, mainly due to the presence of associated polyphenolic compounds (98.3 mg/g), and it can be incorporated into various functional foods to improve their health-promoting properties.

A number of researchers have investigated and developed suitable extraction, refining, and estimation techniques for various bioactive compounds from mango fruit constituents (Geerkens *et al.* 2015; Jiang *et al.* 2010; Ruiz-Montanez *et al.* 2014). The ethyl gallate and penta-O-galloyl-glucoside were the two major antioxidants in mango peel that showed a strong scavenging property against hydroxyl radicals (•OH), superoxide (O_2^-), and singlet oxygen ($^1O_2$), and showed synergistic effects (Jiang *et al.* 2010). Ruiz-Montanez *et al.* (2014) evaluated various emerging extraction methods (sonication, microwave, and hydrostatic pressures) to determine two bioactive compounds (mangiferin and lupeol) from mango peel, and found sonication to give increased yield within less time. Mango peel has been recommended as a source of carbohydrates replacement (maximum of 33% replacement level) for corn to raise tilapia fish without adversely affecting the growth performance and carcass chemical composition (Souza *et al.* 2013).

11.8 Future Research Needs

With a global increase in public health issues, there is an ever-increasing demand for natural foods that may offer remedies for diseases, without worrying about the side effects of allopathic drugs. The importance of all parts of the mango tree must be emphasized, which warrants a need to study the composition, including bioactive compounds, of all botanical parts of the tree to provide more information about their medical, nutritional, and industrial applications for the benefit of consumers. This will help in the proper utilization of mango fruits to improve nutrition and health of the consumers, as well as to supplement our understanding of the defense mechanisms of plants using these bioactive compounds.

More research is needed on how the perceived differences in flavor and texture of different mango cultivars could arise when they are processed into pulp and other products. This information would also help in developing/selecting cultivars that would become successful when exported to target markets around the world. The high pressure processing is another area that needs to be further investigated. Because of the negligible rise in temperature during the HPP, it has the ability to preserve the freshness of mango flavor and maximum retention of health promoting bioactive compounds in addition to achieving required microbial load reduction for ensuring food safety. Fresh-cut mangoes are another area of research that needs more focus to fully exploit the potential of this flavorsome and nutrient-rich fruit.

References

Abdalla, A.E.M., Darwish, S.M., Ayad, E.H.E. and El-Hamahmy, R.M. (2007) Egyptian mango by-product. 1: Compositional quality of mango seed kernel. *Food Chemistry*, **103**, 1141–1152.

Abdul Aziz, N.A., Wong, L.M., Bhat, R. and Cheng, L.H. (2011a) Evaluation of processed green and ripe mango peel and pulp flours (*Mangifera indica* var. Chokanan) in terms of chemical composition, antioxidant compounds and functional properties. *Journal of the Science of Food and Agriculture*, **92**, 557–563.

Abdul Aziz, N.A., Wong, L.M. and Bhat, R. (2011b) Nutritional and sensory quality evaluation of sponge cake prepared by incorporation of high dietary fiber containing mango (*Mangifera indica* var. Chokanan) pulp and peel flours. *International Journal of Food Science and Nutrition*, **62**, 559–567.

Ajila, C.M. and Prasada Rao, U.J.S. (2008) Protection against hydrogen peroxide induced oxidative damage in rat erythrocytes by *Mangifera indica* L. peel extract. *Food Chemistry and Toxicology*, **46**, 303–309.

Ajila, C.M. and Prasada Rao, U.J.S. (2013) Mango peel dietary fiber: Composition and associated bound phenolics. *Journal of Functional Foods*, **5**, 444–450.

Ajila, C.M., Leelavathi, K. and Prasada Rao, U.J.S. (2008) Improvement of dietary fiber content and antioxidant properties in soft dough biscuits with the incorporation of mango peel powder. *Journal of Cereal Science*, **48**, 319–326.

Ajila, C.M., Aalami, M., Leelavathi, K. and Prasada Rao, U.J.S. (2010) Mango peel powder: A potential source of antioxidant and dietary fiber in macaroni preparations. *Innovative Food Science and Emerging Technologies*, **11**, 219–224.

Ali, S., Haq, C.A. and Hussain, S. (1992) Physicochemical studies of some varieties of mango grown at Shujabad. *Pakistan Journal of Agricultural Research*, **13**, 350–356.

Bebarta, B., Jhansi, M., Kotasthane, P. and Sunkireddy, Y.R. (2013) Medium chain and behenic acid incorporated structured lipids from sal, mango and kokum fats by lipase acidolysis. *Food Chemistry*, **136**, 889–894.

Bicanic, D., Dimitrovski, D., Luterotti, S., Twisk, C., Buijnsters, J.G. and Doka, O. (2010) Estimating rapidly and precisely the concentration of beta carotene in mango homogenates by measuring the amplitude of optothermal signals, chromaticity indices and the intensities of Raman peaks. *Food Chemistry*, **121**, 832–838.

Boon, C.S., McClements, D.J., Weiss, J. and Decker, E.A. (2010) Factors influencing the chemical stability of carotenoids in foods. *Critical Reviews in Food Science and Nutrition*, **50**, 515–532.

Charles, F., Vidal, V., Olive, F. Filgueiras, H. and Sallanon, H. (2013) Pulsed light treatment as new method to maintain physical and nutritional quality of fresh-cut mangoes. *Innovative Food Science and Emerging Technologies*, **18**, 190–195.

Cheema, S. and Sommerhalter, M. (2015) Characterization of polyphenol oxidase activity in Ataulfo mango. *Food Chemistry*, **171**, 382–387.

Damiani, C., Vilas Boas, E.V.B., Soares, Jr, M.S., Caliari, M. and de Paula, M.L. (2009) Chemical evaluation of mango (*Mangifera indica* L.) jams formulated with different levels of peels in substitution to pulp. *Ciencia e Agrotechnologia*, **33**, 177–184.

Dorta, E., Gonzalez, M., Lobo, M.G., Sanchez-Moreno, C. and de Ancos, B. (2014) Screening of phenolic compounds in by-product extracts from mangoes (*Mangifera indica* L.) by HPLC-ESI-QTOF-MS and multivariate analysis for use as a food ingredient. *Food Research International*, **57**, 51–60.

Dzamic, A.M., Marin, P.D. Gbolade, A.A. and Ristic, M.S. (2010) Chemical composition of *Mangifera indica* essential oil from Nigeria. *Journal of Essential Oils Research*, **22**, 123–125.

Friel, S., Barosh, L. and Lawrence, M. (2014) Towards healthy and sustainable food consumption: An Australian case study. *Public Health Nutrition*, **17**, 1156–1166.

Garcia-Magana, M. de L., Garcia, H.S., Bello-Perez, L.A., Sayago-Ayerdi, S.G. and Montes de Oca, M.M. (2013) Functional properties and dietary fiber characterization of mango processing by-products (*Mangifera indica* L., cv *Ataulfo* and *Tommy Atkins*). *Plant Foods for Human Nutrition*, **68**, 254–258.

Garg, N. and Tandon, D. K. (1997) Amylase activity of *A. oryzae* grown on mango kernel after certain pretreatments and aeration. *Indian Food Packer*, **51**, 26–29.

Garrido, G., Gonzalez, D., Romay, C., Nunez-Selles, A.J. and Delgado, R. (2008) Scavenger effect of a mango (*Mangifera indica* L.) food supplement's active ingredient on free radicals produced by human polymorphonuclear cells and hypoxanthine-xanthine oxidase chemiluminescence systems. *Food Chemistry*, **107**, 1008–1014.

Geerkens, C.H., Matejka, A.E., Carle, R. and Schweiggert, R.M. (2015) Development and validation of an HPLC method for the determination of alk(en)ylresorcinols using rapid ultrasound-assisted extraction of mango peels and rye grains. *Food Chemistry*, **169**, 261–269.

Gondi, M, Basha, S.A., Bhaskar, J.J., Salimath, P.V. and Prasada Rao, U.J.S. (2015) Anti-diabetic effect of dietary mango (*Mangifera indica* L.) peel in streptozotocin-induced diabetic rats. *Journal of the Science of Food and Agriculture*, **95**, 991–999.

Gopalan, C., Rama Shastri, B.V. and Balasubramanian, S.C. (eds) (1977) *Nutritive Value of Indian Foods*. National Institute of Nutrition, Hyderabad, India, 156 p.

Hymavathi, T.V. and Khader, V. (2005) Carotene, ascorbic acid and sugar content of vacuum dehydrated ripe mango powders stored in flexible packaging material. *Journal of Food Composition and Analysis*, **18**, 181–192.

Jiang, L.Y., He, S., Pan, Y.J. and Sun, C.R. (2010) Bioassay-guided isolation and EPR-assisted antioxidant evaluation of two valuable compounds from mango peels. *Food Chemistry*, **119**, 1285–1292.

Kadam, D.M., Wilson, R.A. and Kaur, S. (2010) Determination of biochemical properties of foam-mat dried mango powder. *International Journal of Food Science and Technology*, **45**, 1626–1632.

Karunasawat, K. and Anprung, P. (2010) Effect of depolymerized mango pulp as a stabilizer in, oil-in-water emulsion containing sodium caseinate. *Food Bioproducts Processing*, **88**, 202–208.

Kaur, H. and Kumar, P. (2008) Mango soy fortified probiotic yogurt: Effect of culture addition and temperature on textural characteristics. *Journal of Texture Studies*, **39**, 673–694.

Kaur, M., Singh, N., Sandhu, K.S. and Guraya, H.S. (2004) Physicochemical, morphological, thermal and rheological properties of starches separated from kernels of some Indian mango cultivars (*Mangifera indica* L.). *Food Chemistry*, **85**, 131–140.

Kaushik, N., Kaur, B.P., Srinivasa, Rao, P. and Mishra, H.N. (2014) Effects of high pressure processing on color, biochemical and microbiological characteristics of mango pulp (*Mangifera indica* cv. Amrapali). *Innovative Food Science and Emerging Technologies*, **22**, 40–50.

Kermani, Z.J., Shpigelman, A., Pham, H.T.T., Loey, A.M.V. and Hendricks, M.E. (2015) Functional properties of citric acid extracted mango peel pectin as related to its chemical structure. *Food Hydrocolloids*, **44**, 424–434.

Kienzle, S., Carle, R., Sruamsiri, P., Tosta, C. and Neidhart, S. (2013) Occurrence of alk(en) ylresorcinols in the fruits of two mango (*Mangifera indica* L.) cultivars during on-tree maturation and postharvest storage. *Journal of Agricultural and Food Chemistry*, **62**, 28–40.

Kim, Y., Brecht, J.K. and Talcott, S.T. (2007) Antioxidant phytochemical and fruit quality changes in mango (*Mangifera indica* L.) following hot water immersion and controlled atmosphere storage. *Food Chemistry*, **105**, 1327–1334.

Kim, Y., Lounds-Singleton, A.J. and Talcott, S.T. (2009) Antioxidant phytochemical and quality changes associated with hot water immersion treatment of mangoes (*Mangifera indica* L.). *Food Chemistry*, **115**, 989–993.

Kim, H., Moon, J.Y., Kim, H., Lee, D.S., Cho, M. *et al.* (2010) Antioxidant and antiproliferative activities of mango (*Mangifera indica* L.) flesh and peel. *Food Chemistry*, **121**, 429–436.

Kim, H., Kim, H., Mosaddik, A., Gyawali, R., Ahn, K.S. and Cho, S.K. (2012) Induction of apoptosis by ethanolic extract of mango peel and comparative analysis of the chemical constituents of mango peel and flesh. *Food Chemistry*, **133**, 416–422.

Kumar, C., Mythily, R. and Chandraju, S. (2012) Utilization of mango peels (*Mangifera indica*) for the extraction of sugars. *Der Pharma Chemica*, **4**, 2422–2426.

Lalel, H.J.D., Singh, Z. and Tan, S.C. (2003a) Aroma volatile production during ripening of "Kensington Pride" mango. *Postharvest Biology and Technology*, **27**, 323–336.

Lalel, H.J.D., Singh, Z. and Tan, S.C. (2003b) Glycosidically-bound aroma volatile compounds in the skin and pulp of "Kensington Pride" mango fruit at different stages of maturity. *Postharvest Biology and Technology*, **29**, 205–218.

Lemmens, L., Tchuenche, E.S., Van Loey, A.M. and Hendricks, M.E. (2013) Beta-carotene isomerization in mango puree as influenced by thermal processing and high-pressure homogenization. *European Food Research and Technology*, **236**, 155–163.

Lima, G.P.P., Vianello, F., Correa, C.R., Campos, R.A.S. and Borguini, M.G. (2014) Polyphenols in fruits and vegetables and its effect on human health. *Food and Nutrition Sciences*, **5**, 1065–1082.

Liu, F.X., Fu, S.F., Bi, X.F., Liao, X.J., Hu, X.S. and Wu, J.H. (2013) Physicochemical and antioxidant properties of four mango (*Mangifera indica* L.) cultivars in China. *Food Chemistry*, **138**, 396–405.

Liu, F.X., Wang, Y., Li, R., Bi, X. and Liao, X. (2014) Effect of high hydrostatic pressure and high temperature short time on antioxidant activity, antioxidant compounds and color of mango nectars. *Innovative Food Science and Emerging Technologies*, **21**, 35–43.

Low, D.Y., D'Arcy, B. and Gidley, M.J. (2015) Mastication effects on carotenoid bioavailability from mango fruit tissues. *Food Research International*, **67**, 238–246.

Mahto, R. and Das, M. (2013) Effect of gamma irradiation on the physicochemical and visual properties of mango (*Mangifera indica* L.) cv. "Dushehri" and "Fazli" stored at 20°C. *Postharvest Biology and Technology*, **86**, 447–455.

Masibo, M. and He, Q. (2008) Major mango polyphenols and their potential significance to human health. *Comprehensive Reviews in Food Science and Food Safety*, **7**, 309–319.

Masibo, M. and He, Q. (2009) Mango bioactive compounds and related nutraceutical properties: A review. *Food Reviews International*, **25**, 346–370.

Medlicott, A.P. and Thompson, A.K. (1985) Analysis of sugars and organic acids in ripening of mango fruit (*Mangifera indica* L. var Keitt) by high-performance liquid chromatography. *Journal of the Science of Food and Agriculture*, **36**, 561–566.

Narayana, N.M.N.K. and Gupta, V.K. (2013) Effect of total milk solid content adjusted by adding ultra-filtered milk retentate on quality of set mango yogurt. *International Journal of Dairy Technology*, **66**, 570–575.

Nunez-Selles, A.J., Villa, D.G. and Rastrelli, L. (2015. Mango polyphenols and its protective effects on diseases associated to oxidative stress. *Current Pharmaceutical Biotechnology*, **16**, 272–280.

Onimawo, I.A., Oteno, F., Orokpo, G. and Akubor, P.I. (2003) Physicochemical and nutrient evaluation of African bush mango (*Irvingia gabonensis*) seeds and pulp. *Plant Foods for Human Nutrition*, **58**, 1–6.

Padda, M.S., Amarante, C.V.T., Garcia, R.M., Slaughter, D.C. and Mitcham, E.J. (2011) Methods to analyze physicochemical changes during mango ripening: A multivariate approach. *Postharvest Biology and Technology*, **62**(3), 267–274.

Pandit, S.S., Chidley, H.G., Kulkarni, R.S., Pujari, K.H., Giri, A.P. and Gupta, V.S. (2009) Cultivar relationships in mango based on fruit volatile profiles. *Food Chemistry*, **114**, 363–372.

Paz, M., Gullon, P., Barroso, M.F., Carvalho, A.P., Domingues, V.F., Gomes, A.M. *et al.* (2015) Brazilian fruit pulps as functional foods and additives: Evaluation of bioactive compounds. *Food Chemistry*, **172**, 462–468.

Perera, C.O. and Yen, G.M. (2007) Functional properties of carotenoids in human health. *International Journal of Food Properties*, **10**, 201–230.

Ragab, M. and El-Nemr, S.E. (1990) Chemical composition and volatile constituents of canned mango juice. *Nahrung*, **34**: 591–599.

Reis, L.C.B., de Souza, C.O., da Silva, J.B.A., Martins, A.C., Nunes, I.L. and Druzian, J.I. (2015) Active biocomposites of cassava starch: The effect of yerba mate extract and mango pulp as antioxidant additives on the properties and the stability of a packaged product. *Food Bioproducts Processing*, **94**, 382–391.

Robles-Sanchez, R.M., Rojas-Grau, M.A., Odriozola-Serrano, I., Gonzalez-Aguilar, G. and Martin-Belloso, O. (2009) Effect of minimal processing on bioactive compounds and antioxidant activity of fresh-cut "Kent" mango (*Mangifera indica* L.). *Postharvest Biology and Technology*, **51**, 384–390.

Robles-Sanchez, R.M., Rojas-Grau, M.A., Odriozola-Serrano, I., Gonzalez-Aguilar, G. and Martin-Belloso, O. (2013) Influence of alginate-based edible coating as carrier of anti-browning agents on bioactive compounds and antioxidant activity in fresh-cut Kent mangoes. *LWT-Food Science and Technology*, **50**, 240–246.

Rodriguez-Amaya, D.B., Kimura, M., Godoy, H. and Amaya-Farfan, J. (2008) Updated Brazilian database on food carotenoid: Factors affecting carotenoid composition. *Journal of Food Composition and Analysis*, **21**, 445–463.

Rodriguez-Pleguezuelo, C.R., Durian Zuazo, V.H., Muriel-Fernandez, J.L. and Franco-Tarifa, D. (2012) Physicochemical quality parameters of mango (*Mangifera indica* L.) fruit grown in a Mediterranean subtropical climate (SE Spain). *Journal of Agricultural Science and Technology*, **14**, 365–374.

Ruiz-Montanez, G., Ragazzo-Sanchez, J.A., Calderon-Santoyo, M., Velazquez-de-la Cruz, G., Ramirez de Leon, J.A. and Navarro-Ocana, A. (2014) Evaluation of extraction methods for preparative scale obtention of mangiferin and lupeol from mango peels (*Mangifera indica* L.). *Food Chemistry*, **159**, 267–272.

Rungpichayapichet, P., Mahayothee, B., Khuwijitjaru, P., Nagle, M. and Mueller, J. (2015) Non-destructive determination of β–carotene content in mango by near-infrared spectroscopy compared with colorimetric measurements. *Journal of Food Composition and Analysis*, **38**, 32–41.

Rymbai, H., Srivastav, M., Sharma, R.R., Patel, C.R. and Singh, A.K. (2013) Bioactive compounds in mango (*Mangifera indica* L.) and their role in human health and plant defense: A review. *Journal of Horticultural Science and Biotechnology*, **88**, 369–379.

Saranwong, S., Sornsrivichai, J. and Kawano, S. (2004) Prediction of ripe-stage eating quality of mango fruit from its harvest quality measured non-destructively by near infrared spectroscopy. *Postharvest Biology and Technology*, **31**, 137–145.

Siddiq, M., Sogi, D.S. and Dolan, K.D. (2013) Antioxidant properties, total phenolics, and quality of fresh-cut Tommy Atkins mangoes as affected by different pre-treatments. *LWT-Food Science and Technology*, **53**, 156–162.

Singh, R.N. (1990) *Mango*. Indian Council of Agricultural Research Publication, New Delhi, India. Singh, R. and Dwivedi, U.N. (2008). Effect of Ethrel and 1-methylcyclopropene (1-MCP) on antioxidants in mango (*Mangifera indica var. Dashehari*) during fruit ripening. *Food Chemistry*, **111**, 951–956.

Singh, R.N. and Dwivedi, U.N. (2008) Effect of ethrel and 1-methylcyclopropene (1-MCP) on antioxidants in mango (*Mangifera indica* var. Dashehari) during fruit ripening. *Food Chemistry*, **111**, 951–956.

Singh, S.P. and Saini, M.K. (2014) Postharvest vapor heat treatment as a phytosanitary measure influences the aroma volatiles profile of mango fruit. *Food Chemistry*, **164**, 387–395.

Singh, Z. and Singh, S.P. (2012) Mango. In *Crop post-harvest: Science and Technology, vol. 3: Perishables* (eds D. Rees, G. Farrell and J.E. Orchard), pp. 108–142. Blackwell Publishing, Oxford.

Sivakumar, D., Jiang, Y. and Yahia, E.M. (2011) Maintaining mango (*Mangifera indica* L.) fruit quality during the export chain. *Food Research International*, **44**, 1254–1263.

Sogi, D.S., Siddiq, M., Roidoung, S. and Dolan, K.D. (2012) Total phenolics, carotenoids, ascorbic acid, and antioxidant properties of fresh-cut mango (*Mangifera indica* L., cv. Tommy Atkins) as affected by infrared heat treatment. *Journal of Food Science*, **27**, C1197–C1207.

Souza, R.C., Melo, J.F.B., Nogueira Filho, R.M., Campeche, D.F.B. and Figueiredo, R.A.C.R. (2013) Influence of mango meal on growth and body composition of Nile tilapia. *Archivos de Zootecnia*, **62**, 217–225.

Sun, J. and Chen, P. (2012) High-performance Liquid Chromatography with high-resolution mass spectroscopy analysis of African mango (*Iringia gabonensis*) seeds, extract and related dietary supplements. *Journal of Agricultural and Food Chemistry*, **60**, 8703–8709.

Thanaraj, T., Terry, L.A. and Bessant, C. (2009) Chemometric profiling of pre-climacteric Sri Lankan mango fruit (*Mangifera indica* L.). *Food Chemistry* **112**, 786–794.

Tharanathan, R.N., Yashoda, H.M. and Prabha, T.N. (2006) Mango (*Mangifera indica*, L.), "The King of Fruits" – An overview. *Food Research International*, **22**, 95–123.

USDA [US Dept of Agriculture] (2016) USDA National Nutrient Database for Standard Reference Release 28 [Online]. Available: https://ndb.nal.usda.gov/ndb/search [21 October 2016].

Vega-Vega, V., Silva-Espinoza, B.A., Cruz-Valenzuela, M.R., Bernal-Mercado, A.T., Gonzalez-Aguilar, G.A. *et al.* (2013) Antioxidant enrichment and antimicrobial protection of fresh-cut mango applying bioactive extracts from their seeds by-products. *Food and Nutrition Sciences*, **4**, 197–203.

Vergara-Valencia, N., Granados-Perez, E., Agama-Acevedo, E., Tovar, J., Ruales, J. and Bello-Perez, L.A. (2007) Fiber concentrate from mango fruit: Characterization, associated antioxidant capacity and application as a bakery product ingredient. *LWT-Food Science and Technology*, **40**, 722–729.

Vijayanand, P., Deepu, E. and Kulkarni, S.G. (2015) Physicochemical characterization and the effect of processing on the quality characteristics of Sindura, Mallika and Totapuri mango cultivars. *Journal of Food Science and Technology*, **52**, 1047–1053.

Willis, R., McGlasson, B., Graham, D. and Joyce, D. (eds) (1998) *Postharvest: An Introduction To The Physiology And Handling Of Fruit, Vegetables Ornamentals*, 4th edition, pp. 262. CAB International Wallingford, Oxford.

12

Phytochemical Compounds in Functional Properties of Mangoes

Yearul Kabir[1], Hossain Uddin Shekhar[1] and Jiwan S. Sidhu[2]

[1] Department of Biochemistry and Molecular Biology, University of Dhaka, Dhaka, Bangladesh
[2] Department of Food Science and Nutrition, College of Life Sciences, Kuwait University, Safat, Kuwait

12.1 Introduction

Mango (*Mangifera indica* L.), which belongs to the family Anacardiaceae, is a rich source of various bioactive phytochemical compounds. The polyphenols are the major phytochemicals in mango plant and fruit in terms of antioxidant capacity; these include mangiferin, catechins, quercetin, kaempferol, rhamnetin, anthocyanins, gallic acid, ellagic acid, propyl and methyl gallate, benzoic acid, and protocatechuic acid. The chemical structures of selected phytochemicals found in mango are shown in Figure 12.1. The nutraceutical and pharmaceutical significance of mangiferin, which is the predominant antioxidant in the mango, has been extensively demonstrated and continues to attract much attention, especially in its potential to combat degenerative diseases like heart disease and cancer (Luo *et al.* 2012; Pardo-Andreu *et al.* 2008; Wauthoz *et al.* 2007). The concentration of the different polyphenolic compounds varies in the different parts of mango (pulp, peel, seed, bark, leaf, and flower), with most polyphenols being found in all the parts (Ajila *et al.* 2007; Ali *et al.* 2012; Ashoush and Gadallah 2011). Mango polyphenols, like other polyphenolic compounds, work mainly as antioxidants, a property that enables them to protect human cells against damage due to oxidative stress leading to lipid peroxidation, DNA damage, and many degenerative diseases. Use of pure isolated compounds has been found to be less effective than the use of crude mixtures from the particular mango part, suggesting that synergism of the various mango polyphenols is important for maximum antioxidant activity.

12.2 Phytochemical Compounds in Mango

Mango is a rich source of many phytochemical compounds, which also impart a characteristic color and flavor to the fruit (Grundhöfer *et al.* 2001). The phytochemical compounds of mango are mainly phenolic compounds and these compounds can be obtained from various parts of the plant such as fruit, kernel (stone), leaves, and bark

Handbook of Mango Fruit: Production, Postharvest Science, Processing Technology and Nutrition,
First Edition. Edited by Muhammad Siddiq, Jeffrey K. Brecht and Jiwan S. Sidhu.

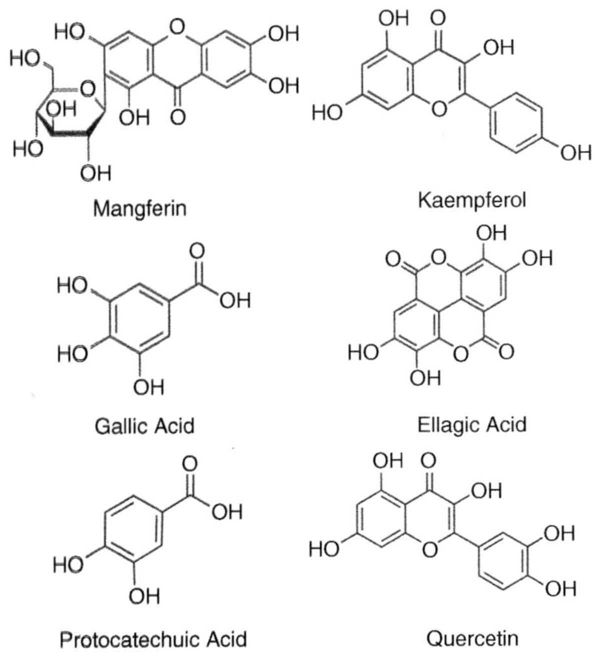

Figure 12.1 Chemical structures of selected phytochemicals found in mango plant and fruit.

(Ajila *et al.* 2007; Dorta *et al.* 2012). These results corroborate previous information that mangoes are a good source of many phytochemicals (Barreto *et al.* 2008).

The quantities and characteristics of different mango phenolics differ in the different plant parts, besides being affected by the geographic location of the plants. The chemical composition of mango varies with the location of cultivation, variety, and stage of maturity (Manthey and Perkins-Veazie 2009). Abdalla *et al.* (2007) characterized the phenolic compounds in Egyptian mango seed kernels. The components included tannins, gallic acid, coumarin, ellagic acid, vanillin, mangiferin, ferulic acid, cinnamic acid, and unknown compounds. Cheema and Sommerhalter (2015) reported that crude extracts of Ataulfo exhibited polyphenol oxidase activity with pyrogallol, 3-methylcatechol, catechol, gallic acid, and protocatechuic acid. During different stages of ripening, besides skin color changes, total phenolic acids and flavonoids concentration of mango fruit also changes, as shown in Figure 12.2.

Early studies showed that the main phenolics found in mango are leucocyanidin, catechin, epicatechin, chlorogenic acid, quercitrin, and quercetin (Schieber *et al.* 2000). Mango also contains significant amounts of pigments, including chlorophylls and carotenoids (Grundhofer *et al.* 2001). Gallic acid is one of the most abundant phenolic compounds detected in mango fruit (Palafox-Carlos *et al.* 2012). Gallic acid has been identified as the major polyphenol present in mango fruits, followed by 6 hydrolysable tannins and 6 minor compounds, for example *p*-hydroxy-benzoic acid, *m*-coumaric acid, *p*-coumaric acid, and ferulic acid (Kim *et al.* 2009). Schieber *et al.* (2000) identified 20 polyphenols in mango puree, and these compounds included gallic acid, caffeic acid, protocatechuic acid, *p*-coumaric acid, mangiferin, quercetin 3-ara-glc (peltatoside), kaempferol (hexose), six quercetin derivatives, gallotannin, and others. The most

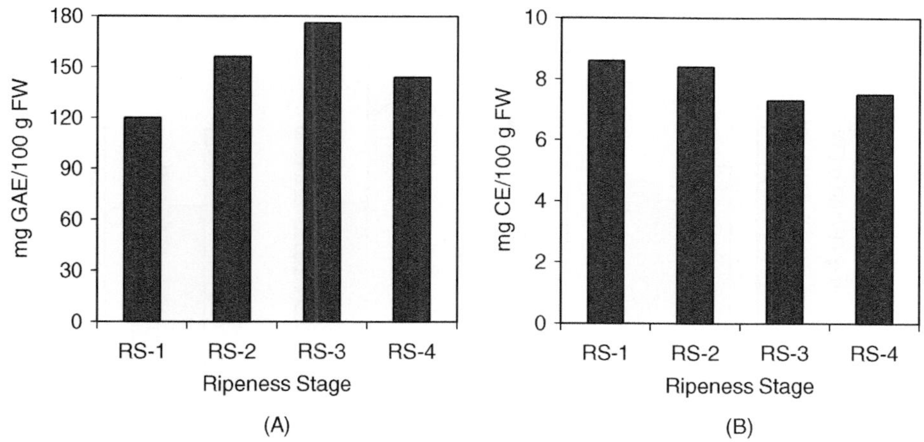

Figure 12.2 Total phenolic acids (A) and flavonoids (B) in Ataulfo mango at four ripening stages, based on skin yellow color of 0–10% (RS-1), 20–30% (RS-2), 70–80% (RS-3) and 100% (RS-4). GAE = Gallic acid equivalent, CE = catechin equivalents, FW = fresh weight basis *Source:* adapted from Palafox-Carlos *et al.* (2012).

abundant phenolics in mango cubes were gallic acid, p-hydroxybenzoic acid, p-coumaric acid, sinapic acid, quercetin, ellagic acid, and catechin (Robles-Sánchez *et al.* 2009a). In another study, it was reported that the major phenolics in mango are leucocyanidin, catechin, epicatechin, chlorogenic acid, and quercetin (Robles-Sánchez *et al.* 2009b).

12.2.1 Mango Pulp

Pulp, a major processed product from mangoes, is rich in a number of phytochemicals. Mangiferin, gallic acids, gallotannins, quercetin, isoquercetin, ellagic acid, and β-glucogallin are among the polyphenolic compounds identified in the mango pulp (Schieber *et al.* 2000). In a polyphenol screening of 20 mango varieties, Saleh and El-Ansari (1975) reported the co-occurrence of mangiferin, iso-mangiferin, and homo-mangiferin in mango fruit pulp. Rastraelli *et al.* (2002) reported the presence of gallic acid, 3,4-dihydroxybenzoic acid, benzoic acid, gallic acid methyl ester, gallic acid propyl ester, and benzoic acid propyl ester in pulp. The phenolic acids are either gallic acid in gallotannins (mango fruit) or other phenolic acids derived from the oxidation of galloyl residues in ellagitannins (Scalbert and Williamson 2000). Besides the previously mentioned phenolic acids, other phenolic acids found in the mango, albeit in small amounts, are caffeic acid, ferulic acid, and cinnamic acid (Ahmed *et al.* 2007). Ramirez *et al.* (2014) identified 21 phenolic compounds in peel and pulp of Pica and Tommy Atkins cultivars of mango in northern Chile. Three procyanidin dimers, seven acid derivatives and four xanthones were identified, mainly mangiferin and mangiferin gallate, which were present in both peel and pulp. Homo-mangiferin was also present in both fruit pulps and dimethyl mangiferin was present only in Tommy Atkins mango pulp. These researchers also reported that the Pica fruits showed better antioxidant capacities and higher polyphenolic content than the Tommy Atkins fruits. The fruit ripening stage typically has a significant effect on the antioxidant capacity (Figure 12.3), for Ataulfo mangoes at different stages of ripening.

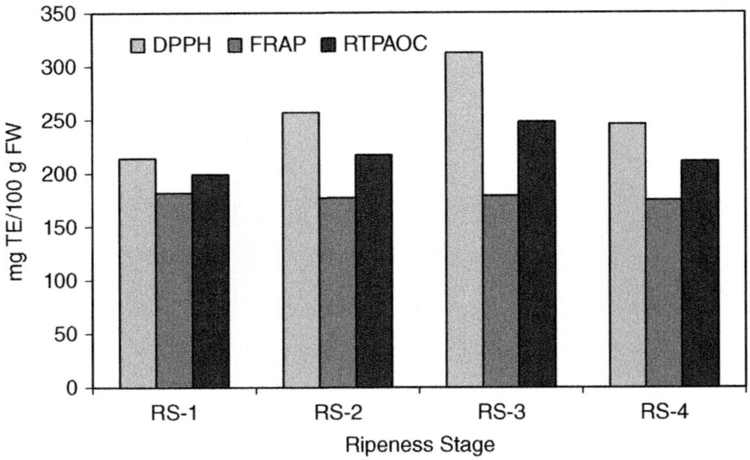

Figure 12.3 Antioxidant capacity of Ataulfo mango at four ripening stages, based on skin yellow color of 0–10% (RS-1), 20–30% (RS-2), 70–80% (RS-3) and 100% (RS-4). RTPAOX = relative total phenolic antioxidant capacity, as mean of 2,2-diphenyl-1-picrylhydrazyl (DPPH) + Ferric reducing ability of plasma (FRAP), = Trolox equivalent, FW = fresh weight basis *Source:* adapted from Palafox-Carlos *et al.* (2012).

12.2.2 Mango Peel

Mango peel is a major by-product of the mango processing industry and constitutes about 15–20% of total weight of mango fruit. Peel has been found to be a good source of polyphenols, carotenoids, dietary fiber, vitamin E, and vitamin C (Ajila *et al.* 2007). Masibo and He (2008) reported that the polyphenolic constituents of mango peel include mangiferin, quercetin, rhamnetin, ellagic acid, kaempferol, and their related conjugates. Luo *et al.* (2012) reported that mango peel and seed kernel were good sources of mangiferin. Ajila *et al.* (2010) reported that acetone extract of mango peel contained gallic, protocatechuic, gentisic, and syringic acids as bound phenolic acids. Relatively high contents of carotenoids and anthocyanins were present in the ripened mango peel, whereas unripened mango peel contained more polyphenol constituents (Ajila *et al.* 2007). Berardini *et al.* (2005a) reported the presence of anthocyanins in the mango peel in the range from 203 to 565 mg/100 g (dry matter), depending on variety and stage of maturity. Mango peels have been shown to be a rich source of flavonol O- and xanthone C-glycosides, gallotanins, and benzophenone derivatives. Berardini *et al.* (2004) detected 18 gallotannins and 5 benzophenone derivatives in mango peel, which were tentatively identified as galloylatedmaclurin and iriflophenone glucosides. In a separate study on mango peel, Berardini *et al.* (2005b) found higher quantities of quercetin and its related glycosides. Ajila and Prasada Rao (2013) reported that among the bound flavonoids identified in mango peel, kaempferol and quercetin were the major flavonoids, and rutin was a minor component. The variations in phytochemical compounds in mango peel extracted by different solvents and at different temperatures are shown in Table 12.1.

Dorta *et al.* (2014) characterized the phenolic compounds of peel and seed extracts of 3 mango varieties, Keitt, Sensation, and Gomera 3, produced in Spain by 3 different extraction systems and identified 30 compounds that belong to 5 phenolic families:

Table 12.1 Phytochemical compounds in mango peel extracts obtained using different solvents and temperatures (mg/100 g, dry basis).

Extraction Temperature	Extraction Solvent	Total Flavonoids[1]	Tannins[2]	Proanthocyanindins[3]
25 °C	Methanol	700	9,400	580
	Ethanol	500	6,000	320
	Acetone	550	1,100	200
	Water	180	3,400	240
	Methanol : Water (1:1)	340	4,100	230
	Ethanol : Water (1:1)	340	8,100	300
	Acetone : Water (1:1)	430	7,100	380
50 °C	Methanol	380	7,700	540
	Ethanol	420	6,000	310
	Acetone	490	2,700	250
	Water	170	3,800	120
	Methanol : Water (1:1)	340	6,000	180
	Ethanol : Water (1:1)	380	7,400	310
	Acetone : Water (1:1)	540	10,000	430
75 °C	Methanol	370	7,800	540
	Ethanol	530	7,500	430
	Acetone	510	3,300	270
	Water	170	2,100	220
	Methanol : Water (1:1)	390	5,700	320
	Ethanol : Water (1:1)	250	8,000	280
	Acetone : Water (1:1)	560	11,000	520

[1] Catechin equivalents.
[2] Tannic acid equivalents.
[3] Leucoanthocyanidin equivalents.

gallates and gallotannins; flavonoids, mainly quercetin derivatives; ellagic acid and derivatives; xanthones, principally mangiferin; and benzophenones and derivatives, such as maclurin derivatives. The contribution of each of these groups depends on the mango variety, the part of fruit studied (peel or seed) and the extraction conditions used. The extraction method had a greater influence on the phenolic content than variety or the mango by-product analyzed. Ribeiro *et al.* (2008) observed that mango peel has higher free radical scavenging activity of 53.3% than that of seed at 24.2%. At all developmental stages, the peel of mango fruit contains higher levels of polyphenols than the flesh.

12.2.3 Mango Kernel

The mango kernel or stone has been shown to be a good source of phenolic compounds (Soong and Barlow 2004), including micro-elements like selenium, copper, and zinc.

Table 12.2 Total phenolic, ascorbic acid, and scavenging activity of mango peel powder (MPP) and mango kernel powder (MKP).

Characteristics	MPP	MKP
Total phenolics (mg GAE/g)[1, 2]	19.06–23.20	23.90–125.35
Ascorbic acid (mg/g)[2]	8.47	6.12
Scavenging activity (%)[1]	93.89	95.08
Scavenging activity, ORAC (μmol TE/g)[2]	611	1600

GAE, Gallic acid equivalent; TE, Trolox equivalent
Source: [1]Ashoush and Gadallah (2011), [2]Sogi *et al.* (2013) *Source:* Kaur *et al.* (2010).

Schieber *et al.* (2003) suggested that the antioxidant effect of the mango kernel was due to its high content of polyphenols, sesquiterpenoids, phytosterols, and microelements – selenium, copper, and zinc. Ahmed *et al.* (2007) quantified various polyphenolic compounds in the mango kernel: tannin, gallic acid, coumarin, caffeic acid, vanillin, mangiferin, ferulic acid, cinnamic acid, and unknown compounds. The extract of mango kernel (Chok-Anan cultivar) contained total phenolics at a concentration of 194 mg GAE/g dry weight of mango kernel, of which 85.7% was identified as methyl gallate (Maisuthisakul and Gordon, 2014). The total phenolics, ascorbic acid, and scavenging activity of mango peel powder and mango kernel powder are shown in Table 12.2.

Kaur *et al.* (2010) summarized the tests that can be used for detecting the phytochemical substances in the crude methanolic mango seed kernel extract:

- *Benedict's test (to determine reducing sugars)*: extract changes to dark brick red color, showing the presence of glycosides;
- *Frothing test for saponins*: formation of frothing is an indication of the presence of saponins;
- *Borntrager's test (anthraquinone derivatives)*: change of extract to greenish color, indicating the absence of anthraquinone derivatives;
- *Flavonoid test*: development of yellow color, showing the presence of flavonoids;
- *Ferric Chloride test for tannins*: extract change into dark blue color, showing the presence of tannins; and
- *Test for alkaloids*: formation of white precipitate in extract, indicating the presence of alkaloids.

12.2.4 Mango Leaves and Bark

Mango leaves as well as the bark contain very high levels of phenolic compounds, mainly mangiferin, that is responsible for the various pharmaceutical activities (Mohan *et al.* 2013). Mangiferin (C 2- *β* –D –glucopyranosyl-1, 3, 6, 7-tetrahydroxyxanthone) was first isolated from mango leaves, while homo-mangiferin (1,6,7-trihydroxy-3-methoxy-2-C-*β*-D-glucopyranosyl-xanthone) was isolated from bark (Masibo and He 2008). Mango bark is reported to contain protocatechic acid, catechin, mangiferin, alanine, glycine, γ-amino-butyric acid, kinic acid, shikimic acid, and the tetra cyclic triterpenoids cycloart-24-en-3p,26-diol,3-keto dammar-24(ES-en-2Os,26-diol,C-24) epimers of cycloart-25 en 3β,24,27-triol and cycloartan-3β, 24,27-triol (Wauthoz *et al.* 2007).

Zhang *et al.* (2011) reported that mango leaves contained significant amounts of polyphenols, including xanthonoids, mangiferin, and gallic acid. In addition, the presence of mangiferin has been reported in mango leaves (Schieber *et al.* 2003). Monzer *et al.* (2013) isolated and identified a phenolic compound pyrogallol (1,2,3-Trihydroxybenzene), considered a toxic compound, from Alphonso mango leaves. The presence of mangiferin has been widely reported and is the main component of the mango bark extract commercialized in Brazil under the name of Vimang (Pardo-Andreu *et al.* 2008). Sánchez *et al.* (2003) isolated and identified phenolic acids (gallic acid, 3,4-dihydroxy benzoic acid, benzoic acid), phenolic esters (gallic acid methyl ester, gallic acid propyl ester, benzoic acid propyl ester), flavan-3-ols (catechin and epicatechin), and the mangiferin from the aqueous extract (10%) of mango stem bark available in Cuba under the brand name of Vimang.

12.3 Flavonoids in Mango

Shivashankara *et al.* (2004) reported that quercetin and catechin were the main flavonoids present in mango. Catechin is a flavonoid from the group of catechins, including (−)-epicatechin, (−)-epigallocatechin, (−)-epicatechingallate, and (+)-gallocatechin. Schieber *et al.* (2000) demonstrated the presence of quercetin and related glycosides in mango pulp, with the predominant flavonol glycoside being quercetin 3-galactoside, followed by quercetin 3-glucoside and quercetin 3-arabinoside. Kanwali *et al.* (2009) isolated five flavonoids: (−)-epicatechin-3-*O*-β-glucopyranoside, 5-hydroxy-3-(4-hydroxylphenyl)pyrano[3,2-*g*]chromene-4(8*H*)-one,6-(*p*-hydroxybenzyl)taxifolin-7-*O*-β-D-glucoside (tricuspid), quercetin-3-*O*-α-glucopyranosyl-(1→2)-β-glucopyranoside, and (−)-epicatechin(2-(3,4-dihydroxyphenyl)-3,4-dihydro-2*H*-chromene-3,5,7-triol), from the leaves of mango. Vilela *et al.* (2013) investigated the chemical composition of the lipophilic extracts of the ripe pulp of mangoes from 12 cultivars of mango from the island of Madeira and reported that the predominant compounds were phytosterols (free and glycosylated) and fatty acids (ω−3 and ω−6), minor amounts of long chain aliphatic alcohols (C14–C30), and α-tocopherol in the extracts.

12.4 Pigments in Mango

The pigment accumulation in mango fruit is cultivar dependent, but in general, mango has a high content of carotenoids in mesocarp tissue responsible for the intense yellow color (Singh *et al.* 2013). Several carotenoids have been identified in fruit of different mango cultivars, but only a few of them occur in significant concentrations (Ornelas-Paz *et al.* 2007). Chen *et al.* (2007) isolated 25 different carotenoids from mango pulp, the densest of which was β-carotene that accounts for the yellow orange pigmentation of most mango cultivars. Ornelas-Paz *et al.* (2007) quantified xanthophyll esters, carotenes, and tocopherols in the fruit of seven Mexican mango cultivars. During fruit ripening, carotenoids accumulate in an exponential manner, especially all-*trans*-β-carotene (Ornelas-Paz *et al.* 2008). Jacxsens *et al.* (2010) reported that Haden and Ataulfo mango varieties have higher β-carotene constituents as compared to Kent and Tommy Atkins. Vásquez-Caicedo *et al.* (2005) determined the total β-carotene content of four bright yellow-orange and five other poor-colored Thai mango cultivars and reported β-carotene content of 6.54–11.25 mg/100 g and 1.02–2.20 mg/100 g dry weight, respectively.

Berardini *et al.* (2005b,c) reported that the total anthocyanin content was higher in ripe mango peel than that of raw mango peels. In addition, they also identified a novel anthocyanin 7-O-methylcyanidin 3-*O*-*β*-D-galactopyranoside in the mango peel of a red-colored Tommy Atkins cultivar. Rozema *et al.* (2002) reported that the amount of anthocyanin changes during growth of mango leaf or epidermal cells, where the polyphenols are mainly accumulated. Zhichang *et al.* (2014) reported that there was no response to fruit coloration, which was affected by anthocyanins. They also reported that the expression of anthocyanidin synthase, the key enzyme in the anthocyanin biosynthesis, was maximum in the green rather than the yellow and red fruit peel, and expression was down-regulated in mature fruit.

12.5 Functional Properties and Health Benefits

Fruits, including mangoes, are a good source of various health promoting bioactive compounds. Polyphenols are ubiquitously present in all the fruits and vegetables and provide antioxidant and protective effects in human body. Postharvest stress-type treatments further enhance the concentration of bioactive compounds. Gonzalez-Aguilar *et al.* (2010) reported that to prevent injuries, the plant cell has developed mechanisms involving some secondary metabolic compounds (flavonoids, lignans, carotenoids, ascorbate, glutathione, etc.) and some enzymes (supeoxide dismutase, catalase, peroxidase, ascorbate peroxidase, glutathione reductase, etc.) that convert reactive oxygen species (ROS) into less toxic products. ROS are involved in signaling pathways for the production of antioxidant molecules. This section provides a brief description of various functional properties and health benefits of bioactive compounds found in mango.

12.5.1 Anti-parasitic Activities

Garrido *et al.* (2005) reported that an aqueous mango stem bark extract (Vimang) exhibited anthelminthic and anti-allergic activities. Rad *et al.* (2011) reported that mango extract displayed a protective action against gentamicin-induced nephrotoxicity in the rat. The aqueous extract of stem bark of mango was shown to prevent the GM-induced free radical formation (Bibu *et al.* 2011). Awe *et al.* (1998) evaluated anti-plasmodial activity of the stem bark extract of mango against *Plasmodium yoeliinigeriensis*; the results showed a marked respiratory activity and schizontocidal effect during early infection. Tona *et al.* (1998) reported *in vitro* anti-amoebic activity of mangiferin extract from mango stem bark using Metronidazole as a reference drug.

12.5.2 Antimicrobial Activity

Hannan *et al.* (2013) reported antibacterial effects of mango leaves extract against antibiotic sensitive and multi-drug resistant *Salmonella typhi*. Subbiya *et al.* (2013) evaluated and compared the antibacterial efficacy of mango kernel extracts with conventional irrigants (5% sodium hypochlorite (NaOCl) and 2% chlorhexidine) against *Enterococcus*

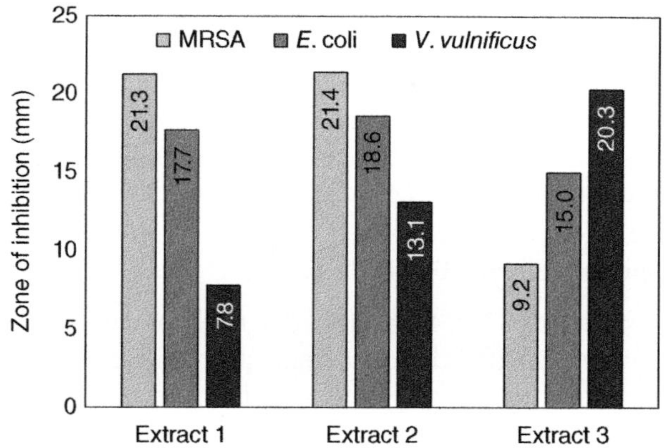

Figure 12.4 Antibacterial activity of crude methanolic extract of mango seed kernel by disc diffusion assay. (MRSA = Methicillin resistant *Staphylococcus aureus*; Extract 1: Single mango species methanolic extract (100 mg/mL); Extract 2: Mixed mango species methanolic extract (100 mg/mL); and Extract 3: Chloramphenicol disc). Source: adapted from Kaur *et al.* (2010)

faecalis dentinal biofilm, and reported comparable antibacterial activity of mango extract to that of conventional irrigants on the biofilm. Kaur *et al.* (2010) also reported the antibacterial activity of the methanolic extract of mango kernel. The antimicrobial spectrum of the mango kernel extract is more effective against Gram-negative bacteria and some pathogenic fungi (Mounica and Subbiah 2014). Zakaria *et al.* (2006) showed antibacterial activity of methanolic and ethanolic extracts of mango leaves against *Cholera diptheriae, Staphylococcus aureus, S. typhi,* and *Pseudomonas aeruginosa*. The crude extracts of kernel and fractions of mango were reported to have significant antimicrobial activity against the enteropathogen, *Shigella dysenteriae*, isolated from diarrhoeal stool specimens (Rajan *et al.* 2011). Bharti (2013) reported that the hexane-ethyl acetate leaf extract of mango exhibited pronounced antimicrobial activity against selected pathogenic bacteria. Talba *et al.* (2014) also reported that the methanol extract of mango kernel at different concentrations inhibited the growth of *Aeromonas caviae*. The antimicrobial activity of methanolic extract of mango seed kernel is presented in Figure 12.4.

12.5.3 Antitumor Activity

Different mango fruit extracts exhibited anticancer properties in several cancer cell lines, including leukemia, lung, breast, cervical, prostate, and colon cells among others (Ali *et al.* 2012; García-Rivera *et al.* 2011). However, the anticancer activity of mango extracts showed cell line specificity and the activity efficacy was affected by many factors, such as cultivar, tissue, or the fruit growth environment. For example, in a screening of the anti-proliferative effect of plant food in Mexico, mango extracts did not show significant inhibitory effects on the breast cancer cell line MCF-7 (Garcia-Solis *et al.* 2009).

Taing *et al.* (2015) evaluated the ability of methanolic extracts of mango peel and flesh from three genetically distinct mango cultivars (Irwin, Nam Doc Mai, and Kensington Pride) to inhibit proliferation of MCF-7 breast cancer cells and found that only the Nam

Doc Mai peel extract significantly inhibited the proliferation of hormone sensitive MCF-7 cells. Noratto *et al.* (2010) demonstrated differential growth inhibitory activities of extracts from diverse mango cultivars in SW-480 colon cancer cells. Hoang *et al.* (2015) reported that the Nam Doc Mai mango peel extract of content had very high bioactivity values against human breast cancer cell lines (both MCF-7 and MDA-MB-231 cells), based on the reduction of cell numbers and increasing cell death.

12.5.4 Anti-hyperlipidemic and Antidiabetic Activity

Recently, Osorio-Esquive *et al.* (2014) reported that the hypercholesterolemic mice supplemented with microwave dehydrated mango powder diet showed a significant decrease in total serum cholesterol and significant increase in HDL-C levels compared to the hypercholesterolemic mice and the mice fed with conventional dehydrated mango powder. Moreover, the mice fed with microwave dehydrated mango powder had significantly lower levels of serum LDL-C compared to hypercholesterolemic mice. Robles-Sánchez *et al.* (2011) reported that the serum triglycerides were significantly reduced after supplementation of whole mango and fresh-cut fruit for 30 days in normolipidemic subjects (37% and 38%, respectively). Shah *et al.* (2010) reported that treatment with aqueous extract of mango leaves significantly decreased total serum cholesterol, triglycerides, low density lipoprotein, very low density lipoprotein, and increased the high density lipoproteins in rats fed a high cholesterol diet. Luka and Mohammed (2012) reported that the aqueous extract of mango leaf showed significant reduction in the levels of glucose, cholesterol, triglyceride, and enzymes activities in alloxan induced diabetic rats, therefore possessing hypoglycaemic and hypolipidaemic properties.

Evans *et al.* (2013) suggested that mango consumption may help lower blood sugar levels in obese adults, which was supported by previously reported findings that adding 1% mango to high-fat diets in mice was effective in reducing body fat accumulation and lowering blood glucose levels. Lucas *et al.* (2012) reported that the freeze-dried mango pulp lowered blood glucose to a similar level as the hypoglycemic drug Rosiglitazone, but did not exert the same negative consequences of Rosiglitazone on bone mass and structural parameters in mice fed a high fat diet.

12.5.5 Analgesic and Anti-inflammatory Activity

Aqueous extract of mango (Vimang), when administered orally at 50–200 mg/kg body weight, was shown to reduce ear edema induced by arachidonic acid and phorbol-myristate acetate in mice (Garrido *et al.* 2006). Oluwole and Esume (2014) investigated the anti-inflammatory effects of mango extract in a carrageenin induced rat paw edema model of acute inflammation and reported a dose-dependent and significant inhibition of acute inflammation when compared with controls, and suggested that mango extract contained active compounds with an anti-inflammatory activity. Mango juice prepared from mango pulp was shown to inhibit free radical production and neoplastic transformation in both BALB/3T3 and HL-60 cell lines (Percival *et al.* 2006).

12.5.6 Immuno-modulatory Effects

Makare *et al* (2001) investigated the immune modulatory activity of alcoholic extract of mango bark and stem for its effect on cell mediated and humoral components of the immune system in mice and reported that the extract possessed immune-modulatory activity. The inhibitory effects of mango extract on rat macrophage functions, including phagocytic activity and the respiratory burst, also has been reported by Leiro *et al.* (2002).

12.5.7 Antipyretic and Anti-allergic Activity

Awe *et al.* (1998) reported antipyretic activity of the mango stem bark extract (mangiferin) in mice. A reduction in yeast induced hyperpyrexia was also produced by the extract. Rivera *et al.* (2006) reported anti-allergic properties of Vimang on an *in-vivo* model of inflammatory allergy mediated by mast cells and thus suggested that this natural extract could be successfully used in the treatment of allergic disorders. Leiro *et al.* (2003) observed the anti-allergic effect of the mango stem bark component mangiferin in mice infected with the nematode, *T. spiralis*.

12.5.8 Hepatoprotective Activity

The cytotoxic effects of mango aqueous extract (Vimang) on rat hepatocytes, possible interactions of the extract with drug-metabolizing enzymes and its effects on GSH levels and lipid peroxidation was investigated by Rodeiro *et al.* (2007). The results showed that the pre-treatment of cells with Vimang (25–200 μg/mL) strongly inhibited the decrease of GSH levels and lipid peroxidation. Tolosa *et al.* (2013) showed that the mango stem bark extract has cytoprotective effects against oxidative damage and mitochondrial toxicity induced by xenobiotics to human hepatic cells. Prasad *et al.* (2007) evaluated the chemopreventive properties of mango pulp extract against 7,12-dimethylbenz[a]anthracene (DMBA) induced alteration in liver of Swiss albino mice and found that the extract was effective in combating oxidative stress induced cellular injury of mouse liver by modulating cell-growth regulators. Karuppanan *et al.* (2014) reported that oral supplementation of ethanolic and methanolic leaf extract of mango remarkably reduced mercuric chloride induced hepatotoxicity in mice, possibly due to its potent radical scavenging property.

El-Makawy *et al.* (2015) investigated the protective role of mango peel or kernel defatted extracts against Pb-acetate adverse effects on oxidant/antioxidant status, liver dysfunction biomarkers, histopathological changes, and genotoxicity in male mice and reported that the extracts prevented the histopathological changes in liver induced by Pb-acetate and decreased the cytotoxicity of lead.

12.6 Future Research Needs

During the last few years, a number of reports addressed the genetic diversity of mango for application in cultivar identification using PCR and other sequencing based techniques. The elucidation of mango transcriptome will be very helpful in identifying genes for expression studies in early and late flowering mangos during fruit ripening, to offset the problem of alternate bearing of fruits by mango plants. A number of terpene and benzenoid metabolism related genes, including geranylgeranyl pyrophosphate synthase, farnesyl pyrophosphate synthase, and isochorismate hydrolase were found in mango unigenes. Terpenoids and benzenoids are important natural products in mango. Characterization of the genome and proteome of this plant would help in the improvement of mango fruit's ripening, eating, storage, and transportation quality.

References

Abdalla, A.E.M., Darwish, S.M., Ayad, E.H.E. and El-Hamahmy, R.M. (2007) Egyptian mango by-product: Compositional quality of mango seed kernel. *Food Chemistry*, **103**, 1134–1140.

Ahmed, A., Saeid, D., Eman, A. and Reham, E. (2007) Egyptian mango by-product. 1: Compositional quality of mango seed kernel. *Food Chemistry*, **103**, 1141–1152.

Ajila, C.M. and Prasada Rao, U.J.S. (2013) Mango peel dietary fiber: Composition and associated bound phenolics. *Journal of Functional Foods*, **5**, 444–450.

Ajila, C.M., Naidu, K.A., Bhat, S.G. and Prasada Rao, U.J.S. (2007) Bioactive compounds and antioxidant potential of mango peel extract. *Food Chemistry*, **105**, 982–988.

Ajila, C.M., Jaganmohan Rao, L. and Prasada Rao, U.J.S. (2010) Characterization of bioactive compounds from raw and ripe *Mangifera indica* L. peel extracts. *Food and Chemical Toxicology*, **48**, 3406–3411.

Ali, M.R., Yong, M.J., Gyawali, R., Mosaddik, A., Ryu, Y.C. and Cho, S.K. (2012) Mango (*Mangifera indica* L.) peel extracts inhibit proliferation of HeLa human cervical carcinoma cell via induction of apoptosis. *Journal of the Korean Society for Applied Biological Chemistry*, **54**, 397–405.

Ashoush, I.S. and Gadallah, M.G.E. (2011) Utilization of mango peels and seed kernels powders as sources of phytochemicals in biscuit. *World Journal of Dairy and Food Sciences*, **6**, 35–42.

Awe, S.O., Olajide, O.A., Oladiran O.O. and Makinde, J.M. (1998) Antiplasmodial and antipyretic screening of *Mangifera indica* extract. *Phytotherapy Research*, **12**, 437–438.

Berardini, N., Carle, R. and Schieber, A. (2004) Characterization of gallotannins and benzophenone derivatives from mango (*Mangifera indica* L. cv. "Tommy Atkins") peels, pulp and kernels by high performance liquid chromatography/electrospray ionization mass spectrometry. *Rapid Communications in Mass Spectrometry*, **18**, 2208–2216.

Berardini, N., Fezer, R., Conrad, J., Beifuss, U., Carle, R. and Schieber, A. (2005a) Screening of mango (*Mangifera indica* L.) cultivars for their contents of flavonol O-and xanthone C glycosides, anthocyanins, and pectin. *Journal of Agricultural and Food Chemistry*, **53**, 1563–1570.

Berardini, N., Knödler, M., Schieber, A. and Carle, R. (2005b) Utilization of mango peels as a source of pectin and polyphenolics. *Innovative Food Science and Emerging Technologies*, **6**, 442–452.

Berardini, N., Schieber, A., Klaiber, I., Beifuss, U., Carlea, R. and Conrad, J. (2005c) 7-O-methylcyanidin 3-O- β-D-galactopyranoside, a novel anthocyanin from mango (*Mangifera indica* L. cv. "Tommy Atkins") peels. *Zeitschrift Fur Naturforschung B*, **60**, 801–804.

Barreto, J.C., Trevisan, M.T.S., Hull, W.E., Erben, G., de Brito, E.S. *et al.* (2008) Characterization and quantitation of polyphenolic compounds in bark, kernel, leaves, and peel of mango (*Mangifera indica* L.). *Journal of Agricultural and Food Chemistry*, **56**, 5599–5610.

Bharti, R.P. (2013) Studies on antimicrobial activity and phytochemical profile of *Mangifera indica* leaf extract. *IOSR Journal of Environmental Science, Toxicology and Food Technology*, **7**, 74–78.

Bibu, K.J., Joy, A.D., Abraham, M.J. and Mercey, K.A. (2011) Evaluation of therapeutic effect of aqueous extract of stem bark of *Mangifera indica* Linn. on gentamicin-induced nephrotoxicity. *Journal of Applied Animal Research*, **39**, 252–256.

Cheema, S. and Sommerhalter, M. (2015) Characterization of polyphenol oxidase activity in Ataulfo mango. *Food Chemistry*, **171**, 382–387.

Chen, J.P., Tai, C.Y. and Chen, B.H. (2007) Effects of different drying treatments on the stability of carotenoids in Taiwanese mango (*Mangifera indica* L.). *Food Chemistry*, **100**, 1005–1010.

Dorta, E., Lobo, M.G. and Gonzalez, M. (2012) Reutilization of mango by-products: Study of the effect of extraction solvent and temperature on their antioxidant properties. *Journal of Food Science*, **71**, 80–88.

Dorta, E., González, M., Lobo, M.G., Sánchez-Moreno, C. and de Ancos, B. (2014) Screening of phenolic compounds in by-product extracts from mangoes (*Mangifera indica* L.) by HPLC-ESI-QTOF-MS and multivariate analysis for use as a food ingredient. *Food Research International*, **57**, 51–60.

El-Makawy, A.I., Ashoush, I.S. and Abd-Elmoneim, O.M. (2015) Evaluation of mango by-product extracts as antioxidant against Pb-acetate-induced oxidative stress and genotoxicity in mice. *Polish Journal of Food and Nutrition Sciences*, **65**(1), 39–47.

Evans, S., Peterson, S., Perkins-Veazie, P., Clarke, S.L., Payton, M. *et al.* (2013) Effects of mango supplementation on body weight and composition and clinical parameters of obese individuals [Poster Abstract]. *The FASEB Journal*, **27**, 1057–1055.

Garcia-Solis, P., Yahia, E.M., Morales-Tlalpan, V. and Diaz-Munoz, M. (2009) Screening of anti-proliferative effect of aqueous extracts of plant foods consumed in Mexico on the breast cancer cell line MCF-7. *International Journal of Food Sciences and Nutrition*, **60**, 32–46.

García-Rivera, D., Delgado, R., Bougarne, N., Haegeman, G., and Berghe, W.V. (2011) Gallic acid indanone and mangiferin xanthone are strong determinants of immunosuppressive anti-tumour effects of *Mangifera indica* L. bark in MDA-MB231 breast cancer cells. *Cancer Letters*, **305**, 21–31.

Garrido, G., Blanco-Molina, M., Sancho, R., Macho, A., Delgado, R. and Munoz, E. (2005) An aqueous stem bark extract of *Mangifera indica* (Vimang) inhibits T-cell proliferation and TNF-induced activation of nuclear transcription factor NF-kappaB. *Phytotherapy Research*, **19**, 211–215.

Garrido, G., González, D., Lemus, Y., Delporte, C. and Delgado, R. (2006) Protective effects of a standard extract of *Mangifera indica* L. (VIMANG®) against mouse ear edemas and its inhibition of eicosanoid production in J774 murine macrophages. *Phytomedicine*, **13**, 412–418.

Gonzalez-Aguilar, G.A., Villa-Rodriguez, J.A., Ayala-Zavala, J.F. and Yahia, E.M. (2010) Improvement of the antioxidant status of tropical fruits as a secondary response to some postharvest treatments. *Trends in Food Science and Technology*, **21**, 475–482.

Grundhofer, P., Niemetz, R., Schilling, G. and Gross, G. (2001) Biosynthesis and subcellular distribution of hydrolyzable tannins. *Phytochemistry.*, **57**, 915–927.

Hannan, A., Asghar, S., Naeem, T., Ullah, M.I., Ahmed, I. *et al.* (2013) Antibacterial effect of mango (*Mangifera indica* Linn.) leaf extract against antibiotic sensitive and multi-drug resistant *Salmonella typhi*. *Pakistan Journal of Pharmaceutical Sciences*, **26**, 715–719.

Hoang, V.L.T., Pierson, J-T., Curry, M.C., Shaw, P.N., Dietzgen, R.G. *et al.* (2015) Polyphenolic contents and the effects of methanol extracts from mango varieties on breast cancer cells. *Food Science Biotechnology*, **24**, 265–271.

Jacxsens, L., Luning, P.A., van der Vorst, J.G.A.J., Devlieghere, F., Leemans, R. and Uyttendaele, M. (2010) Simulation modelling and risk assessment as tools to identify the impact of climate change on microbiological food safety: The case study of fresh produce supply chain. *Food Research International*, **43**, 1925–1935.

Kanwali, Q., Hussain, I., Siddiqui, H.L. and Javaid, A. (2009) Flavonoids from mango leaves with antibacterial activity. *Journal of the Serbian Chemical Society*, **74**, 1389–1399.

Karuppanan, M., Krishnan, M., Padarthi, P. and Namasivayam, E. (2014) Hepatoprotective and antioxidant effect of *Mangifera indica* leaf extracts against mercuric chloride-induced liver toxicity in mine. *Euroasian Journal of Hepato-Gastroenterology*, **4**, 18–24.

Kaur, J., Rathinam, X., Kasi, M., Leng, K.M., Ayyalu, R. Kathiresan, S. and Subramaniam, S. (2010) Preliminary investigation on the antibacterial activity of mango (*Mangifera indica* L: Anacardiaceae) seed kernel. *Asian Pacific Journal of Tropical Medicine*, **3**, 707–710

Kim, Y., Lounds-Singleton, A.J. and Talcott, S.T. (2009) Antioxidant phytochemical and quality changes associated with hot water immersion treatment of mangoes (*Mangifera indica* L.). *Food Chemistry*, **115**, 989–993.

Leiro, J., García, D., Delgado, R. and Ubeira, F.M. (2002) Modulation of rat macrophage function by the *Mangifera indica* L. extracts Vimang and mangiferin. *International Immunology*, **2**, 797–806.

Leiro, J., García, D., Escalante, M., Delgado, R. and Ubeira, F.M. (2003) Anthelminthic and anti-allergic activities of *Mangifera indica* L. stem bark components Vimang and Mangiferin. *Phytotheraphy Research*, **17**, 1203–1208.

Lucas, E.A., Brown, A., Li, W., Peterson1, S.K., Wang, Y. et al. (2012) Mango modulates blood glucose similar to rosiglitazone without compromising bone parameters in mice fed high fat diet. *Journal of Pharmacy and Nutrition Sciences*, **2**, 115–126.

Luka, C.D. and Mohammed, A. (2012) Evaluation of the antidiabetic property of aqueous extract of *Mangifera indica* leaf on normal and alloxan-induced diabetic rats. *Journal of Natural Product and Plant Resources*, **2**, 239–243.

Luo, F., Qiang, L., Zhao, Y., Hu, G., Huang, G. *et al.* (2012) Quantification and purification of mangiferin from Chinese mango (*Mangifera indica* L.) cultivars and its protective effect on human umbilical vein endothelial cells under H_2O_2-induced stress. *International Journal of Molecular Sciences*, **13**, 11260–11274.

Maisuthisakul, P. and Gordon, M.H. (2014) Characterization and storage stability of the extract of Thai mango (*Mangifera indica* Linn. Cultivar Chok–Anan) seed kernels. *Journal of Food Science and Technology*, **51**, 1453–1462.

Makare, N., Bodhankar, S. and Rangari, V. (2001) Immunomodulatory activity of alcoholic extract of *Mangifera indica* L. in mice. *Journal of Ethnopharmacology*, **78**, 133–137.

Manthey, J.A. and Perkins-Veazie, P. (2009) Influences of harvest date and location on the levels of beta-carotene, ascorbic acid, total phenols, the *in vitro* antioxidant capacity, and phenolic profiles of five commercial varieties of mango (*Mangifera indica* L.). *Journal of Agricultural and Food Chemistry*, **57**, 10825–10830.

Masibo, M. and He, Q. (2008) Major mango polyphenols and their potential significance to human health. *Comprehensive Reviews in Food Science and Food Safety*, **7**, 309–319.

Mohan, C.G., Deepak, M., Viswanatha, G.L., Savinay, G., Hanumantharaju, V. and Rajendra, C.E. (2013) Anti-oxidant and anti-inflammatory activity of leaf extracts and fractions of *Mangifera indica*. *Asian Pacific Journal of Tropical Medicine*, **6**, 311–314.

Monzer, A.E.M., Srour, H.A. and Abd El-Ghany, A.M. (2013) Isolation and identification of *Kilifia acuminata* (Signoret) (Hemiptera: Coccidae) toxic compounds in Alphonso mango leaves. *World Rural Observations*, **5**, 114–119.

Mounica, V. and Subbiah, V. (2014) Extraction and characterization of mango seed kernel oil and its antimicrobial activity. *International Journal of Advanced Information Science and Technology*, **21**, 25–29.

Noratto, G.D., Bertoldi, M.C., Krenek, K., Talcott, S.T., Stringheta, P.C. and Mertens-Talcott, S.U. (2010) Anticarcinogenic effects of polyphenolics from mango (*Mangifera indica*) varieties. *Journal of Agricultural and Food Chemistry*, **58**, 4104–4112.

Oluwole, O.G and Esume, C. (2014) Anti-inflammatory effects of aqueous extract of *Mangifera indica* in Wistar rats. *Journal of Basic and Clinical Physiology and Pharmacology*, **26**, 313–315.

Ornelas-Paz, J.J, Yahia, E.M. and Gardea-Bejar, A. (2007) Identification and quantification of xanthophyll esters, carotenes, and tocopherols in the fruit of seven Mexican mango cultivars by liquid chromatography-atmospheric pressure chemical ionization-time-of-flight mass spectrometry [LC-(APcI(+))-MS]. *Journal of Agricultural and Food Chemistry*, **55**, 6628–6635.

Ornelas-Paz, J.J., Yahia , E.M. and Gardea, A.A. (2008) Changes in external and internal color during postharvest ripening of "Manila" and "Ataulfo" mango fruit and relationship with carotenoid content determined by liquid chromatography –APcl +-time-of-flight mass spectrometry. *Postharvest Biology and Technology*, **50**, 145–152.

Osorio-Esquive, O., Cortés-Viguri, V., Garduño-Siciliano, L., Ortiz-Moreno, A. and Sánchez-Pardo, M.E. (2014) Hypolipidemic activity of microwave-dehydrated mango (*Mangifera indica* L.) powder in mice fed a hypercholesterolemic diet. *Journal of Biomedical Science and Engineering*, **7**, 809–817.

Palafox-Carlos, H., Yahia, E. Islas-Osuna, M.A., Gutierrez-Martinez, P., Robles- Sánchez, M. and González-Aguilar, G.A. (2012) Effect of ripeness stage of mango fruit (*Mangifera indica* L., cv. Ataulfo) on physiological parameters and antioxidant activity. *Scientia Horticulturae*, **135**, 7–13.

Pardo-Andreu, G.L., Paim, B.A., Castilho, R.F., Velho, J.A., Delgado, R. *et al.* (2008) *Mangifera indica* L. extract (Vimang) and its main polyphenol mangiferin prevent mitochondrial oxidative stress in atherosclerosis-prone hypercholesterolemic mouse. *Pharmacological Research*, **57**, 332–338.

Percival, S.S., Talcott, S.T., Chin, S.T., Mallak, A.C., Lounds-Singleton, A. and Pettit-Moore, J. (2006) Neoplastic transformation of BALB/3 T3 cells and cell cycle of HL-60 cells are inhibited by mango (*Mangifera indica* L.) juice and mango juice extracts. *Journal of Nutrition*, **136**, 1300–1304.

Prasad, S., Kalra, N, and Shukla, Y. (2007) Hepatoprotective effects of lupeol and mango pulp extract of carcinogen induced alteration in Swiss albino mice. *Molecular Nutrition and Food Research*, **51**, 352–359.

Rad, A.K., Ghazi, L., Boroushaki, M.T., Khooei, A., Keshavarzi, Z. *et al.* (2011) Effect of commercial (Vimang) and hydroalcoholic extract of *Mangifera indica* (Mango) on gentamicin-induced nephrotoxicity in rat. *Avicenna Journal of Phytomedicine*, **1**, 98–105.

Rajan, S, Thirunalasundari, T. and Jeeva, S. (2011) Anti-enteric bacterial activity and phytochemical analysis of the seed kernel extract of *Mangifera indica* Linnaeus against *Shigella dysenteriae* (Shiga, corrig.) Castellani and Chalmers. *Asian Pacific Journal of Tropical Medicine*, **4**, 294–300.

Ramirez, J.F., Zambrano, R., Sepúlveda, B. and Simirgiotis, M.J. (2014) Antioxidant properties and hyphenated HPLC-PDA-MS profiling of Chilean pica *m*ango fruits (*Mangifera indica* L. Cv. piqueño). *Molecules*, **19**, 438–458.

Rastraelli, L., Selles, A.J.N., Castro, H.T.V., Aguero-Aguero, J., Gonzalez-Gonzalez, J. *et al.* (2002) Isolation and quantitative analysis of phenolic antioxidants, free sugars, and polyols from mango (*Mangifera indica* L.) stem bark aqueous decoction used in Cuba as a nutritional supplement. *Journal of Agriculture and Food Chemistry*, **50**, 762–766.

Ribeiro, S.M.R., Barbosa, L.C.A., Queiroz, J.H., Knodler, M. and Schieber. A. (2008) Phenolic compounds and antioxidant capacity of Brazilian mango (*Mangifera indica* L.) varieties. *Food Chemistry*, **110**, 620–626.

Rivera, D.G., Balmaseda, I.H., León, A.Á., Hernández, B.C., Montiel, L.M. *et al.* (2006) Anti-allergic properties of *Mangifera indica* L. extract (Vimang) and contribution of its glucosylxanthone mangiferin. *Journal of Pharmacy and Pharmacology*, **58**, 385–392.

Robles-Sánchez, M., Astiazaran-García, H., Martín-Belloso, O., Gorinstein, S., Alvarez-Parrilla, E. *et al.* (2011) Influence of whole and fresh-cut mango intake on plasma lipids and antioxidant capacity of healthy adults. *Food Research International*, **44**, 1386–1391.

Robles-Sánchez, R.M., Islas-Osuna, M.A., Astiazaran-Garcia, H., Vazquez-Ortiz, F.A., Martin-Belloso, O. *et al.* (2009a) Quality index consumer acceptability, bioactive compounds, and antioxidant activity of fresh-cut Ataulfo mangoes (*Mangifera indica* L.) as affected by low-temperature storage. *Journal of Food Science*, **74**, S126–S134.

Robles- Sánchez, R.M., Rojas-Grau, M.A., Odriozola-Serrano, I., Gonzalez-Aguilar, G.A. and Martin-Belloso, O. (2009b) Effect of minimal processing on bioactive compounds and antioxidant activity of fresh-cut "Kent" mango (*Mangifera indica* L.). *Postharvest Biology and Technology*, **51**, 384–390.

Rodeiro, I., Donato, M.T., Jimnenez, N., Garrido, G., Delgado, R. and Gomez-Lechon, M.J. (2007) Effects of *Mangifera indica* L. aqueous extract (Vimang) on primary culture of rat hepatocytes. *Food Chemical Toxicology*, **45**, 2506–2512.

Rozema, J., Bornman, J.F., Gaberscik, A., Haeder, D.P. and Trost, T. (2002) The role of UV-B radiation in aquatic and terrestrial ecosystems-an experimental and functional analysis of the evolution of UV-absorbing compounds. *Journal of Photochemistry and Photobiology*. B: Biology, **66**, 2–12.

Saleh, N.A. and El-Ansari, M.A. (1975) Polyphenolics of twenty local varieties of *Mangifera indica*. *Planta Medica*, **28**, 124–130.

Sánchez. G.M., Rodriguez, M.A., Giuliani. A., Nilncz Sclles, A.J., Rodriguez, N.P. *et al.* (2003) Protective effect of *Mangifera indica* L. extract (Vimang) on the injury associated with hepatic ischaemia reperfusion. *Phytotherapy Research*, **17**, 197–201.

Scalbert, A. and Williamson, G. (2000) Dietary intake and bioavailability of polyphenols. *Journal of Nutrition*, **130**, 2073S–2085S.

Schieber, A., Ullrich, W. and Carle, R. (2000) Characterization of polyphenols in mango puree concentrate by HPLC with diode array and mass spectrometric detection. *Innovative Food Science and Emerging Technologies*, **1**, 161–166.

Schieber, A., Berardini, N. and Carle, R. (2003) Identification of flavonol and xanthone glycosides from mango (*Mangifera indica* L. cv. "Tommy Atkins") peels by high-performance liquid chromatography-electrospray ionization mass spectrometry. *Journal of Agricultural and Food Chemistry* **51**, 5006–5011.

Shah, K.A., Patel, M.B., Shah, S.S., Chauhan, K.N., Parmar P.K. and Patel, N.M. (2010) Antihyperlipidemic activity of *Mangifera indica* l. leaf extract on rats fed with high cholesterol diet. *Der Pharmacia Sinica,* **1**, 156–161.

Shivashankara, K., Isobe, S., Al-Haq, M., Takenaka, M. and Shiina, T. (2004) Fruit antioxidant activity, ascorbic acid, total phenol, quercetin, and carotene of Irwin mango fruits stored at low temperature after high electric field pretreatment. *Journal of Agricultural and Food Chemistry,* **52**, 128–1286.

Singh, Z., Singh, R.K., Sane, V.A. and Nath, P. (2013) Mango-postharvest biology and biotechnology. *CRC Critical Reviews in Plant Sciences* **32**, 217–236.

Sogi, D.S., Siddiq, M., Greiby, I. and Dolan, K.D. (2013)Total phenolics, antioxidant activity, and functional properties of "Tommy Atkins" mango peel and kernel as affected by drying methods. *Food Chemistry,* **141**, 2649–2655.

Soong, Y.Y. and Barlow, J.P. (2004) Antioxidant activity and phenolic content of selected fruit seeds. *Food Chemistry,* **88**, 411–417.

Subbiya, A., Mahalakshmi, K., Pushpangadan, S., Padmavathy, K., Vivekanandan, P. and Sukumaran, V.G. (2013) Antibacterial efficacy of *Mangifera indica* L. kernel and *Ocimum sanctum* L. leaves against *Enterococcus faecalis* dentinal biofilm. *Journal of Conservative Dentistry,* **16**, 454–457.

Taing M.W, Pierson, J.T., Shaw, P.N., Dietzgen, R.G., Roberts-Thomson, S.J. *et al.* (2015) Mango fruit extracts differentially affect proliferation and intracellular calcium signaling in MCF-7 human breast cancer cells. *Journal of Chemistry, Article ID* **613268**, 10 pages.

Talba, A.M., Suleiman, M.M., Raji, M.A. and Oniye, S.J. (2014) Phytochemical screening and *in-vitro* antibacterial activity of *Mangifera indica* (Mango) kernel on *Aeromonas caviae. IOSR Journal of Pharmacy,* **4**, 45–50.

Tolosa, L., Rodeiro, I., Donato, M.T., Herrera, J.A., Delgado, R. *et al.* (2013) Multiparametric evaluation of the cytoprotective effect of the *Mangifera indica* L. stem bark extract and mangiferin in HepG2 cells. *Journal of Pharmacy and Pharmacology,* **65**, 1073–1082.

Tona, L., Kambu, K., Ngimbi, N., Cimanga, K. and Vlietinck, A.J. (1998) Antiamoebic and phytochemical screening of some Congolese medicinal plants. *Journal of Ethnopharmacology,* **61**, 57–65.

Vásquez-Caicedo, A.L., Sruamsiri, P., Carle, R. and Neidhart, S. (2005) Accumulation of all-trans-β-carotene and its 9-cis and 13-cis stereoisomers during postharvest ripening of nine Thai mango cultivars. *Journal of Agricultural and Food Chemistry,* **53**, 4827–4835.

Vilela, C., Santos, S.A.O., Oliveira, L., Camacho, J.F., Cordeiro, N., Freire, C.S.R. and Armando, J.D. (2013) Silvestre. The ripe pulp of *Mangifera indica* L.: A rich source of phytosterols and other lipophilic phytochemicals. *Food Research International,* **54**, 1535–1540.

Wauthoz, N., Aliou Balde, A., Balde, E.S., van Damme, M. and Duez, P. (2007) Ethnopharmacology of *Mangifera indica* L. bark and pharmacological studies of its main C-glucosylxanthone, Mangiferin. *International Journal of Biomedical and Pharmaceutical Sciences,* **1**, 112–119.

Zakaria, Z.A., Mat Jais, A.M., Sulaiman, M.R., Mohamed Isa, S.S.P. and Riffin, S. (2006) The *in vitro* antibacterial activity of methanol and ethanol extracts of *Carica papaya* flowers and *Magnifera indica* leaves. *Journal of Pharmacology and Toxicology*, **1**, 278–283.

Zhang, Y., Qian, Q., Ge, D., Li. Y., Wang, X. *et al.* (2011) Identification of benzophenone C-glucosides from mango tree leaves and their inhibitory effect on triglyceride accumulation in 3T3-L1 adipocytes. *Journal of Agricultural and Food Chemistry*, **59**, 11526–11533.

Zhichang, Z., Yeyuan, C., Aiping, G., Jianfeng, H. and Zhiguo, D. (2014) Cloning and expression of anthocyanidin synthase (ANS) gene from peel of mango (*Mangifera indica* Linn). *African Journal of Plant Science*, **8**, 147–152.

13

Microbiology of Fresh Mangoes and Processed Products

Anu Kalia[1] *and Rajinder P. Gupta*[2]

[1] *Electron Microscopy and Nanoscience Laboratory, College of Agriculture, Punjab Agricultural University, Ludhiana, Punjab, India*
[2] *BIS Institute of Science and Technology, Gagra, Moga, Punjab, India*

13.1 Introduction

Mango (*Mangifera indica* L.), a member of the family Ariocardiaceae, is one of the most popular tropical fruits due to the unique flavor of the edible fleshy mesocarp of this drupe (Jacobi *et al.* 2001). The fleshy mesocarp is rich in organic acids, such as oxalic, malic, succinic, pyruvic, glucuronic, and galacturonic acids, with the most abundant acid being citric acid and phytochemicals such as ascorbic acid, carotenoids, polyphenols, mangiferin, coumaric acid, gallic acid, and many more (Sivakumar *et al.* 2011). The presence of these antioxidants means mango can be included in the daily diet for several health benefits, having anti-activities/properties against cancer and viruses, apart from improving cardiac health, and decreasing the risk of cardiac disease. Mango is a nutrient-rich fruit with high content of total soluble solids and sugar – an environment suitable for the growth of certain microbes on the surface and inside the mango tissue. This chapter reviews the microbiology of fresh and minimally processed mangoes, and methods used to evaluate and maintain the quality of fresh and processed mangoes.

13.2 Microbiology of Fresh and Minimally processed Mango

The mango fruit surface microflora is comprised of a myriad of microorganisms including bacteria, yeast, and molds that are unintentionally inoculated on the surface of the produce during pre- and postharvest operations (Nguyen and Carlin 1994). The normal microflora of the outer surface of fresh fruits can generally vary according to the microbiologically safe conditions/practices followed during handling, processing, storage, and distribution (Scheper *et al.* 2007).

13.2.1 Microflora of Whole Fresh Mangoes

Although fresh fruit skin is an inhospitable environment for most microbes, microbes that do cling to fruit surfaces may exist as vegetative cells or spores

Handbook of Mango Fruit: Production, Postharvest Science, Processing Technology and Nutrition, First Edition. Edited by Muhammad Siddiq, Jeffrey K. Brecht and Jiwan S. Sidhu.

(exo- or endospores, cysts, other bodies). However, the fruit surfaces mainly contain non-pathogenic epiphytic microbes, which may include bacteria, yeast, molds, and viruses (Hanklin and Lacy 1992). The mango fruit surface has certain physico-chemical attributes such as hydrophobicity, which influences hydration (Vacheethasanee *et al.* 1998), fruit surface because its uneven features result in the creation of micro-niches that can be inhabited by microbes (Wang *et al.* 2009). It also has an electrical charge that depends on chemical functional groups on the surface, and the roughness that may result in enhanced adherence of human pathogenic microbes (Hanning *et al.* 2009). Since the microbial adhesion is a complex process involving a plethora of physical and chemical properties of the inoculated/contaminating bacterial cell surface as well as the food contact surfaces (Araújo *et al.* 2009; Fernández *et al.* 2014), the mango fruit surface is a dynamic niche for diverse microbial genera.

The general bacteria associated with the outer surfaces include both the Gram-positive and Gram-negative bacteria, consisting of varied morphological structures as bacillary rods to spherical coccoid cells. The predominant Gram-negative bacteria include the genera belonging to the family Pseudomonadaceae *viz., Pseudomonas, Xanthomonas*; Enterobacteriaceae *viz., Enterobacter, Aerobacter, Escherichia, Erwinia*, Gram-positive rods belonging to the family Bacillaceae, *Bacillus* and *Lactobacillus*; while the coccoid bacterial genera include members of Micrococcaceae, *Micrococcus*; Stapylococcaceae, *Staphylococcus*; Streptococcaceae, *Streptococcus equines, S. salviarus, S. faecalis, S. mitis*, and sarcinae. Normal fungal microflora of fruits is as diverse as the bacterial forms and includes hyphal fungi or molds such as *Rhizopus, Aspergillus*, and *Penicillum*, and the single-cell yeasts such as *Saccharomyces, Zygosaccharomyces, Hanseniaspora, Candida, Debaryomyces*, and *Pichia* sp. The spoilage molds (certain strains of *Aspergillus*) are known to produce mycotoxins, which may affect vital organs of human body, in particular the liver, kidney, and lungs.

13.2.2 Microflora of Minimally processed Mango Fruit

Fresh-cut or minimally processed mangoes are highly perishable, since cutting/slicing accelerates biochemical, physiological, and microbial changes, and so require refrigerated storage (Garcia and Barret 2002). The inner tissues of mango fruit usually do not contain any microbes; however, the high sugar content and other nutrients present may invite certain osmophilic molds and yeasts (Abadias *et al.* 2008). The cut surfaces also provide an increased surface area for contamination and growth and allow microbial infiltration of the tissues. These microbes may cling to the bruised surfaces and enter newer tissue niches and may persist there as harmless commensals (Kalia and Gupta 2012). The quality of water used for washing and rinsing of the whole mango fruit may contaminate it with an array of pathogenic microbes such as *E. coli 0157:H7, Enterobacter, Shigella, Salmonella* sp., *Vibrio chloreae*, Rota viruses and other causative agents of foodborne illnesses (Sivapalasingam *et al.* 2003). Poubol and Izumi (2005) reported the presence of *Klebsiella* and *Pantoea* in fresh-cut cubes of Nam Dok Mai mangoes.

As the fresh-cut fruits are kept under low temperature conditions, a study by Sriwimon and Boonsupthip (2011) suggested an alternative technique of enhancing the textural and organoleptic characteristics of the frozen cut mangoes. Since ice crystallization causes greater textural damage during the freezing of the ripened mango, use of partially ripened mango, vacuum infused with mango juice and sugar (natural flavor), avoids the textural and flavor losses in comparison to fully ripened mangoes. Interestingly, a by-product of the fresh-cut mango fruit industry could provide certain compounds to help curb the growth of certain opportunistic microbes. Vega-Vega *et al.* (2013) have reported that the ethanolic extract of mango seed curbed the growth of mesophilic aerobic microorganisms, total fungi, and yeast. Moreover, they also observed that this by-product from minimally processed mango enhanced the total phenolic content and antioxidant capacity of the treated mango flesh.

13.2.3 Microflora of Processed Mango Products

The processed mango fruits and fruit products exhibit the occurrence of different microbial genera and species (Nour *et al.* 2011). Largely, processed mango products are dominated by temperature tolerant (high or low) microbes, particularly bacterial endo/exospores and fungal spores. Mango jam, jelly, and preserves have reduced water activity (a_w) due to sugar addition that preferentially helps the growth of osmo-tolerant bacteria and yeast, as well as xero-tolerant fungi. Mango juice and purees are susceptible to microbial growth; Warnasuriya *et al.* (1985) reported the presence of *S. italicus* in mango juice.

The heating of some processed mango products, such as mango juice pasteurization, kills the mesophilic bacteria and vegetative fungal hyphae; however, it is not enough to kill heat-resistant ascospores or sclerotia producing *Paecilomyces* sp., *Aspergillus* sp., and *Penicillum* sp. (Splittstoesser 1991). The incidences of contamination of mango juice by *Alicyclobacillus acidoterrestris*, most commonly isolated heat-tolerant, endospore forming spoilage bacteria in canned commercially pasteurized juices (Silva and Gibbs 2001), have been reported. Danyluk *et al.* (2010) performed a gas chromatography (GC) study to ascertain the potential compound, guaiacol, responsible for the production of a typical medicinal off odor during growth by inoculation of a five-strain cocktail of *A. acidoterrestris* in mango and pineapple juices. However, Durak *et al.* (2010) reported the isolation of *A. acidocaldarius* from the mango samples and suggested that the contamination by these bacteria may be occurring during processing, transportation, or handling.

The presence of significant amounts of soluble solids make the flow characteristics and rheological properties of the mango juice dependent on temperature. For example, Dak *et al.* (2007) observed pseudo-plasticity, that is the shear-thinning nature of mango juice. Thus the possibility of preferred growth of largely pectinolytic spoilage microbes is speculated in mango juice and puree preparations. The microbial diversity of the mango purees is also determined by the processing protocol and pre-processing treatments, particularly faulty pasteurization followed for the making of raw puree. O'Connor-Shaw *et al.* (1995) observed the presence of opportunistic pathogenic *K. pneumoniae* in the mango cheeks-a-puree.

13.3 Factors Affecting Microbial Quality and Safety

Mango fruits are composed of pulp that contains pectin, sugars, vitamins, antioxidant compounds, and some minerals, which function as reservoirs or substrates for microbial growth. There occurs development of distaste on microbial contamination followed by growth in mango, as the flesh taste largely depends on the balance between organic acids (citric and malic acids) and soluble sugars (sucrose, fructose, and glucose) (Léchaudel and Joas 2007). Poubol and Izumi (2005) reported that general browning discoloration and water-soaked appearance of fresh-cut mango within 2 days of storage at 5 °C was due to the degradation of several compounds by oxidation or microbial activity. Fresh whole mango fruits have a mixture of varied microbial population, and the growth rate of each microbe type would depend upon an array of factors, as discussed below.

13.3.1 Pre-harvest Factors, Postharvest Handling, and Processing

Pre-harvest factors basically involve production practices that have an effect on the microbial quality of mangoes. The mango fruit quality is affected by a variety of pre-harvest factors, including different cultural practices, such as tree pruning for enhanced availability of light and temperature, and fruit thinning to increase the carbon availability and irrigation management to provide optimum water. The visual quality and fruit storage potential is also affected by the carbon availability during fruit development, such an increased leaf-to-fruit ratio decreased shelf life, as well as increased incidence and severity of disease (Léchaudel and Joas 2007). Similarly, Shrestha (2002) reported significant reduction in the spoilage loss of fruits during storage by foliar application of mineral nutrients (1–2% urea, 0.5–1% borax) and growth regulators (gibberellic acid) just after fruit set and at 10 days before harvesting of mango (cv Langra). Léchaudel and Joas (2007) also reported that the supply of certain minerals such as potassium, calcium, and magnesium during fruit development also affects the shelf life of the fruit. The appearance of storage disorders is not only affected by variable mineral content but also by the ratio of calcium content to other minerals. Besides environmental pre-harvest factors, general agricultural practices like foliar application of fungicides also alter the shelf life during storage. Subramanyam *et al.* (1972) reported that pre-harvest application of Captan, Bordeaux mixture, and copper oxychloride fungicides could curb the incipient fungal infections, particularly in grafted varieties in mango during storage.

Being a climacteric fruit, the mango exhibits high respiration rates during ripening that result in a shortened shelf life. A decrease in the acidity of the mango fruit occurs as it matures. The two major acids that contribute to the acidity are citric and malic acids (Younis *et al.* 2011). Therefore, mango based products that contain high soluble solids and near neutral pH could be preserved well by the addition of certain chemical additives (Doreyappy-Gowda and Huddar 2001), such as sodium benzoate, benzoic acid, potassium sorbate, potassium metabisulfite, and organic acids such as citric or malic acid and many more (Younis *et al.* 2011). The addition of these chemicals may alter the sugar content, pH, and fat content in preserved mango products. A study by Younis *et al.* (2011) reported a decline in the fat content of mango pulp by addition of SO_2. However, benzoic acid application decreased the acidity. Similarly, apart from the ash content, the fat and protein content also decreased in pulp with time during storage (Germain *et al.* 2003). The soluble solids acidity and reducing sugars increased in a whey-based mango beverage, while its pH and ascorbic acid content decreased with time during storage (Ismail *et al.* 2011). The alteration in these compounds also changes the microbial flora of the mango product.

It is imperative to obtain information on the microbial ecology of mango fruit, such that the mechanism of susceptibility for microbial spoilage among different mango varieties could be deciphered. There may be occurrence of major components, structures, and compounds that would support climax community of microorganisms, which may be responsible for spoilage of mango fruit (Jha *et al.* 2010). The use of partially treated manure, irrigation with livestock-used pond water, or storage near roosting birds contaminates mangoes with opportunistic human pathogenic microbes (Kalia and Gupta 2006). Postharvest management practices can affect product quality, since stressed produce or mechanical injuries permit microbial contamination. The increase in water loss during storage may influence the structure and/or composition of the cuticle and/or lenticels. Improper or harsh handling of mangoes may lead to bruised skin that invites greater microbial damage. Postharvest rots are most prevalent in damaged or bruised fruits (Bachmann and Earles 2000). Ismail *et al.* (2011) concluded that mango fruit juice contamination primarily occurs due to use of unwashed instruments and utensils, unhygienic water for dilution, dressing with contaminated ice, prolonged preservation without refrigeration, and unhygienic surroundings (presence of houseflies, fruit flies, dust, etc.). These researchers further reported the presence of both fecal and non-fecal coliforms in a whey-based mango beverage.

13.3.2 Mango Fruit Spoilage

The postharvest losses of mango have been reported to be as high as 27% in Asian countries. These losses are mainly due to the microbial spoilage causing fruit decay (Prabakar *et al.* 2005). Microbial spoilage affects quality of fruits and renders them inedible. Moreover, senescence and spoilage depend on cultivar type, abiotic factors, and microbes involved in the deterioration process. A common approach is to name the spoilage type by symptoms/appearance, such as soft rot or black rot. However, this definitely results in discrepancy in ascertaining the causal pathogen of spoilage and this ambiguity could be overruled by classifying on the basis of the causal microbe. The common diseases that affect mangoes include anthracnose (*Colletotrichum gloeosporioides*) (Figure 13.1), diplodia stem-end rot (*Lasiodiplodia theobromae*), and soft rots including *Alternaria* rot (*Alternaria* sp.) (Figure 13.2) and *Aspergillus* rot (*Aspergillus* sp.) (Figure 13.3), bacterial canker (*X. campestris* pv. *mangiferaeindicae*), and die back (*Botryodiplodia* (*Lasiodiplodia*) *theobromae*) (Table 13.1). These pathogens may also cause mixed infections leading to increased spoilage and decay of mango fruits, such as *Colletotrichum-Penicillium-Alternaria* and *Penicillium-Mucor* mixed infections (Figure 13.4).

Fresh mangoes possess an effective defense, including a thicker epidermal tissue and relatively higher concentration of antimicrobial organic acids. However, the higher water activity, higher sugar content, and acidic pH (<4.4) of fresh fruit favor the growth of xerotolerant fungi or osmophilic yeasts. Lamikanra *et al.* (2000) reported bacterial spoilage in neutral pH fruits like mango. Al-Najada and Al-Suabeyl (2014) reported entrance of pathogenic fungi via injured or weakened mango fruit flesh and isolated *A. niger* from brown circular spots with depressions/darker lesions, *A. flavus* and *Penicillium* spp., from infected mangoes. The mangoes also exhibit variation in their extent of spoilage at different stages of marketing, such that the spoilage may vary at wholesale and consumer levels. Similarly, *Botryodiplodia theobromae*, a pycnidia-based endogenous spore-forming fungus that causes fruit rot was isolated by Mascarenhas *et al.* (1995) as one the major postharvest pathogens of mango. The fresh-cut mangoes may become spoiled by bacteria such as *P. agglomerans*, *Aeromonas enteropelogenes*, *M. lylae*, and *Spingobacterium*

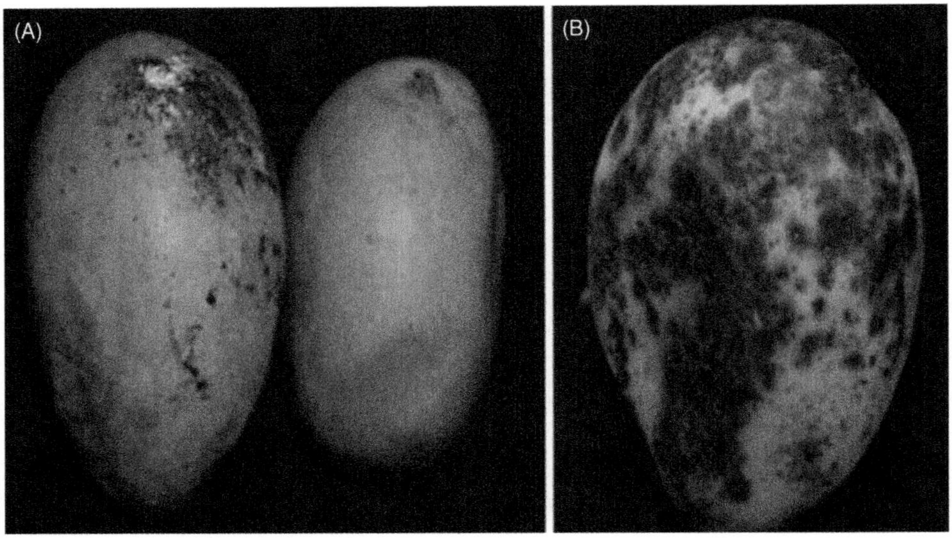

Figure 13.1 Anthracnose affected (incipient stage) and normal mango fruit (A), and anthracnose affected mango showing presence of averculi leads to rotting of the whole fruit (B).

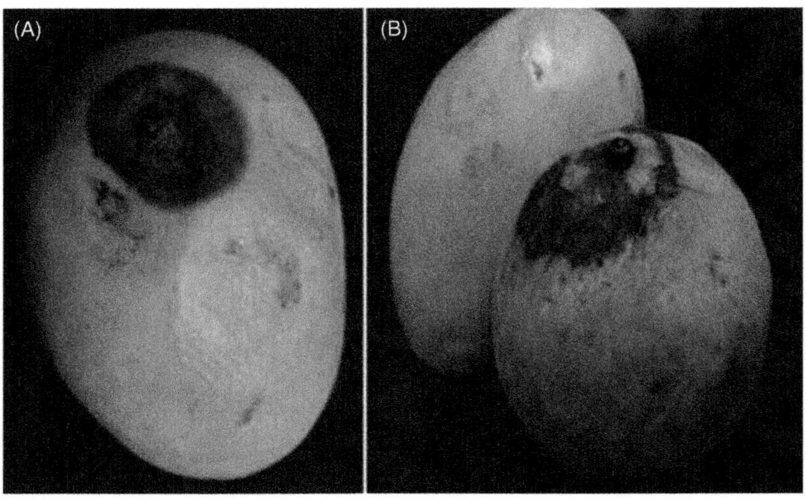

Figure 13.2 Stem end rot showing circular darkening of mango epicarp around the base of the peduncle (A), and soft rot caused by *A. alternata* in mango (B).

spiritovorum, molds such as *Alternaria* sp., *Penicillum* sp., *Aspergillus* sp., and *Fusarium* sp., and several yeasts (Ngarmsak *et al.* 2006).

Unlike the normal microflora of other fruits, mango microflora is less diverse and is predominated by spore forming fungi, with less bacteria, such as *Pseudomonas, Erwinia, Enterobacter,* and *Lactobacillus* sp. (Pao and Petracek 1997). These microbes remain adhered to the outer skin of the fruit. They can be inoculated from several sources such as air, soil, compost, and insect infestation. The major insects infesting the mango fruits

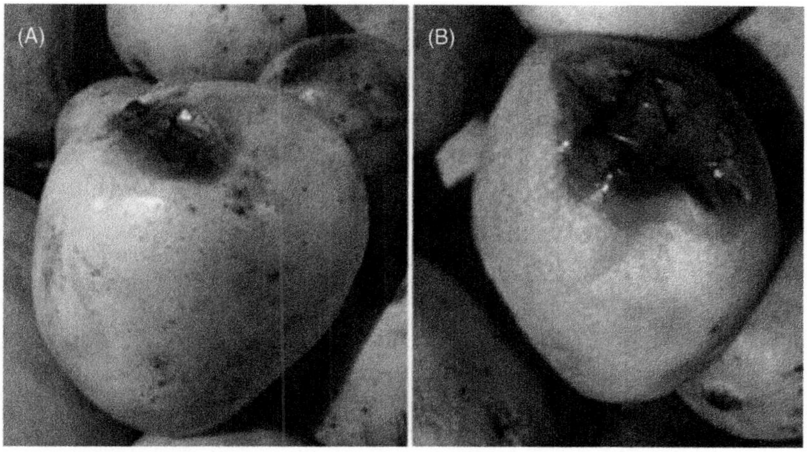

Figure 13.3 *A. niger* infection (black mold rot) in mango characterized by appearance of sunken pale brown lesions giving symptom of water soaked fruit (A, early stage; B, advance stage).

are fruit flies (*Drosophila melanogaster*), such as tephritid fruit flies, Queensland fruit fly (*Bactrocera tryoni*), Mediterranean fruit fly (*Ceratitis capitata*), and the Papaya fruit fly (*B. papayae*) (Jacobi *et al.* 2001). The source and quality of water dictates the potential for human pathogen contamination upon contact with the harvested mangoes.

Lund and Snowdon (2000) reported certain common molds, such as *Penicillum* sp., *Aspergillus* sp., *Eurotium* sp., *Alternaria* sp., *Cladosporium* sp., and *Botrytis cinerea*, involved in spoilage of fresh and dried fruits. Molds producing heat-resistant ascospores or sclerotia such as *P. fulvus*, *P. niveus*, *A. fischeri*, *P. vermiculatum*, and *P. dangeardii* were observed to cause spoilage of thermally processed fruits or the fruit products. These products had off flavors, visible mold growth, and texture breakdown (Beuchat and Pitt 1992). Apart from these changes, the mango fruits infested with *B. theobromae* exhibit enhanced respiratory rates, thus rendering them unable to attain a climacteric peak or full ripeness (Mascarenhas *et al.* 1995).

Fruit safety risks could be increased by certain spoilage types that create micro-environments suitable for the growth of human pathogens, as the primary spoilage by one group of phytopathogens produces substances required for nurturing growth and development of human pathogens. Wade and Beuchat (2003) reported on the role of proteolytic fungi and the associated changes in pH of the decayed and damaged raw fruits on survival and growth of various foodborne pathogens. *Botrytis* or *Rhizopus* spoilage of fruits can help create a suitable environment for proliferation of *Salmonella enterica* serovar *typhimurium* (Wells and Butterfield 1997).

13.3.3 Mango Fruit Safety Issues and HACCP

Several human pathogens can be transmitted by consumption of contaminated mango fruit or its products. Interestingly, the mango fruit surface and the neutral pH and high sugar content of the mango pulp can be contaminated by human pathogenic bacteria such as the common enteric bacteria *Salmonella*, notoriously associated with foodborne outbreaks of salmonellosis worldwide and particularly in the USA

Table 13.1 Mango fruit pathogens causing characteristic spoilage or rots and their physical, chemical, and biological control measures.

Disease	Causative agent	Physical	Chemical	Biological
Preharvest:				
Anthracnose	*Colletotrichum gloeosporioides*	–	Copper fungicides (Johnson and Hofman 2009); Dithiocarbamet, mancozeb, febran (Akem 2006)	*Bacillus licheniformis* (Silimela and Korsten 2007)
Stem end rot	*Lasiodiplodia theobromae, Homopsis mangifera, Dithiorella dominicana, C. gloeosporioides*	Planting disease resistant or tolerant cultivars; reduction of potential wounds; covering fruit with polyethylene caps	Preharvest fungicidal sprays	*B.licheniformis* (Govender and Korsten 2005)
Alternaria rot	*Alternaria alternata*	Hot water spray and fruit brushing treatment for 15–20s with 225 ppm prochloraz	Dithiocarbamate suppresses latent infection; 0.1% Azoxystrobin	–
Black mold rot	*Aspergillus niger A. flavus*	Keeping low temperature during harvesting	–	–
Penicillum (Blue mold) rot	*Penicillum expansum, P. cyclopium, P. digitatum*	–	–	–
Sooty mold	*Capnodium ramosum*	Pruning of infested plant parts; application of sticky compounds around the stem; trimming of limbs touching ground or buildings	Spray of Nottasul + Metacin + gumacasea (0.2% + 0.1% + 0.3%)	–

Table 13.1 (Continued)

Disease	Causative agent	Physical	Chemical	Biological
Postharvest:				
Anthracnose	*C. gloeosporioides*	Hot water treatment (50–55°C for 2–5 minutes) (Alemu 2014)	Thiabendazole benomyl; Prochloraz, imazilil, strobilurin	*B. subtilis* (Alemu 2014); bacteria and yeast antagonists (Kefialew and Ayalew 2008)
Stem end rot	*L. theobromae, H. mangifera, D. dominicana, C. gloeosporioides*	Increased CO_2 at low temp. storage (Alemu 2014)	Benomyl (500 ppm, 52 °C, 5 mins (Iram and Ahmad 2013); Hot benomyl followed by Prochloraz (0.025%, 31 °C); captan, mancozeb, copper oxychloride; geranium, mint, thyme oils (Tripathi and Shukla 2009)	Warm water dip of *B. lichiforms*, followed by prochloraz treatment (Alemu 2014)
Alternaria rot	*A. alternata*	–	Halfabarr, ginger, avocado, cinnamon, laurel extracts, neem leaf extract (5%); Strobulin (0.1%)	–
Black mold rot	*A. niger* *A. flavus*	Washing with mild hot water; low temperature (10–15°C) Storage; avoid sun exposure of fruits	Hot water containing fungicides	*Debaromyces hansenii* + 0.5% chitosan (Tripathi *et al.* 2013)
Penicillum (Blue mold) rot	*P. expansum, P. cyclopium, P. digitatum*	–	–	*Cryptococcus laurentii* + 100 ppm Indole acetic acid
Sooty mold	*C. ramosum*	5% starch sprayed on fruits followed by rubbing to remove soot	Carbaryl (0.2%), Phosphomidon (0.03%)	Garlic extracts, *Artemisia, Aloe vera, Ocium basilicum,* Rebolledo-Martínez *et al.* 2013)

(Hanning *et al.* 2009). The mango fruit contamination with *Salmonella* occurs by using inadequately treated wash water. These bacteria can become internalized and contaminate the internal tissues of the mango fruit and can be easily prevented by use of wash water disinfected by proper chlorination or application of copper salt (Hanning *et al.* 2009). Interestingly, Bordini *et al.* (2007) reported that the hot water

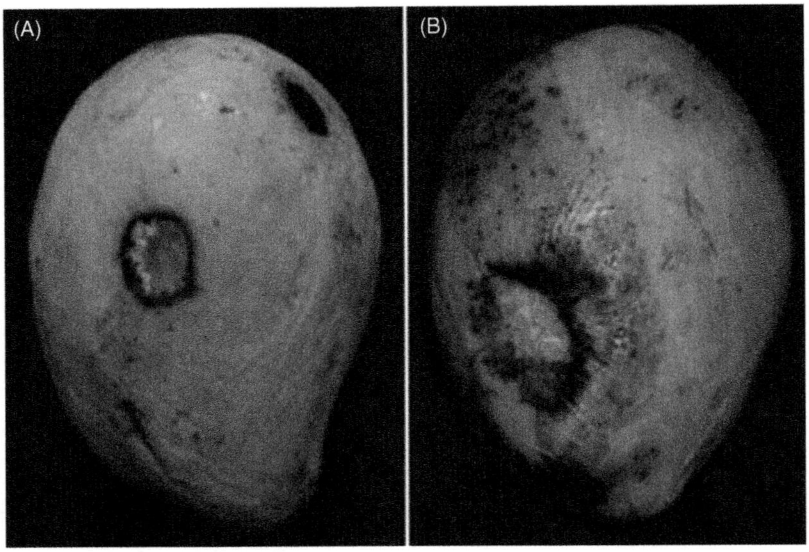

Figure 13.4 Mixed infection of *C. gloeosporioides, P. expansum,* and *A. alternata* (A); and *P. cyclopium* and *M. circinelloides* (B) in mango fruit.

immersion or hydrothermal treatment utilized for fly larvae control caused contamination of mangoes with coliforms, which migrated to the interior of mangoes, survived, and grew in all fruit portions.

In order to achieve microbiological quality, good agricultural production (GAP) and distribution practices are important. The food safety issues have resulted in identification of certain Critical Control Points (CCP), which can be easily modulated to reduce or eliminate the risks of physical, chemical, or biological hazards. This approach has been based on utilizing the preventive measures to decrease or eliminate growth of spoilage or foodborne pathogens in foods.

The fresh-cut or minimally processed fruit industry requires the stringent implementation of this approach, though it is being used voluntarily in this sector in spite of mandatory Hazard Analysis and Critical Control Points (HACCP) programs for the fruit juice industry by the FDA. The whole raw fruits, diced/sliced/skinned/shredded or minimally processed fruit products and unpasteurized fruit juices have been associated predominately with millions of cases of foodborne illnesses annually, particularly the gastrointestinal illnesses of bacterial etiology followed by protozoal parasite-caused outbreaks. The determination of a CCP in the HACCP system can be facilitated using a decision tree, as shown in Figure 13.5.

Tapia *et al.* (2009) reviewed the implementation of HACCP strategies for production of safe fresh-cut fruits and vegetables in particular. The HACCP or similar certification like EurepGap for international practice and standards have to be obtained by the mango growers that export mangoes, which would usher the marketing and storage of good quality and microbiologically safe mangoes as well as minimizing the occurrence of hazards in the supply chain and to ensure correct product handling (Sivakumar *et al.* 2011).

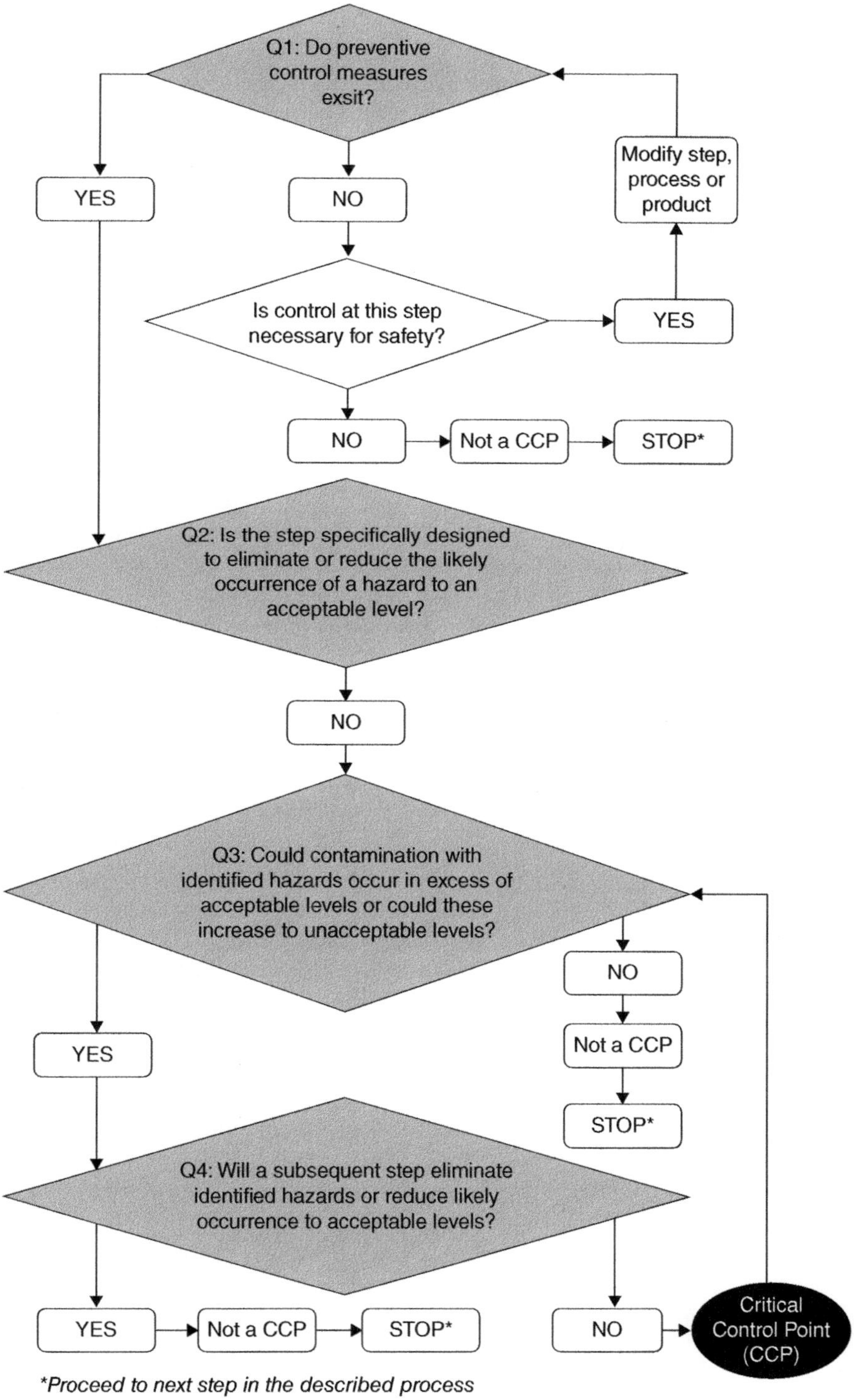

Figure 13.5 Application of decision tree toward critical control point (CCP) identification in HACCP. *Source:* FDA (1997)

13.4 Maintaining Microbial Quality

Fruit variety, and abiotic or environmental factors such as soil type, temperature, frost, and wet weather at harvest may adversely affect the storage life and quality of produce. Fresh produce that has been stressed by too much or too little water, high rates of nitrogen application, or mechanical injury (scrapes, bruises, abrasions) is particularly susceptible to postharvest diseases. Storage temperature and water activity play an important role during storage and transportation of the mango fruit. It is noteworthy that the extent of disease development and microbial contamination is governed by the developmental stage of the mango fruit and the environmental conditions (Jha *et al.* 2010). This section reviews methods used for maintaining microbial quality of mango fruit and its processed products, including fresh-cut mangoes.

13.4.1 Temperature

Fruit quality after harvest is primarily affected by the temperature conditions maintained during storage and transportation of mango fruits. A temperature-based postharvest technique to control the microbial spoilage is the heat disinfestation of mango fruits to get rid of fruit flies, which involves heating the fruit by conductive heat transfer from heated high humidity vapors (Vapor heat treatment), by hot air heating (Forced hot-air treatment), and by immersing in hot water (Hot water immersion treatment) (Jacobi *et al.* 2001).

Since mango is a climacteric fruit, the refrigerated storage retards the aging due to ripening, softening, and textural/color changes, undesirable metabolic changes and respiratory heat production, moisture loss/wilting, and spoilage due to invasion by bacteria/fungi/yeasts. Refrigeration controls the respiration rate of the crop, as this generates heat due to oxidation of sugars, fats, and proteins in the cells resulting in loss of these stored food reserves leading to decreased food value, loss of flavor, loss of saleable weight, and more rapid deterioration. A study on two mango varieties by Devanesan *et al.* (2011) showed that the respiration rate in mango decreased with a decrease in temperature from 40–12°C, and with a decrease in O_2 concentration from 21–1% in the storage glass bottle micro-environment. The respiration rate varied between the two mango varieties Banganapalli and Thothapuri, which was faster in the former than the later.

However, during refrigerated storage, mango fruit containing higher water contents can undergo chilling injury. The effects of chilling injury may be cumulative in mango and include a grayish appearance with chilling symptoms becoming evident as pitting or other skin blemishes, internal discoloration, poor color development, uneven ripening or failure to ripen, and poor flavor. Sometimes over-chilled mangoes become vulnerable or more susceptible to decay by fungal genera (Chaplin *et al.* 1989). There exists a critical time limit for the appearance of symptoms of chilling injury in mango when it is exposed to an injurious, low temperature to initiate irreversible injury. Thus a threshold temperature, that is lowest temperature at which a susceptible fruit or vegetable can be held with no symptoms of chilling injury, must be deduced for a particular fruit. Mangoes are best stored at 45–55°F.

13.4.2 Relative Humidity

Transpiration rates are determined by the moisture content of the air, which is usually expressed as relative humidity (RH). Water loss at low RH levels can severely degrade quality, since sugar-rich perishable fruits, such as mango, exhibit decreased aesthetic value of the product as well as saleable weight loss culminating in reduced profits. Thus, the RH of the storage unit directly influences water loss in fresh produce. The high levels of RH can result in condensation on the surface of the fruit, which can potentially provide a favorable environment for microbial growth.

13.4.3 Control of Respiration and Ethylene Production

Ethylene, a natural phytohormone, is produced by mango that promotes additional ripening of fruit. Over-ripening in mango decreases the quality and reduces shelf life due to appearance of specific symptoms of injury. The quality and shelf life of the fresh-cut mangoes are greatly affected by respiration induced ethylene production, which is enhanced by wounding during fresh-cut processing and causes softening and development of an off flavor (Vilas-Boas and Kader 2007). Over ripening can result in the collapse of fruit skin, especially in the case of thin-skin mango cultivars, which in turn can make fruit vulnerable to greater microbial proliferation.

13.4.4 Irradiation

The irradiation treatment of the mango has manifold benefits, including control of fruit flies (Bustos *et al.* 2004). Irradiation of mangoes with low doses of 250 Gy gamma rays may cause delay in ripening of about a week, even when stored at ambient temperatures of 25–30 °C (Sharma 2008). A dose of 150 Gy of Cobalt-60 gamma radiation is a general quarantine treatment to inhibit the adult emergence from third instar of four types of fruit fly exhibiting potential infestation of the exported mangoes in Mexico. An exposure with UV-C irradiation and storage at 5 °C may reduce deterioration of sliced Tommy Atkins mangoes (Gonzalez-Aguilar *et. al.* 2007).

13.4.5 Packaging

Appropriate packaging of mangoes is crucial to prevent contamination by microbes, as it avoids inward movement of light and air, thus keeping the produce dry/moist, and this prevents any changes in textural integrity of produce along with convenient division of the produce in suitable portions needed for transportation, handling, and sale. Modified atmosphere packaging (MAP) is useful for mangoes as it reduces the population of yeasts and molds on cut mangoes, thereby increasing the shelf life during storage (Martinez-Ferrer *et al.* 2002). Similarly, Kumar *et al.* (2008) reported better retention of beta-carotene, ascorbic acid, sugar, and hence better shelf life of 6 months at low temperature storage (at 7 °C) of osmo-vac dehydrated slices of Amrapali mangoes packaged in COEX nitrogen.

The use of chemicals such as vanillin and antimicrobial coatings (edible or non-edible) are some of the new techniques for enhancing the shelf life of mangoes. These treatments are beneficial to curb the microbial spoilage of minimally processed mango fruits such as fresh-cut mango slices. Low concentrations of 80 mM with minimum inhibitory concentrations (MICs) of 5–13 mM of vanillin could curb the growth of aerobic Gram-positive and Gram-negative bacteria, spore forming molds, as well as unicellular yeasts (Ngarmsak *et. al.* 2006). The application of edible chitosan (at 2%, w/v) as a antimicrobial coating on sliced mangoes delayed the microbial growth as compared to untreated slices during storage at 6 °C (Chien *et al.* 2007).

13.4.6 Application of Biocontrol Agents

Mangoes generally exhibit very high postharvest losses, especially in developing countries. One factor contributing to such losses is the fungal contamination causing fruit decay during storage, which can be controlled by treatment of fruits with synthetic fungicides (Sharma *et al.* 2009). However, with the evolution of resistant strains and species, as well as the growing concern for the avoidance of application of pesticides, the use of obsolete physical and chemical processes of microbial eradication have to be either revised or fresh environmentally benign alternatives, such as the use of biological control agents have to be explored further (Mari *et al.* 2007).

Several microbes have been reported to act as potential postharvest microbial control agents that decrease or even eliminate the postharvest decay of fruits by antagonism, either active, such as curbing the growth of other microbial pathogens due to secretion of antimicrobial compounds like antibiotics, cell wall degrading enzymes and induction of host resistance, or passive by competing for space and nutrition with pathogens (Sharma *et al.* 2009). Kefialew and Ayalew (2008) observed effective control of anthracnose in mango fruits by treating them with biocontrol agents *Brevundimonas diminuta*, *Stenotrophomonas maltophilia*, and *C. membranifaciens*. Similarly, application of a common dominant phylloplane and fructoplane colonizer *Bacillus* spp. can curb latent infections by *Colletotrichum* spp. on mango (Janisiewicz and Korsten 2002). Interestingly, even pre-harvest aerial spray application with the biocontrol agent *P. fluorescence* significantly reduces the pre- and postharvest anthracnose incidences in mango (Vivekananthan *et al.* 2004).

13.4.7 Nanomaterial-based Processing, Preservation, and Packaging

The use of nanomaterials, having 1–100 nm dimensions, in fruit preservation, processing, and packaging is being explored increasingly (Kalia and Parshad 2015). Nanomaterials are expected to exhibit novel physical and chemical properties in comparison to their bulky counterparts. Nanoparticles of silver, gold, zinc oxide, and titanium dioxide have known antimicrobial properties. Amendment of zinc oxide (ZnO) nanoparticles will be beneficial, as zinc is not only an essential micronutrient for maintaining good health in humans, but also the antimicrobial action of ZnO nanoparticles will render the product free from plant as well as human pathogens. Firouzabadi *et al.* (2014) reported the antimicrobial activity against known human pathogens, for example *Listeria monocytogenes*, *E. coli*, *S. aureus*, and *B. cereus* of ZnO nanoparticle suspensions containing 0.3% citric acid on supplementation in mango juice. As mangoes have near neutral pH, the addition of low amount of organic acid enhanced the antimicrobial activity by reducing pH of the juice, thereby decreasing the metabolic activity and multiplication of the microbes, whereas the ZnO nanoparticles may destroy the bacterial cell membrane (bind to lipids and proteins), causing leakage of intracellular contents and hence death of bacterial cells (Huang *et al.* 2008). Moreover, like other nanomaterials, ZnO nanoparticles lead to the generation of reactive oxygen species (ROS), apart from the Zn^{2+} ions that exhibit antibacterial action.

Nano-based edible encapsulation systems are offering potential solutions for the protection, enhanced bioavailability, and sustained or targeted release of functional compounds (Kalia and Parshad 2015). Nanoencapsulation (using synthetic or natural organic polymers to cover/coat/embed active ingredient and functional compound), nanoemulsions (nanoformulating water insoluble compounds to generate nanoparticulate size of active compound in solvent phase) (Silva *et al.* 2012), and nanoparticle adsorption (using porous biocompatible inorganic non-metal nanoparticles like silica nanoparticles), have their pros and cons and any of the above three techniques could be used for creating functional food.

13.5 Methods to Evaluate Microbial Quality

Fruit quality and safety are ensured by analysis for presence of microbes, and such microbial analyses are routinely performed as quarantine/regulatory procedures. The methods employed for adjudging the quality of food include an array of microbiological to biochemical assays to ascertain the acceptability or unacceptability of a food product for human consumption or a processing/handling practice that needs to be followed. Thus, enumerating the microbial load of the produce could help in determining the quality as well as the related safety aspects of the product and effectiveness of the processing technique employed to kill spoilage microbes.

Microbiological methods for pathogen identification primarily involve conventional cultural techniques of growing microbes on culture media and observing the ability to form viable countable colonies showing characteristic growth on such media, as well as the direct microscopic methods for various groups of microbes. The routine culturing techniques require a longer time to obtain results. To overcome this hurdle, use of indicator organisms that provide rapid, simple, and reliable information without isolation and identification of specific pathogens is performed. However, such tests can be used as a presumptive test with the confirmation provided by a battery of biochemical tests, and may include specialized serological typing (Swaminathan and Feng 1994).

The physical characteristics of food can cause a non-uniform distribution of microbes and thus a (non-uniform) sample preparation may not provide consistent and reproducible microbiological data. Rapid methods shorten the assay time by a simple modification of conventional methods or may involve an array of molecular assay formats and diverse technologies that are specific and more sensitive (Mermelstein *et al.* 2002). Some of the assays that have been used for the rapid enumeration of pathogens in mango fruits and its products are PCR-based assays and spectroscopy/spectrometric techniques.

13.5.1 PCR-based Assays

A DNA-based probe has been developed for detection of most foodborne pathogens, which detects the target gene sequence specific to a particular pathogenic microbe in a sample with unique sensitivity and reproducibility (Guo *et al.* 2000). DNA-DNA hybridization, DNA-RNA hybridization, and DNA-protein hybridization are the types of DNA-based probing techniques.

Duvenage *et al.* (2007) utilized the PCR-based denaturing gradient gel electrophoresis technique (PCR-DGGE) for rapid identification of thermophilic acidophilic bacteria (TAB) in several pasteurized juices including mango juice. Researchers were able to identify the presence of even uncultured bacteria close to genera *Bacillus* and *Alcaligenes faecalis* in pasteurized fruit juices. Kamle *et al.* (2013) used sensitive nested PCR using designed primer pairs for the identification of *C. gloeosporioides* causing anthracnose in mango fruit. Casey *et al.* (2004) reported using real-time PCR to detect the presence of spoilage yeast in fruit juice. Haggag *et al.* (2011) suggested using molecular characterization by RAPD-PCR for quick and reliable identification of causative agent of mango malformation disease, *Fusarium* spp.

Gouws *et al.* (2005) performed phylogenetic profiling by 16S rDNA PCR amplification of ultra-high temperature (UHT) mango juice and a carbonated fruit juice blend to identify temperature tolerant *A. acidocaldarius* as the only spoilage microbe. The 16s rRNA gene PCR followed by the DGGE analysis of variable region V3 could decipher bacterial diversity of UV-C radiated Ataulfo mango surface (Fernández-Suárez *et al.* 2013). The UV-C irradiation reduced the microbial load on the surface of mangoes immediately after treatment and the structure of bacterial communities was modified during storage, that is, the enterobacteriaceae member *Enterobacter* was observed in treated mangoes, while *Lactococcus* and *Pantoea* were only detected on the surface of non-treated mangoes.

13.5.2 Spectroscopy/Spectrometric Techniques

Mango fruit contains certain volatile aroma compounds that may be synthesized or accumulated during the ripening stage and hence defines the physiological age of the fruit. Generally the internal quality of mango fruit includes determining the appearance, firmness, and presence of volatiles. Besides PCR-based techniques, other methods for identification of microbes involve qualitative or quantitative occurrence of volatile compounds, signature peptides, glycoproteins, glycan, etc., by Infra-red spectroscopy (IRS), magnetic resonance, chromatography, and mass spectrometry techniques, which are useful to differentiate healthy/diseased or unripe/ripened whole fruits, fresh-cut fruits, and processed fruit products.

13.5.2.1 Fourier Transform–Infra-red Spectroscopy (FT-IRS)

The whole and fresh-cut fruit industry now places emphasis on non-invasive, rapid, reliable/accurate, real-time in-line determination of the fruit quality. IRS is one such non-invasive technique with variants such as FT-IR or Near IR spectroscopy (NIRS), which requires negligible or no-sample processing and avoid wastage of sample produce or use of expensive and environmental corrosive chemicals required in conventional wet chemistry procedures for estimation of volatile and other compounds. The strength of IRS provides both qualitative as well as quantitative determination of an array of chemical constituents, apart from pH, acidity, firmness, and TSS of fresh fruits. Thus, IRS can evaluate the overall sample composition to provide a characteristic fingerprint of the product, thereby eliminating the need to perform different tests to evaluate individual components and characteristics.

Jha and Gunasekaran (2010) reported the benefits of application of FT-IR attenuated total reflection (ATR) spectroscopy to determine added sugar content, TSS, and real juice content in fresh and commercial mango juice. An algorithm were performed based on multivariate statistical analyses, that is partial least squares (PLS) discrimination and principal component analysis (PCA) to classify the samples with or without added sugar. Likewise, Al-Qadiri *et al.* (2006) reported application of FT-IRS to correctly identify the pure as well as mixed cultures of several spoilage-causing *Alicyclobacillus* spp. and human pathogenic *E. coli* microbes in fruit juice samples, on the basis of unique spectral features of various components of the microbial cells.

13.5.2.2 Liquid/Gas Chromatography-Mass Spectrometry (LC/GC-MS)

The LC/GC-MS techniques exploit the separation power of chromatography and the detection power of mass spectrometry, which involves the production and subsequent separation and identification of charged species according to their mass to charge (m/z) ratio (Hajslova and Cajka 2007). The combining of chromatography and spectrometry techniques is very useful in screening/identification/quantification of many groups of non-polar and/or semi-polar food toxicants (Abian 1999), as well as the volatile compounds present in the fruit products. The characteristic flavor of the mango fruit is due to the presence of several volatile compounds and is critical for consumer acceptability. The mango juice quality, including the freshness of the juice, depends on the presence of diverse terpenes and guaiadiene. Shang *et al.* (2002) performed GC-MS of mango juice to identify volatile compounds; 19 compounds were quantified with concentrations varying from 4 to 8,000 μg/liter.

Deme and Upadhyula (2014) developed a novel ultra-performance liquid chromatography-atmospheric pressure photo-ionization high resolution mass spectrometry (UPLC-APPI-HRMS) method for determination of multiple pesticide residues, including organochlorine, synthetic pyrethroid, organophosphate, and carbamate in mango and grapefruit juices. Besides pesticide residues, GC-MS of a volatile compound profile could be performed for early detection of *A. acidoterrestris* in spoiled juice (Bianchi *et al.* 2010). Similarly, Orr *et al.* (2000) detected the presence of *A. acidoterrestris* in apple juice by sensory and chromatographic analysis of compounds produced by bacteria.

13.6 Predictive Microbiology

Predictive microbiology is a sub-discipline of microbiology dealing with development of models for predicting the responses of microbial populations with respect to environmental factors (McMeeken *et al.* 2008). Predictive microbiology provides mathematical tools in the form of primary, secondary, and tertiary models to estimate the consequences of food handling and processing operations on growth, survival, and inactivation of microbes such as the foodborne pathogens (McMeeken *et al.* 2008), and thus helps to describe the microbial behavior in order to prevent food spoilage and foodborne illnesses. Lopez-Malo and Palou (2000) used the logistic model to predict the storage temperature and the preservative for enhancing the shelf life of mango puree. These researchers reported that a critical temperature of 6.5 °C is required to inhibit the growth of *Z. bailii.*

The predictive microbiology techniques, though versatile and proactive, suffer from certain limitations due to the complexity encountered at two levels, that is the complexity of the microbial behavior as well as of the food matrix. New models in system biology are available, which account for both the complexities and improves the quantitative predictions of microbial behavior in a particular sample by utilizing elaborate sequential structure linked equations to provide a graphical interface (Kalia and Gupta 2012).

13.7 Future Research Needs

The microbiological characteristics of mango surface/internal tissues/processed products are required to be stringently defined, as there are a myriad of beneficial as well as plant and human pathogenic microbes that can inhabit mango or its products. The mango microbial status is governed by innumerable abiotic and biotic factors and is a prime indicator of the quality and safety of the whole fruit or its products. The occurrence of specific indicator microbes can be quantified by deciphering the conventional viable cell counts, by using PCR-based fingerprinting, and advanced GC/LC-MS based techniques with enhanced rapidity, accuracy, and sensitivity.

References

Abadias, M., Usall, J., Anguera, M., Solsona, C. and Vinas, I. (2008) Microbiological quality of fresh, minimally processed fruit and vegetables, and sprouts from retail establishments. *International Journal of Food Microbiology*, **123**, 121–129.

Abian, J. (1999) The coupling of Gas and Liquid Chromatography with Mass Spectrometry. *Journal of Mass Spectrometry*, **34**, 157–168.

Akem, C.N. (2006) Mango anthracnose disease: Present status and future research priorities. *Journal of Plant Pathology* **5**, 266–273.

Alemu, K. (2014) Dynamics and management of major postharvest fungal diseases of Mango fruits. *Journal of Biology, Agriculture and Healthcare*, **4**, 13–21.

Al-Najada, A.R. and Al-Suabeyl, M.S. (2014) Isolation and classification of fungi associated with spoilage of post-harvest mango (*Mangifera indica* L.) in Saudi Arabia. *African Journal of Microbiology Research*, **8**(7), 685–688.

Al-Qadiri, H.M., Lin, M., Cavinato, A.G. and Rasco, B.A. (2006) Fourier transform infrared spectroscopy, detection and identification of *Escherichia coli O157:H7* and *Alicyclobacillus* strains in apple juice. *International Journal of Food Microbiology*, **111**, 73–80.

Araújo, E.A., Bernardes, P.C., Andrade, N.J., Fernandes, P.E. and Sá, J.P.N. (2009) Gibbs free energy of adhesion of *Bacillus cereus* isolated from dairy plants on different food processing surfaces evaluated by the hydrophobicity. *International Journal of Food Science and Technology*, **44**, 2519–2525.

Bachmann, J and Earles, R. (2000) Postharvest Handling of Fruit and Vegetables [Online]. Available: http://attra.ncat.org/attra-pub/PDF/postharvest.pdf [26 April 2016].

Beuchat, L.R. and Pitt, J.I. (1992) Detection and enumeration of heat-resistant molds. In: *Compendium of Methods for the Microbiological Examination of Foods* (eds C. Vanderzant and D.F. Splittoesser), pp. 251–263, American Public Health Association, Washington, DC.

Bianchi, F., Careri, M., Mangia, A., Mattarozzi, M., Musci, M. *et al.* (2010) Characterization of the volatile profile of orange juice contaminated with *Alicyclobacillus acidoterrestris*. *Food Chemistry*, **123**, 653–658.

Bordini, M.E.B., Ristori, C.A., Jakabi, M. and Gelli, D.S. (2007) Incidence, internalization and behavior of *Salmonella* in mangoes, var. Tommy Atkins. *Food Control*, **18**, 1002–1007.

Bustos, M.E., Enkerlin, W., Reyes, J. and Toledo, J. (2004) Irradiation of mangoes as a postharvest quarantine treatment for fruit flies (Diptera: Tephritidae). *Journal of Economic Entomology*, **97**, 286–292.

Casey, G.D. and Dobson, A.D.W. (2004) Potential of using real-time PCR-based detection of spoilage yeast in fruit juice: A preliminary study. *International Journal of Food Microbiology*, **91**, 327–335.

Chaplin, G.R., Cole, S.P., Landrigan, M., Nuevo, P.A., Lam, P.F. and Graham, D. (1989) Chilling injury and storage of mango (*Mangifera indica* L.) fruit held under low temperatures. In: *III International Mango Symposium*, pp. 461–471. NT Department of Primary Industry and Fisheries, Darwin, Australia

Chien, P.J., Sheu, F. and Yang, F.H. (2007) Effects of edible chitosan coating on quality and shelf life of sliced mango fruit. *Journal of Food Engineering* **78**, 225–229.

Dak, M., Verma, R.C. and Jaaffrey, S.N.A. (2007) Effect of temperature and concentration on rheological properties of Kesar mango juice. *Journal of Food Engineering*, **80**, 1011–1015.

Danyluk, M.D., Friedrich, L.M., Jouquand, C., Goodrich-Schneider, R., Parish, M.E. and Rouseff, R. (2010) Prevalence, concentration, spoilage, and mitigation of *Alicyclobacillus* spp. in tropical and subtropical fruit juice concentrates. *Food Microbiology*, **28**(3), 472–477.

Deme, P. and Upadhyula, V. (2014) Ultra-performance liquid chromatography atmospheric pressure photo-ionization high resolution mass spectrometric method for determination of multiclass pesticide residues in grape and mango juices. *Food Chemistry*, **173**, 1142–1149.

Devanesan, J.N., Karuppiah, A. and Abirami, C.V.K. (2011) Effect of storage temperatures, O_2 concentrations and variety on respiration of mangoes. *Journal of Agrobiology*, **28**, 119–128.

Doreyappy-Gowda, L.N.D. and Huddar, A.G. (2001) Studies on ripening changes in mango (*Mangifera indica* L.) fruits. *Journal of Food Science and Technology Mysore*, **38**, 135–137.

Durak, M.Z., Churey, J.J., Danyluk, M.D. and Worobo, R.W. (2010) Identification and haplotype distribution of *Alicyclobacillus* spp. from different juices and beverages. *International Journal of Food Microbiology*, **142**, 286–291.

Duvenage, W., Gouws, P.A. and Witthuhn, R.C. (2007) PCR-based DGGE identification of bacteria present in pasteurized South African fruit juices. *South African Journal of Enology and Viticulture*, **28**, 56–60.

FDA [Food and Drug Administration] (1997) Hazard Analysis and Critical Control Point Principles and Application Guidelines [Online]. Available: http://www.fda.gov/Food/GuidanceRegulation/HACCP/ucm2006801.htm#app-e [23 March 2016].

Firouzabadi, F.B., Noori, M., Edalatpanah, Y. and Mirhosseini, M. (2014) ZnO nanoparticle suspensions containing citric acid as antimicrobial to control *Listeria monocytogenes*, *Escherichia coli*, *Staphylococcus aureus* and *Bacillus cereus* in mango juice. *Food Control*, **42**, 310–314.

Fernández, P.E., Sao Jose, J.F.B., Zerdas, E.R.M.A., Andrade, N.J., Fernandes, C.M. and Silva, L.D. (2014) Influence of the hydrophobicity and surface roughness of mangoes and tomatoes on the adhesion of *Salmonella enterica* serovar *typhimurium* and evaluation of cleaning procedures using surfactin. *Food Control*, **41**, 21–26.

Fernández-Suárez, R., Ramírez-Villatoro, G., Díaz-Ruiz, G., Eslava, C., Calderón, M. *et al.* (2013) Effect of postharvest UV-C treatment on the bacterial diversity of Ataulfo mangoes by PCR-DGGE, survival of *E. coli* and antimicrobial activity. *Frontiers in Microbiology*, **4**, 134.

Garcia, E. and Barrett, D.M. (2002) Preservative treatments for fresh-cut fruits and vegetables. In: *Fresh-cut Fruits and Vegetables: Science, technology and market*, (ed. O. Lamikanra), pp. 267–303. CRC Press, Boca Raton, FL.

Germain, K., Benoi, B.K. and Israel, M.L. (2003) Effect of ripening on the composition and the suitability for jam processing of different varieties of mango (*Mangifera indica*). *African Journal of Biotechnology*, **2**, 301–306.

Gonzalez-Aguilar, G.A., Villegas-Ochoa, M.A., Martinez-Tellez, M.A., Gardea, A.A. and Ayala-Zavala, J.F. (2007) Improving antioxidant capacity of fresh-cut mangoes treated with UV-C. *Journal of Food Science*, **72**, S197–S202.

Gouws, P.A., Gie, L., Pretorius, A. and Dhansay, N. (2005) Isolation and identification of *Alicyclobacillus acidocaldarius* by 16 S rDNA from mango juice and concentrate. *International Journal of Food Science and Technology*, **40**, 789–792.

Govender, V., Korsten, L. and Sivakumar, D. (2005) Semi-commercial evaluation of *Bacillus licheniformis* to control mango postharvest diseases in South Africa. *Postharvest Biology and Technology*, **38**, 57–65.

Guo, X., Chen, J., Beuchat, L.R. and Brackett, R.E. (2000) PCR detection of *Salmonella enterica* serotype Montevideo in and on tomatoes using primers derived from *hil*A. *Applied and Environmental Microbiology*, **66**, 5248–5252.

Haggag, W.M., Hazza, M., Saker, M. and El-Wahab, M.A. (2011) Molecular identification of mango malformation pathogens in Egypt. *Advances in Bioscience and Biotechnology*, **2**, 226–232.

Hajslova, J and Cajka, T. (2007) Gas chromatography-mass spectrometry (GC-MS). In: *Food Toxicants Analysis* (ed. Y. Picó), pp. 419–473. Elsevier BV, Amsterdam.

Hanklin, L. and Lacy, G.H. (1992) Pectinolytic microorganisms. In: *Compendium of Methods for the Microbiological Examination of Foods* (eds C. Vanderzant and D.F. Splittstoesser), pp. 176–183, American Public Health Association, Washington, DC.

Hanning, I.B., Nutt, J.D. and Ricke, S.C. (2009) Salmonellosis outbreaks in the United States due to fresh produce: Sources and potential intervention measures. *Foodborne Pathogens Disease*, **6**, 635–648.

Huang, Z., Zheng, X., Yan, D., and Yin, G. (2008) Toxicological effect of ZnO nanoparticles based on bacteria. *Langmuir*, **24**, 4140–4144.

Iram, S. and Ahmad, H.M.I. (2013) Major postharvest diseases of mango and their management. *International Journal of Agronomy and Plant Production*, **4**, 3470–3484.

Ismail, A.E., Abdelgader, M.O. and Ali, A.A. (2011) Microbial and chemical evaluation of whey-based mango beverage. *Advance Journal of Food Science and Technology*, **3**, 250–253.

Jacobi, K.K., MacRae, E.A. and Hetherington, S.A. (2001) Postharvest heat disinfestation treatments of mango fruit. *Scientia Horticulturae*, **89**, 171–193.

Janisiewicz, W.J. and Korsten, L. (2002) Biological control of postharvest disease of fruits. *Annual Review of Phytopathology*, **40**, 411–441.

Jha, S.N. and Gunasekaran, S. (2010) Authentication of sweetness of mango juice using Fourier transform infrared-attenuated total reflection spectroscopy. *Journal of Food Engineering*. **101**, 337–342.

Jha, S.N., Jaiswal, P., Narsaiah, K., Bhardwaj, R., Sharma, R. *et al.* (2010) Post-harvest micro-flora on major cultivars of Indian mangoes. *Scientia Horticulturae*, **125**, 617–621.

Johnson, G.I. and Hofman, P.J. (2009). Postharvest technology and quarantine treatments. In: *The Mango: Botany, Production and Uses* (ed. R.E. Litz), pp. 529–605. CAB International Publishing, Wallingford, UK.

Kalia, A. and Gupta, R.P. (2006) Fruit microbiology. In: *Handbook of Fruits and Fruit Processing* (ed. Y.H. Hui), pp. pp. 3–28. Blackwell Publishing, Ames, IA.

Kalia, A. and Gupta, R.P. (2012) Fruit microbiology: New prospects. *In*: *Handbook of Fruits and Fruit Processing* (eds N.K. Sinha, J.S. Sidhu, J. Barta, J.S.B. Wu and M.P. Cano), pp. 51–72. John Wiley & Sons, Ames, IA.

Kalia, A. and Parshad, V.R. (2015) Novel trends to revolutionize preservation and packaging of fruits/fruit products: Microbiological and nanotechnological perspectives. *Critical Reviews in Food Science and Nutrition*, **55**, 159–182.

Kamle, M., Pandey, B.K., Kumar, P. and Muthu Kumar, M. (2013) A species specific PCR based assay for rapid detection of mango anthracnose pathogen *Colletotrichum gloeosporioides* Penz. and Sacc. *Journal of Plant Pathology and Microbiology*, **4**, 1–5.

Kefialew, Y. and Ayalew, A. (2008) Postharvest biological control of anthracnose (*Colletotrichum gloeosporioides*) on mango (*Mangifera indica*). *Postharvest Biology and Technology*, **50**, 8–11.

Kumar, P.S. and Sagar, V.R. (2008) Quality of osmo-vac dehydrated ripe mango slices influenced by packaging material and storage temperature. *Journal of Scientific and Industrial Research*, **67**, 1108–1114.

Lamikanra, O., Chen, J.C., Banks, D. and Hunter, P.A. (2000) Biochemical and microbial changes during storage of minimally processed cantaloupe. *Journal of Agriculture and Food Chemistry*, **48**, 5955–5961.

Léchaudel, M and Joas, J. (2007) An overview of pre-harvest factors influencing mango fruit growth, quality and postharvest behavior. *Brazilian Journal of Plant Physiology*, **19**, 287–298.

Lopez-Malo, A. and Palou, E. (2000) Modeling the growth/no-growth interface of *Zygosaccharomyces bailii* in mango puree. *Journal of Food Science*, **65**, 516–520.

Lund, B.M. and Snowdon, A.L. (2000) Fresh and processed fruits. In: *The Microbiological Safety and Quality of Food* (eds B.M. Lund, T.C. Baird-Parker and G.W. Gould), vol. I, pp. 738–758. Aspen Publication, Gaithersburg, MD.

Mari, M., Neri, F. and Bertolini, P. (2007) Novel approaches to prevent and control postharvest diseases of fruit. *Stewart Postharvest Review*, **3**(6), Article 4.

Martinez-Ferrer, M., Harper, C., Perez-Muroz, F. and Chaparro, M. (2002) Modified atmosphere packaging of minimally processed mango and pineapple fruits. *Journal of Food Science*, **67**, 3365–3371.

Mascarenhas, P., Behere, A., Sharma A. and Padwal-Desai, S.R. (1995) Postharvest spoilage of mango *(Mangifera indica)* by *Botryodiplodia theobromae*. *Mycology Research*, **100**, 27–30.

McMeekin, T., Bowman, J., McQuestin, O., Mellefont, L., Ross, T. and Tamplin, M. (2008) The future of predictive microbiology: Strategic research, innovative applications and great expectations. *International Journal of Food Microbiology*, **128**, 2–9.

Mermelstein, N.H., Fennema, O.R., Batt, C.A., Goff, H.D., Griffiths, M.W. *et al.* (2002) Food research trends-2003 and beyond. *Food Technology*, **56**(12), 30.

Ngarmsak, M., Delaquis, P., Toivonen, P., Ngarmsak, T., Ooraikul, B and Mazza, G. (2006) Antimicrobial activity of vanillin against spoilage microorganisms in stored fresh-cut mangoes. *Journal of Food Protection*, **69**(7), 1724–1727.

Nguyen, C. and Carlin, F. (1994) The microbiology of minimally processed fresh fruits and vegetables. *Critical Reviews in Food Science and Nutrition*, **34**, 371–401.

Nour, A.A.M, Khalid, K.S.M. and Osman, G.A.M. (2011) Suitability of some Sudanese mango varieties for jam making. *American Journal of Scientific and Industrial Research*, **2**, 17–23.

O'Connor-Shaw, R.E, Guthrie, J.A., Dunlop, K.J. and Roberts, R. (1995) Coliforms in processed mango: Significance and control. *International Journal of Food Microbiology*, **25**, 51–61.

Orr, R.V., Shewfelt, R.L., Huang, C.J., Tefera, S. and Beuchat, L.R. (2000) Detection of guaiacol produced by *Alicyclobacillus acidoterrestris* in apple juice by sensory and chromatographic analyses, and comparison with spore and vegetative cell populations. *Journal of Food Protection*, **63**, 1517–1522.

Pao, S. and Petracek, P.D. (1997) Shelf life extension of peeled oranges by citric acid treatment. *Food Microbiology*, **14**, 485–491.

Poubol, J. and Izumi, H. (2005) Shelf life and microbial quality of fresh-cut mango cubes stored in high CO_2 atmospheres. *Journal of Food Science*, **70**, M69–M74.

Prabakar, K., Raguchander, T., Parthiban, V.K., Muthulakshmi, P. and Prakasam, V. (2005) Postharvest fungal spoilage in mango at different levels marketing. *The Madras Agricultural Journal*, **92**, 42–48.

Rebolledo-Martínez, A., del Angel-Pérez, A.L., Peralta-Antonio, N. and Díaz-Padilla, G. (2013) Sooty mold control (*Capnodium mangiferae* Cooke and Brown) with biofungicides in leaves and fruits of mango "Manila". *Tropical and Subtropical Agroecosystems*, **16**, 355–362.

Scheper, R.W.A., Rogers, D.J., Walker, J.T.S., Manning, M.A. and Wood, P.N. (2007) The incidence of storage rots after postharvest apple washing. *New Zealand Plant Protection*, **60**, 7–14.

Shang, C., Deng, C., Zhang, X., Chen, Z. and Hu, Y. (2002) Headspace Solid-Phase Microextraction and Gas Chromatography-Mass Spectrometry analysis of free volatile compounds in mango. *Chromatographia*, **55**, 737–741.

Sharma, A. (2008) Radiation technology-enabled market access to Indian mango: Journey from Deogarh to DC. *BARC Newsletter*, **296**, 1–8.

Sharma, R.R., Singh, D. and Singh, R. (2009) Biological control of postharvest diseases of fruits and vegetables by microbial antagonists: A review. *Biological Control*, **50**, 205–221.

Shrestha, A.K. (2002) Reduction in the spoilage loss of mango fruits through the pre-harvest application of nutrients and growth substances. *Journal of Institute of Agriculture and Animal Science*, **23**, 77–79.

Silimela, M. and Korsten, L. (2007) Evaluation of pre-harvest *Bacillus licheniformis* sprays to control mango fruit diseases. *Crop Protection*, **26**, 1474–1481.

Silva, H.D., Cerqueira, M.A. and Vicente, A.A. (2012) Nanoemulsions for food applications: Development and characterization. *Food and Bioprocess Technology*, **5**, 854–867.

Silva, F.V.M. and Gibbs, P. (2001) *Alicyclobacillus acidoterrestris* spores in fruit products and design of pasteurization processes. *Trends in Food Science and Technology*, **12**, 68–74.
Sivakumar, D., Jiang, Y. and Yahia, E.M. (2011) Maintaining mango (*Mangifera indica* L.) fruit quality during the export chain. *Food Research International*, **44**, 1254–1263.
Sivapalasingam, S., Barrett, E., Kimura. A., Van Duyne, S., De Witt, W. *et al.* (2003) A multistate outbreak of *Salmonella enterica* Serotype Newport infection linked to mango consumption: Impact of water-dip disinfestations technology. *Clinical and Infectious Diseases*, **37**, 1585–1590.
Splittstoesser, D.F. (1991) Fungi of importance in processed fruits, In: *Handbook of Applied Mycology*, (eds D.K. Arora, K.G. Mukerji and E.H. Marth), pp. 201–219. Marcel Dekker, Inc., New York.
Sriwimon, W. and Boonsupthip, W. (2011) Utilization of partially ripe mangoes for freezing preservation by impregnation of mango juice and sugars. *LWT-Food Science and Technology*, **44**, 375–383.
Subramanyam, H., Narayana Moorthy, N.V., Lakshiminarayana, S. and Dalal, V.B. (1972) Control of fungal spoilage in Alphonso mangoes by pre-harvest application of fungicides. *Acta Horticulture (ISHS)*, **24**, 226–228.
Swaminathan, B. and Feng, P. (1994) Rapid detection of food-borne pathogenic bacteria. *Annual Review of Microbiology*, **48**, 401–426.
Tapia, M.S., Gomez-Lopez, V.M. and Cristina, O. (2009) HACCP implementation in the production of fresh-cut fruits and vegetables. *Stewart Postharvest Review*, **5**(4), 1–7.
Tripathi, P and Shukla, A.K. (2009) Application of essential oils for postharvest control of stem end rot of mango fruits during storage. *International Journal of Postharvest Technology and Innovation*, **1**, 405–415.
Vacheethasanee, K., Temenoff, J.S., Higashi, J.M., Gary, A., Anderson, J.M. *et al.* (1998) Bacterial surface properties of clinically isolated staphylococcus epidermidis strains determine adhesion on polyethylene. *Journal of Material Research*, **42**, 425–432.
Vega-Vega, V., Silva-Espinoza, B.A., Cruz-Valenzuela, M.R., Bernal-Mercado, A.T., González-Aguilar, G.A. *et al.* (2013) Antioxidant enrichment and antimicrobial protection of fresh-cut mango applying bioactive extracts from their seeds by-products. *Food and Nutrition Sciences*, **4**, 197–203.
Vilas-Boas E.V. de B. and Kader, A.A. (2007) Effect of 1-methylcyclopropene (1-MCP) on softening of fresh-cut kiwifruit, mango and persimmon slices. *Postharvest Biology and Technology*, **43**, 238–244.
Vivekananthan, R., Ravi, M., Saravanakumar, D., Kumar, N., Prakasam, V and Samiyappan, R. (2004) Microbially induced defense related proteins against postharvest anthracnose infection in mango. *Crop Protection*, **23**, 1061–1067.
Wade, W.I.N. and Beuchat, L.R. (2003) Proteolytic fungi isolated from decayed and damaged raw tomatoes and implications associated with changes in pericarp pH favorable for survival and growth of food borne pathogens. *Journal of Food Protection*, **66**, 911–917.
Wang, H., Feng, H., Liang, W., Luo, Y. and Malyarchuk, V. (2009) Effect of surface roughness on retention and removal of *Escherichia coli O157:H7* on surfaces of selected fruits. *Journal of Food Science*, **74**, e8–e15.

Warnasuriya, D., Liyanage, A.W., Weerawansa, G.G., Athauda, P.K. and Jayatissa, P.M. (1985) Isolation and characterization of yeasts of some fruits and fruit products of Sri Lanka. *Journal of National Science Council Sri Lanka*, **13**, 71–75.

Wells, J.M. and Butterfield, J.E. (1997) *Salmonella* contamination associated with bacterial soft rot of fresh fruits and vegetables in the marketplace. *Plant Disease*, **81**, 867–872.

Younis, M.S., Butt, M.S., Sharif, M.K., Sulerai, H.A.R. and Hameed, F. (2011) Effect of preservatives on physicochemical, microbial and sensory attributes of mangoes. *Internet Journal of Food Safety*, **13**, 246–263.

14

Value-added Processing and Utilization of Mango By-products

Poonam Aggarwal, Amarjeet Kaur and Suresh Bhise

Department of Food Science and Technology, Punjab Agricultural University, Ludhiana, Punjab, India

14.1 Introduction

Mango (*Mangifera indica* L.) is one of the world's major tropical fruits and its global production ranks fifth among fruit crops. Mango is a seasonal fruit that is consumed mostly at the ripe stage, and is greatly relished for its succulence, exotic flavor, and delicious taste in most countries of the world (Narayana *et al.* 2012). Apart from its delicacy, it is a nutritionally important fruit, being a good source of vitamins A, B, C, and minerals.

Worldwide production of mango in 2014 was 45.23 million metric tons (FAO 2016). In some cases, over 50% of the fruits are knocked off, right from the flowering stage until ripening. The losses, however, can be minimized to a great extent by utilizing the dropped fruits. A number of commercially processed products are prepared from mangoes at various stages of maturity, which include pulp, juice, nectar, drinks, pickles, chutney, preserve, jam, canned slices, dried powder, and many other products. In recent years, fresh-cut mangoes are gaining popularity among consumers, especially in developed countries. During processing of these products, about 30–50% of the fruit ends up as waste. This chapter covers topics related to commercial waste generation, its implications for the environment, and value-added processing and utilization.

14.1.1 Waste Generation and Impact on the Environment

Depending on the cultivar, mango consists of 33–85% edible pulp, 9–40% inedible kernel, and 7–24% inedible peel. Given the high proportion of kernels and peels (16–64%), a significant amount of waste is generated during mango processing. Figure 14.1 shows mango waste generation streams during typical commercial processing operations. Such waste or by-products present serious problems of disposal, therefore ways to process mango waste for value-added products are being explored. Many of the bioactive compounds found in mangoes are higher in the peel and kernel than in the edible flesh/pulp. Mango peel has been reported to be rich in polyphenolic compounds (Arumugam and Manikandan 2011).

Handbook of Mango Fruit: Production, Postharvest Science, Processing Technology and Nutrition,
First Edition. Edited by Muhammad Siddiq, Jeffrey K. Brecht and Jiwan S. Sidhu.

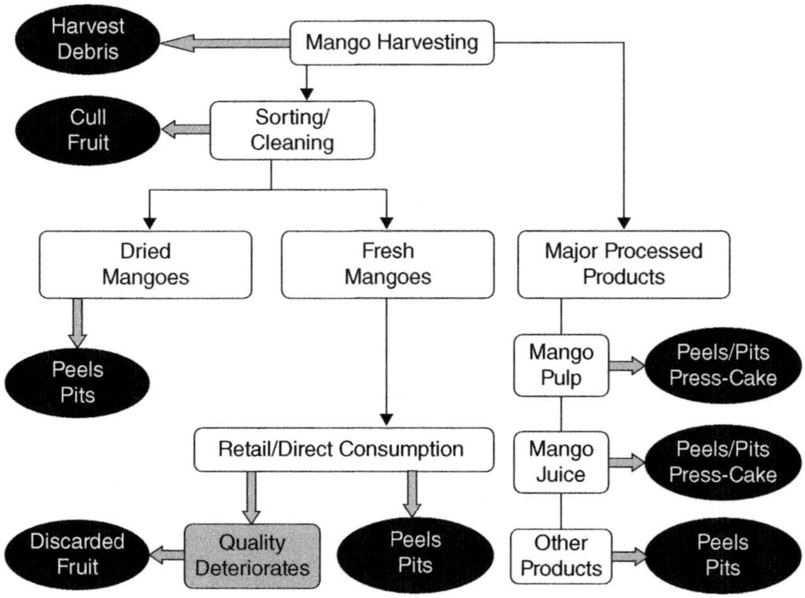

Figure 14.1 Waste generation at various steps of value-added mango processing and fresh fruit marketing.

Disposal of wastes from fruit canning industries has been a problem, due to high transportation costs and limited availability of landfills. Improper disposal of mango kernel and peel appreciably increase environmental pollution. In this regards, a higher level of biological oxygen demand (BOD) and chemical oxygen demand (COD) in mango solid waste creates a further problem in disposal. To a certain extent, use of vermi composting technology could help with the disposal of mango industrial waste in a safe, economic, and useful manner. However, valorization through different techniques will certainly eliminate the disposal problem.

14.2 Value-added Processing of Mango Waste

14.2.1 Mango Peel

Fresh mango peel contains significant amount of moisture. Mango peel is rich in pectin, cellulose, hemicelluloses, lipids, proteins, polyphenols, and carotenoids (Ajila *et al.* 2007; Kim *et al.* 2010). In general, the level of reducing sugars, non-reducing sugars, protein, and cellulose varies, depending on the cultivar. Dried mango peel powder (MPP) offers many functional properties for application in diverse foods. The composition of fresh and dried mango peel, as reported by Puligundla *et al.* (2014), is given in Figure 14.2.

Ajila *et al.* (2010) incorporated mango peel into macaroni at 2.5, 5.0, and 7.5% levels and studied its effect on the cooking properties, firmness, nutraceutical and sensory attributes of the macaroni. The total dietary fiber (DF) content in macaroni increased from 8.6–17.8%. The content of polyphenols increased from 0.46 to 1.80 mg/g and

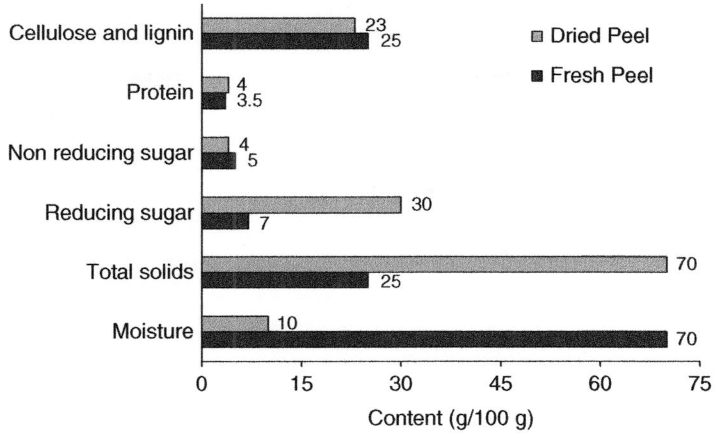

Figure 14.2 Composition of fresh and dried mango peel. *Source:* adapted from Puligundla *et al.* (2014)

carotenoids content increased from 5–84 μg/g of macaroni with 7.5% incorporation of MPP. The macaroni products incorporated with mango peel exhibited improved antioxidant properties. Larrauri *et al.* (1996) developed a method for producing acceptable quality jam from mango peel. Inactivation of polyphenol oxidase was done by immersion of peel in boiling water, followed by extraction of peel soluble compounds at 80 °C, and addition of sugar, citric acid, and pectin as required. The mixture was then concentrated, cooled, and packed. The effects of some technological parameters on preparation of mango peel DF were studied (Larrauri *et al.* 1996). Fresh mango peel was wet milled, washed with water, and dried. Results showed that mango peel is a good source of tropical fruit fiber.

Mango peel flour can be used as a functional ingredient in developing healthy food products such as noodles, bread, biscuits, sponge cakes, and other bakery products (Aziz *et al.* 2012). As mango peel is rich in polyphenols, carotenoids, and vitamins with different health-promoting properties, there is enormous potential for developing functional foods based on mango peel. The solubility, water, and oil absorption values of MPPs play a key role for their utilization in foods. Different drying methods influence the functionality of dried mango peel (Table 14.1), as shown by Sogi *et al.* (2013). Cabinet drying was found to be superior to other drying methods in MPP preparation, which were intended for utilization in several food applications. In a study, yogurt made with supplementation of 10% of MPP showed a good texture, flavor, and color characteristics and exhibited a 30-day shelf life without adding preservatives (Ruiz *et al.* 2011).

14.2.2 Mango Seed Kernel

Mango fruit contains a single oblong seed, which has a fibrous hairy surface that varies from cultivar to cultivar. The seed coat is 1–2 mm thick, and the embryo/kernel is 4–7 cm long, 3–4 cm wide and 1–1.5 cm thick. The seed content of different mango cultivars ranges from 9 to 23% of fruit weight (Palaniswamy *et al.* 1974). Mango kernel is a good source of starch and fat. Maisuthisakul and Gordon (2009) reported that the seed

Table 14.1 Effect of different drying methods on ascorbic acid, total carotenoids, and total phenolics content of mango peel and kernel (mg/100 g, dry wt).

	Ascorbic Acid		Carotenoids		Phenolics	
Processing method	**Peel**	**Kernel**	**Peel**	**Kernel**	**Peel**	**Kernel**
Freeze dried	75.5	65.7	4.05	0.79	3,185	20,035
Cabinet dried	84.7	61.2	2.7	0.48	2,320	12,535
Vacuum dried	68.5	74.5	2.59	0.51	2,032	15,022
Infra-red dried	71.4	69.9	1.88	0.37	3,049	11,228

Source: adapted from Sogi *et al.* (2013).

represents 20–60% of the whole fruit weight, depending on the mango variety and the kernel inside the seed, which represents from 45–75% of the whole seed. The major components of mango seed are starch, fat, and protein. The oil of mango seed kernel consist of about 44–48% saturated (majority stearic) and 52–56% unsaturated fatty acids. Mango seed kernels have a low protein content, but they contain the most of the essential amino acids, with highest values of leucine, valine, and lysine. Mango seed kernels were shown to be a good source of polyphenols, phytosterols such as campesterol, sitosterol, and tocopherols. In addition, mango seed kernel could be used as a potential source for functional food ingredients, and as an antimicrobial compounds. It can be used in cosmetics, due to its high quality of fat and protein, as well as high levels of natural antioxidants. The mango stone obtained after decortication of mango seed can also be utilized as an adsorbent (Kiittiphoom 2012).

Table 14.2 shows the proximate analysis of mango seed kernel in terms of crude protein, oil, ash, crude fiber, and carbohydrate. Variation in chemical composition may be due to the differences in cultivar, climate, ripening stage, the harvesting time of the fruit, and the extraction method used. Mango seed kernel has a high content of potassium, magnesium, phosphorus, calcium, and sodium (Table 14.3). Ashoush and Gadallah (2011) evaluated the addition of MPP and mango kernel powder (MKP) into biscuit (cookie) formulation. The biscuits containing MPP and MKP exhibited an improvement in their antioxidant properties. Acceptable cookies with mango flavor

Table 14.2 Proximate analysis of mango seed kernel from selected cultivars (g/100g).

Content	Chausa[1]	Dusheri[1]	Kibangou[2]	Kaew[3]
Moisture	38.55	50.98	45.20	40.50
Crude protein	5.34	5.25	6.36	1.43
Fats/oils	7.82	6.98	13.0	4.92
Crude fiber	1.75	1.65	2.02	3.96
Ash	2.75	2.45	3.20	0.83
Total carbohydrates	–*	–	32.24	48.19

*not reported
Source: [1]Dingra and Kapoor (1985), [2]Nzikou *et al.* (2010), [3]Changso (2008).

Table 14.3 Mineral elemental composition of mango seed kernels.

Minerals	Content (mg/100 g)
Calcium	10.21[1]–111.30[2]
Magnesium	22.34[1]–94.80[2]
Potassium	22.30[2]–158.0
Sodium	2.70[1]–21.00[2]
Phosphorus	20.0[1]
Iron	11.90[2]
Zinc	1.10[2]
Manganese	0.04[2]

Source: [1]Nzikou *et al.* (2010), [2]Fowomola (2010).

were obtained by incorporating up to 10% MPP and 40% MKP. Considering the color, appearance, texture, and flavor attributes, a 20% level of incorporation of MPP or MKP were optimum. At higher levels of addition, cookies had a slight bitter taste, which may be due to the high polyphenolic content. MPP and MKP would also impart health benefits due to its fiber content in cookies (Bandyopadhyay *et al.* 2014). In another study, Arogba (2002) also prepared acceptable quality biscuits from equal amounts of wheat flour and MKP.

14.2.3 Mango Seed Kernel Oil

Another valuable product which can be extracted from mango seed kernel is pale yellow oil, which is found to be more stable than many other vegetable oils rich in unsaturated fatty acids (Jahurul *et al.* 2013; Moharram and Moustafa 1982). The major saturated fatty acids in mango seed kernel oil are stearic and palmitic acids, and the main unsaturated fatty acids are oleic and linoleic acids (Table 14.4). The comparison of

Table 14.4 Composition of fatty acid in seed kernel oil from different mango cultivars (g/100 g).

Fatty acid	Chausa[1]	Dusheri[1]	Kibangou[2]	Ikankpo[3]
Palmitic acid (C16:0)	7.18	7.50	6.48	10.50
Palmitoleic acid (16:1)	–*	–	–	21.90
Stearic acid (C18:0)	38.90	38.70	37.94	27.00
Oleic acid (C18:1)	42.60	43.20	45.76	48.30
Linoleic acid (C18:2)	5.70	6.20	7.45	14.00
Linolenic acid (C18:3)	5.30	2.90	2.37	–
Saturated	46.08	46.20	44.42	37.50
Unsaturated	53.60	52.30	55.58	62.30

*not reported
Source: [1]Dingra and Kapoor (1985), [2]Nzikou *et al.* (2010), [3]Arogba (1997).

the composition in fatty acids of mango seed kernel oil with that of vegetable oils indicates that this plant is rich in stearic acid and oleic acid.

Mango kernel fat alone or when blended with other nuts fats, for example, almond or palm stearin, can act as a cocoa butter replacer (Jahurul *et al.* 2013; Solís-Fuentes and Durán-de-Bazúa 2004). The cocoa butter replacer has good triglyceride content and thermal properties. Moharram and Moustafa (1982) showed that triglycerides, fatty acids, and phospholipids from mango kernel were comparable to those of cotton seed oil and cocoa butter. They also incorporated mango seed kernel oil in butterscotch toffees and found that mango kernel oil and cocoa butter were almost identical in several of their triglycerides and fatty acids, as well as its effect on taste, odor, and texture of the toffee.

Kaphueakngam *et al.* (2009) reported that oil obtained from the mango seed kernels could be an alternative source of edible oil. With the right proportion, an oil blend between mango seed kernel oil and palm oil could be used as cocoa butter. Fatty acid compositions of mango seed kernel oil, mixture of mango seed kernel oil/palm oil, and cocoa butter are given in Table 14.5. Their results showed that, like cocoa butter, the mixture of mango seed kernel oil/palm oil was also composed mainly of palmitic acid, stearic acid, and oleic acid. Hence, mango seed kernel oil is a good source of stearic acid and oleic acid, whereas palm oil is a source of palmitic acid and oleic acid. The mango seed kernel oil and palm oil blend (80:20) mainly consisted of three fatty acids that were also the main fatty acid components of cocoa butter, whose melting behavior and slip melting point were closest to that of cocoa butter.

14.2.4 Fiber Extraction

Dietary fibers (DF) from fruits generally have a better nutritional quality than those found in cereals, mainly because of their more balanced composition, that is greater soluble dietary fiber (SDF) to insoluble dietary fiber (IDF) ratio, higher water and fat holding capacity, lower metabolic energy value, phytic acid content, and bioactive compounds such as flavonoids and carotenoids. Mango peels have been shown to be a good

Table 14.5 Profile of fatty acids in the mango kernel oil, palm oil, a commercial sample of cocoa and 80:20 blend of mango seed kernel oil and palm oil.

Fatty Acid	Mango kernel oil[1]	Palm oil[2]	Cocoa butter[1]	80:20 blend*
Palmitic acid (C16:0)	9.29	51.69	24.27	16.26
Stearic acid (C18:0)	39.07	4.21	35.10	37.25
Oleic acid (C18:1)	40.81	35.63	36.47	39.60
Linoleic acid (C18:2)	6.06	–**	2.85	–
Linolenic acid (C18:3)	0.64	–	0.30	–
Arachidic acid (C20:0)	2.48	–	1.01	–
Behenic acid (C22:0)	0.64	–	–	–
Lignoceric acid (C24:0)	0.49	–	–	–

*mango kernel oil (80%) and palm oil (20%)
**not reported,
Source: [1]Solís-Fuentes and Durán-de-Bazúa (2004), [2]Kaphueakngam *et al.* (2009).

source of high-quality DF, because of its higher starch, cellulose, hemicelluloses, lignin, and pectin contents (56–66%), and lower lipid content and low cost of the fruits (Berardini *et al.* 2005).

Ajila *et al.* (2010) reported that MPP incorporation into soft dough biscuits increased SDF content, polyphenols, and carotenoids. Sensory evaluation showed that these fiber enriched biscuits were acceptable and showed improved antioxidant properties. *In vitro* studies by Vergara-Valencia *et al.* (2007) on starch have shown low glycemic responses from mango peel fiber.

14.2.5 Modified Cellulose and Hemicellulose

Methylcellulose is being extensively used in the pharmaceutical, food, petrochemical, construction, and other industries. Its use is directly related to the degree of substitution (DS) that influences its physical properties, such as water solubility. Methylcellulose is methyl ether that can be produced by a reaction between cellulose in an alkaline medium, with a methylation agent, such as methyl iodide, methyl chloride, or dimethyl sulfate (Ye and Farriol 2007). Different DS can be obtained by altering the synthesis conditions, such as the reaction time or the methylation agent.

Methylcellulose was produced from the fibers of Uba mango seeds by Vieira *et al.* (2012). Their study showed that mango seeds are a good source of cellulose for producing methylcellulose. The differences in methylation processes led to the production of methylcellulose, with different degrees of crystallinity and molecular structure modification. The methylcellulose produced exhibited good characteristics when applied as mortar additive, both in the fresh state, increasing the consistency index (CI) and, after curing, improving the results of bond tensile strength (TS). The results obtained are higher than the minimum values required for mortar coatings in different applications, justifying the applicability of the polymers produced from this biomass.

14.2.6 Pectin

Mango peel is a good source of pectin, at around 10–15%. A maximum pectin yield of 21% was obtained by soaking finely ground defatted mango peels in a sulfuric acid solution (pH 2.5 at 80 °C for 120 min) (Rehman *et al.* 2004). A pectin level of around 18% has been reported in Totapuri mango peel (Kumar *et al.* 2012a). Mango peel pectin exhibited better gelling capacity than commercial citrus pectin (Koubala *et al.* 2012). Highest pectin yield was obtained at pH 3, bleaching with alcohol, and extraction time of 60 minutes (Pedroza *et al.* 1994). Tandon *et al.* (1991) extracted pectin from dry peel and kernel of immature and ripe Dashehari mangoes using 0.05N NaOH, 0.05N HCl, and 0.3% sodium hexametaphosphate (SHMP). A maximum yield of 12.79% pectin was obtained in ripe mango peel with SHMP extraction. The kernel contained very little amounts of pectin. The water soluble pectin was higher in peel of ripe fruit, while alkali soluble fraction was more in the peel of immature fruit. A combined recovery method of pectin and phenolic compounds from mango peels is shown in Figure 14.3 (Berardini *et al.* 2005).

14.2.7 Starch

The kernel obtained after decortication of mango stone can be utilized as a supplement to wheat flour or for extraction of edible oils (Tandon and Kalra 1989). Besides its use

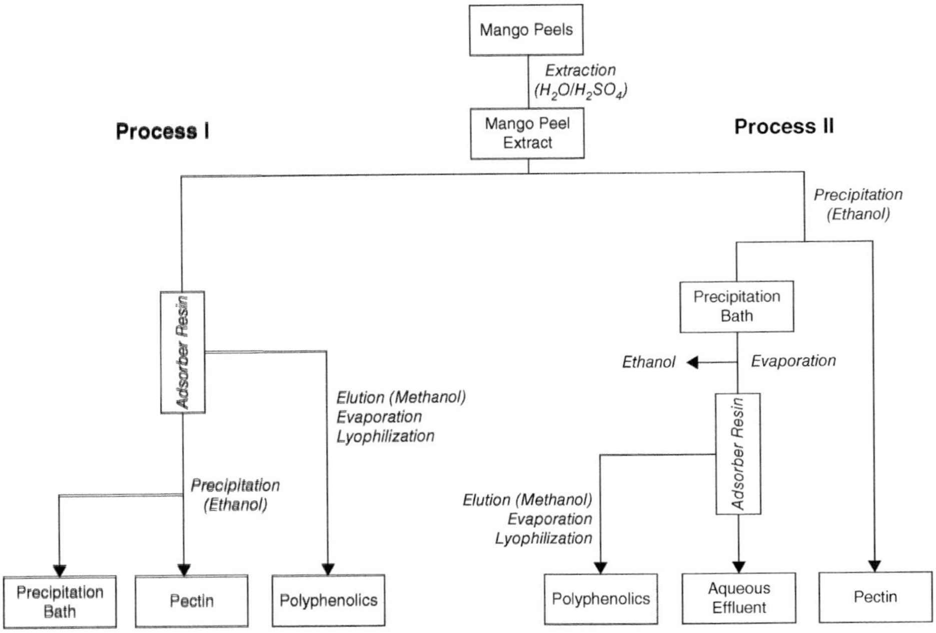

Figure 14.3 A combined recovery method of pectin and phenolic compounds from mango peels. *Source:* Berardini *et al.* (2005).

in animal feed, mango kernel flour can be utilized for edible purposes (Patel *et al.* 1971). Kaur *et al.* (2004) have reported that the amylose content of mango kernel starches was lower than that of corn and potato starches. Mango kernel starch showed oval to elliptical shaped granules, similar to those of legume starch granules. Various properties of mango kernel starch are comparable with the starches from corn, wheat, rice, and potato and thus could be effectively utilized as a source of starch.

14.2.8 Enzymes Production

14.2.8.1 Cellulolytic Enzymes

Mango peel at the level of 7% (w/v) has been used for carboxymethyl cellulase production, using *Paenibacillus polymyxa* (Kumar *et al.* 2012a). Mango peel has also been used for the production of cellulase enzyme by Saravanan *et al.* (2012), with a maximum cellulase activity of 7.8 IU/mL under optimized conditions.

14.2.8.2 Pectic Enzymes

Pectinase/polygalacturonase has been prepared using *Aspergillus awamori* MTCC by the solid state fermentation process using fruit waste and mango peel (Padma *et al.* 2012). The highest enzyme production was obtained with jack fruit rind and mango peel having 65% moisture content, at 28 °C, with a pH of 5.2, and using 10^6 spores/g for jack fruit rind and 10^8 spores/g for mango peel, after 96 hours incubation. Pectinase production using mango peel as a substrate has also been reported by Kumar *et al.* (2012b). As mango peel is rich in pectin, microorganisms (especially filamentous fungi), which are capable of degrading pectin, can be used for fermentative production of pectinases. *Aspergillus foetidus* was employed to produce polygalacturonase and pectin lyase using Totapuri mango peel and the highest productivity was obtained by them with solid state fermentation.

14.2.9 Single-cell Protein (SCP)

Single-cell proteins (SCP) refer to the dried cells of microorganisms. SCPs are used as protein sources in human foods or animal feeds. Many raw materials have been considered as carbon and energy sources for SCP production. In many cases, these raw materials have been hydrolyzed by physical, chemical, and enzymatic methods before use (Jhojaosadati 1999). Mango peel extract has been used for SCP production using *Pichia pinus* yeast. In this study, a maximum yield of 6.2 g/L was observed on the third day of growth and cells contained 62.2% crude protein, 39% protein, and 12.9% nucleic acids (Rashad *et al.* 1990).

14.2.10 Antioxidants and Bioactive Compounds

Mango peel is a rich source of many phenolic compounds, which exhibit antioxidant activity (Ajila *et al.* 2007; Dorta *et al.* 2012). In addition, phenolic compounds also have antiviral, antibacterial, and anti-inflammatory activity (Zgorka and Kawka 2001). The major phenolic compounds present in mango peel extract are reported to be syringic acid, quercetin, mangiferin pentoside, and ellagic acid (Ajila *et al.* 2010). Peels are a good source of mangiferin (C-glucosyl xanthone), a heat-stable and pharmacologically active phytochemical. Mangiferin possesses various bioactivities, such as anti-inflammation, anti-diabetic, immunomodulatory, antitumor, and antioxidant activities (Luo *et al.* 2012). It showed high DPPH· free radical scavenging capacity and thus has potential in the prevention of oxidative stress-associated diseases. Flavonol and xanthone glycosides content in mango peel pectin obtained before and after adsorptive removal of phenolic compounds, as reported by Berardini *et al.* (2005), are shown in Table 14.6.

Table 14.6 Flavonol and xanthone glycosides content in mango peel pectin obtained before and after adsorptive removal of phenolic compounds (mg/kg)

Compound	Before adsorption	After adsorption
Mangiferin	216.0	36.9
Isomangiferin	10.4	1.5
Mangiferin gallate	27.4	0.8
Isomangiferin gallate	8.0	0.2
Quercetin 3-O-glucoside	5.8	Nd*
Rhamnetin 3-O-galactoside/glucoside	0.6	nd
Quercetin	2.7	nd
Total	270.9	39.4

*not detected
Source: Berardini *et al.* (2005).

Polyphenol-rich fractions of peel extract could be used as natural antioxidants and functional food or feed supplements (Berardini *et al.* 2005). Mango seed kernel has also been shown to contain stigmasterol and tocopherols. The antioxidant effect of the mango seed kernel is due to its high content of polyphenols, sesquiterpenoids, and tocopherols. It is also rich in phytosterols and microelements such as selenium, copper, and zinc (Nunez-Selles 2005; Schiber *et al.* 2003).

14.2.11 Ethanol Production

The increasing concern of depleting non-renewable fossil fuels, as well as the escalating social and industrial awareness, is leading us to explore cleaner renewable fuels (Razif *et al.* 2009). Therefore, bioethanol produced from renewable energy sources or non-edible feed stocks, such as sugar molasses and starchy materials, is believed to be one of the better options, and it is concurrently being practiced in waste management (Reddy and Reddy 2005).

Saifuddin *et al.* (2014) studied the feasibility of the utilizing mango processing waste to produce bioethanol through fermentation using *Saccharomyces cerevisiae.* The 15% and 11% yields of bioethanol could be obtained from mango pulp and peel, respectively, using the yeast concentration of 3 g/L at 30 °C, whereas yields at 23 °C or 35 °C were significantly lower than that at 30 °C. Reddy *et al.* (2011) reported that direct fermentation, without nutrient addition, yielded 5.13% ethanol; however, upon supplementation with yeast extract, peptone, and wheat bran, the final ethanol yield was increased by nearly 40%. Arumugam and Manikandan (2011) obtained 9.68% ethanol yield using mango peel; however, the yield and productivity were relatively low compared with mango pulp ethanol fermentation. By using thermo-tolerant and alcohol tolerant yeasts to ferment mango peel at 40 °C for 120 hours, a maximum ethanol yield of 13 g/L was obtained by Somda *et al.* (2011).

14.2.12 Other Products

14.2.12.1 Lactic Acid Production

Lactic acid is used as a preservative and acidulant in foods, and one of the main obstacles in the large-scale production of this substance is the cost of the raw material (Reddy *et al.* 2008). Mango peel as a low-cost substrate for the production of lactic acid has been investigated by Jawad *et al.* (2013). Mango peel was directly fermented using bacteria having both amylolytic and lactic acid producing capabilities. A maximum production of 17.48 g/L lactic acid was obtained through statistical optimization of fermenting conditions. In another study, a lactic acid concentration of 63.33 g/L was obtained from mango peel fermentation using *Lactobacillus casei* (Mudaliyar *et al.* 2012).

14.12.2.2 Alcoholic Fermentation

Mango wine fermentation, using a yeast and mango peel immobilized biocatalyst system, has been reported by Varakumar *et al.* (2012). The operational stability of the biocatalyst system was found to be good with better ethanol productivities (1.53–3.29 g/L/hr) and it can also be suitable for low temperature wine-making. The overall improved quality of wines has been noted using this system.

14.12.2.3 Biogas Production

Organic industrial wastes and agricultural wastes including fruit processing wastes can be used for biogas production by anaerobic digestion. Mango peel has been used for biogas production with a yield of 0.33 m^3 biogas/kg total solids with 53% methane content at a hydraulic retention time of 15 days (Somayaji *et al.* 2001). Ensilage of mango peel for 6 months has been shown to aid in the better conversion of major components of its carbohydrates into volatile fatty acids. This pretreatment effect resulted in 58% more gas production as compared to the control (Madhukara *et al.* 1993). Microbiological pretreatment of mango peel for 6 days was shown to increase the total gas production 8-fold (Devi and Nand 1989).

14.12.2.4 Antimicrobial Compounds

Phenolic compounds, tannins, and flavonoids in mango seed are shown to exhibit antimicrobial properties. They prevent microorganisms by inhibiting extracellular microbial growth and by reducing oxidative phosphorylation (Schiber *et al.* 2003). Mango seed kernel oil has been used for killing abdominal worms and was also used as a cure for vomiting, diarrhea, and hyperacidity (Vatsayan 2002).

14.12.2.5 Activated Carbon

Mango seed shell can also be used as an adsorbent for the removal of methylene blue from its aqueous solution. Itodo *et al.* (2011) have shown that the mango seed shell is potentially a low-cost substrate for generation of activated biosorbent, giving up to 60% dye removal. The sorption capacity of mango seed shell determined in their study was comparable to some of the low-cost adsorbents previously reported for the uptake of methylene blue from aqueous solutions.

14.3 Summary

Research on mango peel and seeds has shown that they can potentially impart health benefits, because of their higher contents of dietary fiber and antioxidant compounds. Once discarded waste products of mango processing factories could now become valuable food ingredients or nutraceuticals for the development of healthy food products. In recent years, mango peel dietary fiber has received a great deal of attention as a functional food ingredient. In addition to fiber, the associated phenolic compounds provide valuable antioxidant benefits. Therefore, it is capable of offering low-cost nutritional dietary supplement for lower income masses. More research is needed in refining technologies for the use of mango peel and kernels to develop functional foods for obtaining enhanced health benefits. Thus, through biotechnological applications, environmentally polluting by-products of the mango processing industry could be converted into products with a great economic importance.

References

Ajila, C.M., Naidu, K.A., Bhat, S.G., and Prasada Rao, U.J.S. (2007) Bioactive compounds and antioxidant potential of mango peel extract. *Food Chemistry*, **105**, 982–988.

Ajila, C.M., Aalami, M., Leelavathi, K. and Prasada Rao, U.J.S. (2010).Mango peel powder: A potential source of antioxidant and dietary fiber in macaroni preparations. *Innovation in Food Science and Emerging Technology*, **11**, 219–224.

Arogba, S.S. (1997) Physical, chemical and functional properties of Nigerian mango (*Mangifera indica*) kernel and its processed flour. *Journal of the Science of Food and Agriculture*, **73**, 321–328.

Arogba, S.S. (2002) Quality characteristics of a model biscuit containing processed mango (*Mangifera indica*) kernel flour. *International Journal of Food Properties*, **5**, 249–260.

Arumugam, R. and Manikandan, M. (2011) Fermentation of pretreated hydrolyzates of banana and mango fruit wastes for ethanol production. *Asian Journal of Experimental Biological Sciences*, **2**, 246–256.

Ashoush, I.S. and Gadallah, M.G.E. (2011) Utilization of mango peels and seed kernels powders as sources of phytochemicals in biscuit. *World Journal of Dairy and Food Sciences*, **6**, 35–42.

Aziz, N.A.A., Wong, L.M., Bhat, R. and Cheng, L.H. (2012) Evaluation of processed green and ripe mango peel and pulp flours (*Mangifera indica* var. Chokanan) in terms of chemical composition, antioxidant compounds and functional properties. *Journal of the Science of Food and Agriculture*, **92**, 557–563.

Bandyopadhyay, K., Chakraborty, C. and Bhattacharyya, S. (2014) Fortification of mango peel and kernel powder in cookies formulation. *Journal of Academia and Industrial Research*, **2**, 661–664.

Berardini, N., Knodler, M., Schieber, A. and Carle, R. (2005) Utilization of mango peels as a source of pectin and polyphenolics. *Innovative Food Science and Emerging*, **6**, 442–452.

Changso, C. (2008) Study of extraction process, chemical and physic properties of mango seed almond fat (cv. Kaew). Bangkok, Thailand: Silpakorn University, MSc thesis. (in Thai)

Devi, S.S. and Nand, K. (1989) Microbial pretreatment of mango peel for biogas production. *Journal of Microbiological Biotechnology*, **4**, 110–115.

Dhingra, S. and Kapoor, A.C. (1985) Nutritive value of mango seed kernel. *Journal of the Science of Food and Agriculture*, **6**, 752–756.

Dorta, E., Lobo, M.G. and González, M. (2012) Using drying treatments to stabilize mango peel and seed: Effect on antioxidant activity. *LWT-Food Science and Technology*, **45**, 261–268.

FAO (2016) Crops Production and Trade Statistics [Online]. Available: http://faostat3.fao.org/download/Q/QC/E [26 July 2016].

Fowomola, M.A. (2010) Some nutrients and antinutrients contents of mango (*Magnifera indica*) seed. *African Journal of Food Science*, **4**, 472–476.

Itodo, A.U. (2011) Submitted to the Department of Pure and Applied Chemistry, Usmanu Danfodiyo, Sokoto, PhD thesis.

Jahurul, M.H.A., Zaidul, I.S.M., Norulaini, N.A.N., Sahena, F., Abedin, M.Z. *et al.* (2013) Hard cocoa butter replacers from mango seed fat and palm stearin. *Food Chemistry*, **154**, 323–329.

Jawad, A.H., Alkarkhi, A.F.M., Jason, O.C., Easa, A.M. and Norulaini, N.A.N. (2013) Production of lactic acid from mango peel waste-factorial experiment. *Journal of King Saud University Science*, **25**, 39–45.

Jhojaosadati, S.A., Rasoul, K., Abbas, J. and Hamid, R.S. (1999) Bioconversion of molasses stillage to protein as an economic treatment of this effluent. *Resources Conservation Recycling*, **27**, 125–138.

Kaphueakngam, P., Flood, A. and Sonwai, S. (2009) Production of cocoa butter equivalent from mango seed almond fat and palm oil – mid fraction. *Asian Journal of Food Agro-Industry*, **2**, 441–447.

Kaur, M., Singh, N., Singh, K., Harmeet, S. and Guraya, S. (2004) Physicochemical, morphological, thermal and rheological properties of starches separated from kernels of some Indian mango cultivars (*Mangifera indica* L.). *Food Chemistry*, **85**, 131–140.

Kiittiphoom, S. (2012) Utilization of mango seed. *International Food Research Journal*, **19**, 1325–1335.

Kim, H.J.Y.H., Moon, D., Kim. M., Lee, H., Cho, Y.S. *et al.* (2010) Antioxidant and antiproliferative activities of mango (*Mangifera indica L.*) flesh and peel. *Food Chemistry*, **121**, 429–436.

Koubala, B.B., Kansci, G., Garnier, C., Ralet, M.C. and Thibault, J.F. (2012) Mango (*Mangifera indica*) and Ambarella (*Spondias cytherea*) peel extracted pectins improve viscoelastic properties of derived jams. *African Journal of Food Agriculture and Nutrition Development*, **12**, 6200–6212.

Kumar, D., Ashfaque, M., Muthukumar, M., Singh, M. and Garg, N. (2012) Production and characterization of carboxymethyl cellulase from *Paenibacillus polymyxa* using mango peel as substrate. *Journal of Environmental Biology*, **33**, 81–84.

Kumar, Y.S., Kumar, P.V. and Reddy, O.V.S. (2012) Pectinase production from mango peel using *Aspergillus foetidus* and its application in processing of mango juice. *Food Biotechnology*, **26**, 107–123.

Larrauri, J.A., Ruperez, P., Borroto, B. and Saura-Calixto, F. (1996) Mango peels as a new tropical fiber: Preparation and characterization. *Lebensmittel-Wissenschaft und - Technologie*, **29**, 729–733.

Luo, F., Lv, Q., Zhao, Y., Hu, G., Huang, G. *et al.* (2012) Quantification and purification of mangiferin from Chinese mango (*Mangifera indica* L.) cultivars and its protective effect on human umbilical vein endothelial cells under H $2O_2$-induced stress. *International Journal of Molecular Sciences*, **13**, 11260–11274.

Madhukara, K., Nand, K., Raju, N.R. and Srilatha, H.R. (1993) Ensilage of mango peel for methane generation. *Process Biochemistry*, **28**, 119–123.

Maisuthisakul, P. and Gordon, M.H. (2009) Antioxidant and tyrosinase inhibitory activity of mango seed kernel by product. *Food Chemistry*, **117**, 332–341.

Moharram, Y.G. and Moustafa, A.M. (1982) Utilization of mango seed kernel (*Mangifera indica*) as a source of oil. *Food Chemistry*, **8**, 269–276.

Mudaliyar, P., Sharma, L. and Kulkarni, C. (2012) Food waste management-Lactic acid production by *Lactobacillus* species. *International Journal of Advanced Biological Research*, **2**, 34–38.

Narayana CK, Rao DS, Roy SK. (2012) Mango production, postharvest physiology and storage. In: *Tropical and Subtropical Fruits: Postharvest Physiology, Processing and Packaging* (ed, Siddiq M). Ames, IA: John Wiley & Sons. pp. 259–276.

Nunez-Selles, A.J. (2005) Antioxidant therapy: Myth or reality? *Journal of the Brazillian Chemical Society*, **6**, 699–710.

Nzikou, J.M., Kimbonguila, A., Matos, L., Loumouamou, B., Pambou-Tobi, N.P.G. *et al.* (2010) Extraction and characteristics of seed kernel oil from mango (*Mangifera indica*). *Research Journal Environmental Earth Science*, **2**, 31–35.

Padma, P.N., Anuradha, K., Nagaraju, B., Kumar, V.S. and Reddy, G. (2012) Use of pectin rich fruit wastes for polygalacturonase production by *Aspergillus awamori* MTCC 9166 in solid state fermentation. *Journal of Bioprocessing and Biotechniques*, **2**, 116.

Palaniswamy, K.P., Muthukrishna, C.R. and Shanmugavelu, K.G. (1974) Physico-chemical characteristics of some varieties of mango. *Indian Food Packer*, **28**, 12–18.

Patel, B.M., Shukla, P.C. and Patel, C.A. (1971) Nutritive value of mango seed kernels. *Food Farm and Agriculture*, **3**, 19–20.

Pedroza-Islas, R., and Aguilar-Esperanza, E. (1994) Obtaining pectins from solid wastes derived from mango (*Mangifera indica*) processing. *AIChE Symposium Series*, **300**, 36–41.

Puligundla, P., Obulam, V.S.R., Oh, S.E. and Mok, C. (2014) Biotechnological potentialities and valorization of mango peel waste: A review. *Sains Malaysiana*, **43**, 1901–1906.

Rashad, M.M., Moharib, S.A. and Jwanny, E.W. (1990) Yeast conversion of mango waste or methanol to single-cell protein and other metabolites. *Biological Waste*, **32**, 277–284.

Razif, H., Michael, K. and Danquah, M.F. (2009) Microalgal biomass as a fermentation feedstock for bioethanol production. *Journal of Chemical Technology and Biotechnology*, **85**, 199–203.

Reddy, G., Altaf, M., Naveena, B.J., Venkateshwar, M. and Vijay Kumar, E. (2008) Amylolytic bacterial lactic acid fermentation-a review. *Biotechnoligical Advances*, **26**, 22–34.

Reddy, L.V.A. and Reddy, O.V.S. (2005) Production and characterization of wine from mango fruits (*Mangifera indica* Linn.). *World Journal of Microbiology and Biotechnology*, **21**, 1345–1350.

Reddy, L.V.A,, Reddy, O.V.S. and Wee, Y.J. (2011) Production of ethanol from mango (*Mangifera indica* L.) peel by *Saccharomyces cerevisiae* CFTRI101. *African Journal of Biotechnology*, **10**, 4183–4189.

Rehman, Z.U., Salariya, A.M., Habib, F. and Shah, W.H. (2004) Utilization of mango peels as a source of pectin. *Journal of the Chemical Society of Pakistan*, **26**, 73–76.

Ruiz, C., Ramirez, C., Gutiérrez de Piñeres, Ángulo, M. and Hedreira, J. (2011) Obtaining and Characterization of Mango Peel Powder and its Use as a Source Of Fiber and a Functional Ingredient in Natural Yogurt [Online]. Available: http://www.icef11.org/content/papers/fpe/FPE1195.pdf [14 August 2015].

Saifuddin, Md., Khandaker M.Md., Hossain, A.B.M.S., Sarwar Jahan, Md., Mat, N.B. and Boyce, A.N. (2014) Bioethanol production from mango waste (*Mangifera indica* L. cv chokanan): Biomass as renewable energy. *Australian Journal of Basic and Applied Sciences*, **8**, 229–237.

Saravanan, P., Muthuvelayudham, R. and Viruthagiri, T. (2012) Application of statistical design for the production of cellulose by *Trichoderma reesei* using mango peel. *Enzyme Research*, 2012: Article ID 157643.

Schiber, A., Berardini, N. and Carle, R. (2003) Identification of flavonol and xanthol glycosides from mango peels by HPLC. *Journal of Agricultural and Food Chemistry*, **51**, 5006–5011.

Sogi, D.S., Siddiq, M., Greiby, I. and Dolan, K.D. (2013) Total phenolics, antioxidant activity, and functional properties of "Tommy Atkins" mango peel and kernel as affected by drying methods. *Food Chemistry*, **141**, 2649–2655.

Solis-Fuentes, J.A. and Durán-de-Bazúa, M.C. (2004) Mango seed uses: Thermal behavior of mango seed almond fat and its mixtures with cocoa butter. *Bioresource Technology*, **92**, 71–78.

Somayaji, D., Padshetty, N.S. and Nand, K. (2001) Recycling of mango-peel waste for biogas production. *Asian Journal of Microbiology Biotechnology and Environmental Sciences*, **3**, 339–341.

Somda, M.K., Savadogo, A., Quattara, C.A.T., Quattara, A.S. and Traore, A.S. (2011) Thermotolerant and alcohol-tolerant yeasts targeted to optimize hydrolyzation from mango peel for high bioethanol production. *Asian Journal of Biotechnology*, **3**, 77–83.

Tandon, D.K. and Kalra, S.K. (1989) Utilization of mango waste. *Beverage Food World*, **16**, 21–22.

Tandon, D.K., Kalra, S.K., Singh, B.P. and Garg, N. (1991) Characterization of pectin from mango fruit waste. *Indian Food Packer*, **45**, 9–12.

Varakumar, S., Naresh, K. and Reddy, O.V.S. (2012) Preparation of mango (*Mangifera indica* L.) wine using a new yeast mango peel immobilized biocatalyst system. *Czech Journal of Food Science*, **30**, 557–566.

Vatsayan, R. (2002) Medicinal plants. In: *Mango, the Fruit Medicine* (ed. J. Hari), pp. 52–59. The Tribune House, Chandigarh, India.

Vergara-Valencia, N., Granados-Perez, E., Agama-Acevedo, E., Tovar, T., Ruales, J. and Bello-Perez, L.A. (2007) Fiber concentrate from mango fruit: Characterization, associated antioxidant capacity and application as a bakery product ingredient. *LWT Food Science Technology*, **40**, 722–729.

Vieira, J.G., Rodrigues Filho, G., Meireles, C.D.S., Faria, F.A., Gomide, D.D. *et al.* (2012) Synthesis and characterization of methylcellulose from cellulose extracted from mango seeds for use as a mortar additive. *Polimeros*, **22**, 80–87.

Ye, D.Y. and Farriol, X. (2007) Preparation and characterization of methylcelluloses from some annual plant pulps. *Indian Crops Production*, **26**, 54.

Zgorka, G. and Kawka, S. (2001) Application of conventional UV, photodiode array (PDA) and fluorescence (FL) detection to analysis of phenolic acids in plant material and pharmaceutical preparation. *Journal Pharmaceutical and Biomedical Analysis*, **24**: 1065–1072.

Index

Handbook of Mango Fruit: Production, Postharvest Science, Processing Technology and Nutrition, First Edition. Edited by Muhammad Siddiq, Jeffrey K. Brecht and Jiwan S. Sidhu.

n

o

Printed and bound by CPI Group (UK) Ltd, Croydon, CR0 4YY

16/09/2026

14963388-0001